More Praise for Great Waters

"Like Rachel Carson before her, Deborah Cramer has captured the mythic proportions of the sea around us. Her voyage of scientific discovery extends our understanding of how intimately connected we are to our watery past, and how deeply we have imperiled our planet's future. There is wonder and beauty and majesty in her writing—and passion in her plea to return Atlantic to wholeness. *Great Waters* is the sequel Carson herself would have wanted to write."

—Linda Lear, author of *Rachel Carson: Witness for Nature*

"In *Great Waters*, Deborah Cramer eloquently portrays the intricate connections between our well-being and that of the earth's seas, and persuasively describes how all of us are putting these vast, life-giving waters at risk. I urge everyone to read it, act on its message, and pass its teachings on so that together we may create a strong, new generation of marine stewards. I can think of no one who wouldn't benefit from reading this inspiring book."

—Al Gore

"[A]n entertaining, instructive, and philosophically perceptive journey . . . . [Cramer's] lucid prose brings to light innumerable facts . . . [and] fascinating fields of study that will enrich any reader's understanding of what's at stake in the depths of the Atlantic."

—*Audubon*

"I was impressed with the level of scholarship evident in the author's research. . . . For the nonspecialist, her prose conveys the essence of the issues without the specialist jargon. . . . I will recommend it to anyone who proposes to be an informed citizen of planet Earth."

—*Science*

"Deborah Cramer makes clear just how small the boundless Atlantic really is. Small in the sweet sense of being a collection of diverse and fascinating neighborhoods; small, too, in the scary sense of no longer being able to absorb the cumulative effects of our carelessness. If you've ever walked the shores of this great ocean and stared off beyond the surf wondering what lay there—well, this fine book will let you know."

—Bill McKibben, author of
Long Distance: A Year of Living Strenuously and The End of Nature

"A wonderful account that reveals an eclectic, comprehensive intelligence . . . . A powerful and provocative synthesis; first-rate science journalism."

—*Kirkus Reviews*

"[Cramer] shows us the Atlantic Ocean in a way we've never seen before; as an inhabitant of planet Earth, as a living entity. . . . Fans of the writings of Rachel Carson, Peter Matthiessen, and Farley Mowat . . . will certainly want to read this provocative work . . . perfect co-curricular reading."

—*Booklist*

"Deborah Cramer has ventured into the sea, nature's least understood frontier, and has written powerfully of the threats to its continuing productivity. . . . She has illuminated the causes of its decline and pointed to the consequences for climate, weather, food supplies, flooding, human life. . . . Read this book and stop eating Patagonian Toothfish (Chilean seabass). It's going fast."

—William K. Reilly, chairman, World Wildlife Fund;
former administrator, U.S. Environmental Protection Agency

"For the general reader, Cramer's chapter 'The Birth of Atlantic' is the best exposition on the subject I've ever read."

—John Teal, *Conservation Matters*;
scientist emeritus at the Woods Hole Oceanographic Institute;
author of *Life and Death of a Salt Marsh*

"[Cramer] has managed to write a book about an entire ocean, but she has an uncanny ability to frame the larger picture in the details. . . . Being a passenger on this marvelous voyage through the ocean was a delight."

—*The* [London] *Times*

"Deborah Cramer's *Great Waters*, as thrilling to read as *The Sea Around Us*, surpasses Rachel Carson's majestic achievement with the fifty years of scientific findings that have increased knowledge since Carson wrote. Ms. Cramer takes full measure of the damage that human actions have lately inflicted on our planet through invasive pollution of air, land, and sea, and by overfishing. Yet, what emerges from her vast research is a vital and eloquent narrative, biblical in scope, recounting cycles of creation and destruction in a long and eventful terrestrial history. This book is an eye-opener. I know of no other like it."

—Peter Davison, author of *Breathing Room: New Poems*

"[Cramer] shines, narrating in language that plies the verge between science and poetry. . . . [A] lament that rings with all the sonorous truth, all the warning persistence, of a buoy bell." —*San Francisco Chronicle*

"Gliding through 'bioluminescent blooms' and sailing alongside the 'broad black backs [of right whales that] barely break the surface,' Deborah Cramer guides us through the physical contours and diverse biological communities of Atlantic Ocean. Her work explores the ocean's role in our evolution and in regulating climate throughout the interwoven harmonics of Earth's orbital cycles; and the role of humans in disturbing its rhythms. A beautifully written portrait of an essential treasure few know so well." —Paul R. Epstein, Center for Health and the Global Environment, Harvard Medical School

"An ocean-going Rachel Carson." —*The Economist*

"Cramer takes us deep into the sea to show us its rhythms, life and role in our lives. She is a fine writer and her story is persuasive."
—*Tampa Tribune*

"Cramer is at her best combining the esoteric data of hydrology with her personal musings on the mysteries of life, achieving originality and poetic grace." —*Publishers Weekly*

"*Great Waters* is a song of the sea, stunning in its breadth and beauty. Deborah Cramer has written a hugely satisfying exploration of the Atlantic Ocean. Not since Rachel Carson has there been such an intimate portrait of a whole sea—its webs of life, its movements, its birth and shaping by human endeavor, and its inevitable death—scientifically accurate but so skillfully written it reads like fiction. If you love the ocean, you will love this book." —Jennifer Ackerman, author of *Chance in the House of Fate: A Natural History of Heredity*

W. W. Norton & Company
New York / London

GREAT WATERS

An Atlantic Passage

DEBORAH CRAMER

Copyright © 2001 by Deborah Cramer

All rights reserved
Printed in the United States of America
First published as a Norton paperback 2002

Illustration by Sarah Landry

Paleogeographic maps by Christopher R. Scotese,
PALEOMAP Project, University of Texas
at Arlington (*www.scotese.com*)

Other maps by William Haxby

Portions of Chapter 2, "A Diminished Thing," appeared
in a different form in *The Atlantic Monthly,* June 1995.

For information about permission to reproduce selections from this book,
write to Permissions, W. W. Norton & Company, Inc.,
500 Fifth Avenue, New York, NY 10110

The text of this book is composed in Fairfield Light
with the display set in Ribbon and Fairfield Medium
Composition by Gina Webster
Manufacturing by The Haddon Craftsmen, Inc.
Book design by JAM Design
Production manager: Andrew Marasia

Library of Congress Cataloging-in-Publication Data

Cramer, Deborah.
Great waters : an Atlantic passage / by Deborah Cramer.
p. cm.
Includes bibliographic references and index.
ISBN 0-393-02019-3
1. Atlantic Ocean. I. Title.

GC481.C73 2001
551,46'1—dc21 2001024005

ISBN 0-393-32334-X pbk.

W. W. Norton & Company, Inc., 500 Fifth Avenue, New York, N.Y. 10110
www.wwnorton.com

W. W. Norton & Company Ltd., Castle House, 75/76 Wells Street, London W1T 3QT

1 2 3 4 5 6 7 8 9 0

To Dan, Abby, and Susannah
with love and many thanks

In memory
of
Ellie Dorsey

They that go down to the sea in ships,
That do business in great waters—
These saw the works of the Lord,
And His wonders in the deep.

Psalm 107:23–24

Passage (păs´ij) *n.* 1. A transit, or journey from one place to another. 2. A motion across time, as in the passage of time. 3. The passing from one condition or stage to another, as in the passage from childhood to adulthood.

Atlantic Ocean

Contents

List of Maps and Illustrations

Introduction

A N Atlantic tide pool is a piece of the ocean, left by a receding sea. Small, its breadth within reach, I can walk its perimeter, see clearly into its depths. Ebbing waters expose the pool, bring its life into view. Limpets make their homes in rocky niches eroded perfectly to shape their shells. Sea urchins feed in slippery seaweed, scraping algae off granite. On the dry rim above the pool, motionless periwinkles, sealed tight against desiccating wind, await the returning water. The pool reveals itself but briefly. Save for those few hours on either side of low tide, this community of life lies, like the rest of the ocean, hidden beneath the waves.

A tide pool is small, close, but the rest of Atlantic, 32 million square miles, 12,000 feet deep, resists our easy or immediate grasp. Man has flown great distances into space and walked on the moon, but no one has set foot on the deep-ocean floor, and few have seen beneath Atlantic's impenetrable surface. From the air, earth's dry contours stand out—craggy mountains, rolling plains, thick forests—but the sea, flat sheet of opaque water, discloses little.

We are drawn to the sea, but only with humility can we begin to fathom its complex immensity. There is reason to know this water. Life emerged from the ocean; its watery realm nourishes us, sustains us, endows us with a benign climate. Our very existence depends on its fullness. By 2020, as much as three quarters of the world's population may be living within forty miles of the seashore, yet most of us have no inkling of the time kept by the tides. One third of the United States' gross domestic product is generated in coastal areas, but most of us cannot see how marshy estuaries clean our water and nurture our fisheries.

What hope is there for the sea if we do not love, nurture, and protect its life-giving waters? In these pages I will plumb Atlantic, peer into her depths, and take her measure, seeking to describe this ocean, one of earth's last frontiers. I will ask to what extent our own well-being is intimately connected to its tiny drifting plants, to the constant rush of waters from the Gulf Stream, to the oxygen levels in the Chesapeake. We are of the sea, and the sea is of us.

Throughout the book, I have chosen to call this sea "Atlantic," rather than the more commonly used phrase "the Atlantic." The article suggests a sense of Atlantic as an object. It creates a distance between people and the sea, corroborates the tendency to regard the ocean as something out there, away. The more I researched this book, the more that distance receded. For me, dropping the article conveys more accurately the idea of a living Atlantic and a wellspring of life.

Oceanography is a fascinating but rather inaccessible field. Like most sciences, it has become increasingly specialized; biological, chemical, and physical oceanographers often speak different and mutually incomprehensible languages. Yet as knowledge grows, the disparate fields reconnect. I sought to understand the relationships and their implications for us, to understand why the sea really matters, over a period of ten years. I read many, many papers across the disciplines of oceanography and listened to physical oceanographers, geologists, biologists, and chemists on both sides of Atlantic explain and reexplain their work.

When mathematical equations and chemical formulae confused rather than clarified, I listened to the sea. I went out on the water with fishermen and watched them comb what were once some of the world's most abundant fisheries for the little that remained. I stood at the upper reaches of an estuary, and as the tide poured in, I watched hundreds of tiny elvers arriving from the distant Sargasso. I hiked along the cold and rainy island of Newfoundland, whose ancient rocks tell the story of Atlantic's birth. As a guest of the Sea Education Association, I sailed from Woods Hole to Barbados, crossing Atlantic's realm—the rough, chilly Gulf of Maine, the intense cobalt of the Gulf Stream, the calm, weedy Sargasso. I wanted immersion in the water I was attempting to describe.

Science, examining the wholeness of our world by breaking it apart, has sometimes led us far from intrinsic truths. If I have learned anything from my effort, it is that science is now sophisticated enough to encompass the larger view, and to bring us home. If I have failed to achieve my purpose, it is due, I believe, not to some weakness or fallacy in the argument, but rather to my own shortcomings in deciphering an exceedingly complex subject and recasting it in language that can speak to us all.

In Robert Frost's poem "For Once, Then, Something," the speaker peers into a well, searching the deep water. He catches a momentary glimpse of something, but then it is gone, obscured by his own reflection. A possible way to understand Atlantic is to see beyond ourselves, to let the lines of our own self-importance fade. Then, perhaps, each of us can hear the water resonate and sing. Hearing the voice of the sea, we may come to understand how we humans are but one among many species, inextricably linked to this mysterious, watery place.

Part 1

Moving Offshore

CHAPTER 1

Waters of Life

EARTH is brimming with water. She holds these stores as liquid, as vapor, and as ice. No other substance appearing here assumes all three forms. The planet harbors no seas of methane, no oceans of ammonia; these more flighty substances evaporate, dissipate into the air. Water is tenacious, holding its own, enduring in generous, abundant liquidity. What keeps it here, what gives it strength and stability, is the architecture of individual water molecules. Two atoms of hydrogen bond to one of oxygen, creating a tiny but powerful magnet. Hydrogen in one molecule attracts and adheres to oxygen in another. Though molecules constantly come together, break apart, then come together again, the overall structure, each molecule linked to others, is strong. It holds together, and at the same time pulls other substances into its sphere, breaking them down. Water is a powerful solvent; the magnetic pull of its molecules rips apart and dissolves the salt in the sea.

Water sticks. It clings to itself in seeming defiance of gravity or pressure. Glistening early-morning dew hanging from a frail spiderweb, swollen raindrops stuck on a windowpane, a glass filled to overflowing:

these are the outward signs of water's inner bond. The bond is tough, and hard to break. Otherwise the sea might have boiled away long ago. Instead, the oceans endure, in tropical heat and raw polar winds. Intense cold cannot fully freeze earth's seas. Other liquids grow heavy as they cool and turn solid, but water follows its own rhythms. Seawater becomes dense, sinking as it chills, until temperatures reach the freezing point. Then water's extraordinary molecules fly apart, buoying the newly forming ice, lifting it to float on the surface. There it insulates the sea below, keeping it warm and wet, able to sustain life throughout winter.

Not so long ago we believed that light was essential to life, but today we know water to be the critical element. Bacteria dwell in scalding hot springs on the seafloor and in damp crevices of granite bedrock deep in the ground, well beyond the sun's reach. Our quest for life beyond earth draws us to other watery spheres, to distant Jupiter's moon Europa, where a living ocean may lie beneath a thick crust of cracked ice, and to Mars, whose meteorites smack of briny beginnings and whose craters, freshly cut with gullies, hint of running water. Liquid water, with strength to wear a mountain into dust and gentleness to cradle and nurture the tiniest, most delicate organism, makes the life we know possible. Water is intrinsic to earth's nature, giving breath and life. While this truth may have slipped outside modern consciousness, where water is taken for granted and its constant and infinite supply assumed, ancient peoples knew to stand in awe. Unaware of the scientific particulars of water, of its chemistry and physics, they understood its essence. Biblical legends tell of a lens of pure water hidden in the earth by God during Creation. Beginning with Hagar and Ishmael, and continuing with Miriam, generation after generation drew from this well whose living waters refresh the body and restore the soul.

Earth, the watery planet. Water, the source of life. Spinning out our lives on dry land, we easily forget that life was conceived in the sea, easily lose touch with a world that has receded from our vision, becoming removed, foreign, and dim. Astronauts provide the wider view, the liquid perspective. From their distance, they beam back photographs of a blue earth, of land swimming in water.

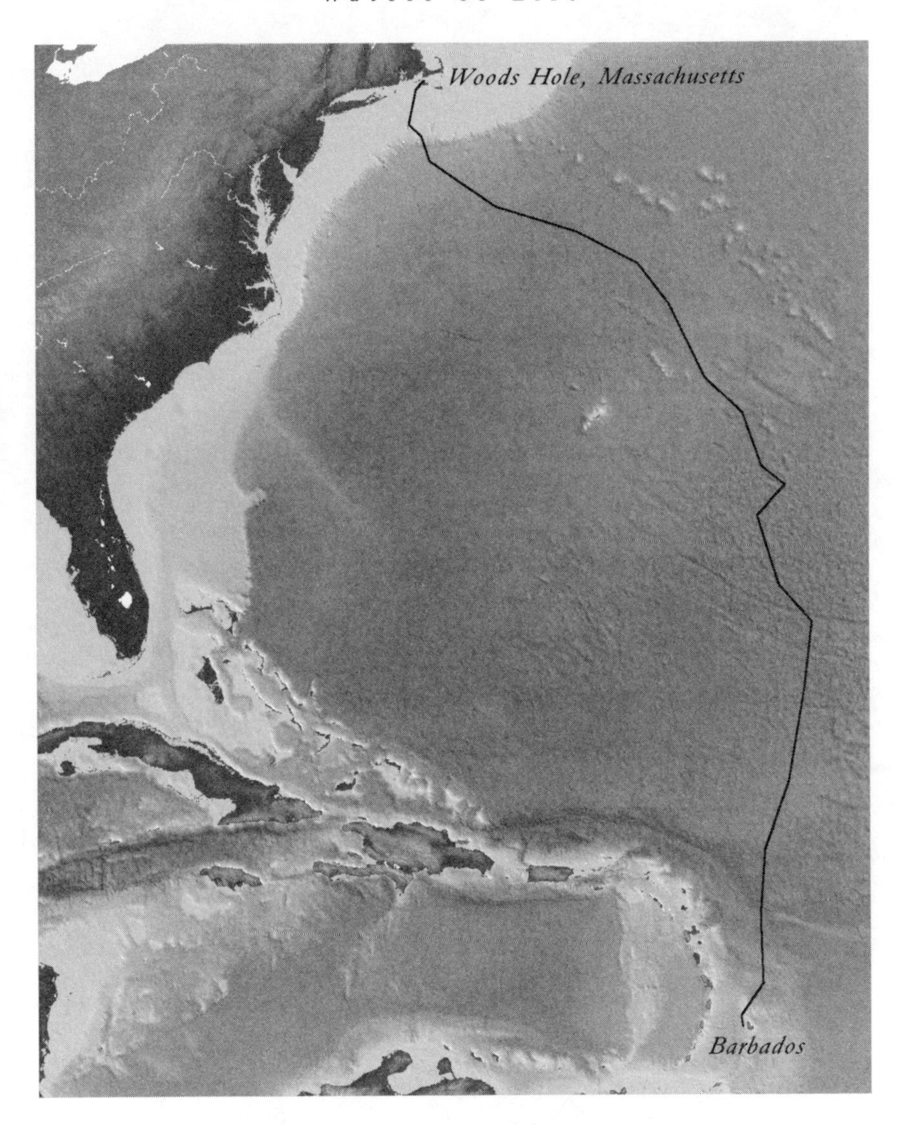

Corwith Cramer *sailing track C–129, first leg*

Water covers two thirds of the planet; land is but a few dry islands rising from a spacious sea. Terrestrial habitat, a thin skin stretched between cracked bedrock and the heights of airy trees, constitutes but a small fraction of the space on earth where life is found. Almost the entire biosphere, 99.5 percent, belongs to the sea. All of it supports life—sunlit surface, dark mid-waters, cold plains of the abyss, and thick lavas lying below the seafloor. Atlantic, earth's second-largest sea, is named for Atlas, the Greek Titan who supported the world on his shoulders. It is a name well deserved; Atlantic lives up to it in ways we could never have imagined, in ways we may never fully understand. Time and time again, she surprises us, nourishing a variety of minute inhabitants we never thought possible, and a web of life whose complexities and intricacies are still unfolding, whose well-being is so intertwined with our own.[1]

Sailing from Woods Hole to Barbados, I can begin to appreciate this richness. Joining scientists and college students aboard the research vessel SSV *Corwith Cramer* (no relation) on the first leg of a fall semester at sea, I am surrounded by more water than I have ever seen. For over three weeks, we are out of sight and reach of land. Occasionally a large tanker looms in the distance, and one day a sailboat bound for Bermuda overtakes us, but by and large the *Cramer,* an elegant 134-foot (41-meter) brigantine, travels alone, surrounded by gray sea and sky. From the deck of this ship, the sea seems without edge. In every direction, as far as the eye can see, there is water. Onshore, we humans, rooted to the land, take that dry place to be the world of consequence, but ours is a myopic view. Out on this boat, hour after hour, day after day, with nothing but water beneath a canopy of sky, we are embraced by the sheer immensity of the sea, by the ubiquity of water.

A touch of the ethereal graces the midwatch, deep in the night, when the sun has long since disappeared and ship and sea are clothed in blackness. From my perch on the bowsprit, harnessed to the rigging, bobbing up and down as the boat rises and falls in the swell, I know peace and quiet and solitude. Watching the waves, one after another, rising and cresting in the empty night, I have an inkling, a vague notion of the marine world usually so distant. Waves break, and tiny unseen organisms, emissaries from another realm, shine in the darkness, glowing with a pure light. Streaks of bioluminescence twinkle at my fingertips, a distant galaxy come close. In this uncomplicated, velvet night where boundaries blur, and edges soften, and time stretches on forever, I, dweller on dry land, greet these tiny inhabitants of the sea. The bustle on deck slips away into the blackness; I am soaked in a watery world.

These bioluminescent lights in the ocean emanate from floating plants and animals, inhabitants of the sea surface, wanderers, captives of current and wind and wave. The plants, central to the marine food web, give the sea life, sustain almost all that dwell in her waters. Their populations soar to astronomical heights, to several million in a pail of seawater, but unlike their mainland counterparts—dense forests, rolling prairies, fields of grain—these meadows of the sea elude the naked eye. To our current knowledge, the atmosphere holds no comparable community: the sea's plants, too heavy to lodge in the lightness of air, thrive suspended in the thick ocean, living out their lives afloat. Size is neither virtue nor necessity in this realm, where plants, buoyed and nourished by salt water, need no roots to reach soil, no stiff trunks or branches to reach light. Land plants, seeking sustenance and light, grew in size and mass, while plants in the sea, bathed in sunlight, surrounded by all that is necessary to sustain life, remained tiny. Theirs is the biology of the small.

Small in stature, great in number, the floating meadows of the sea carry on almost as much photosynthesis as terrestrial fields and forests, with barely two tenths of 1 percent of the biomass. Terrestrial plants grow slowly—fields of grain reach fruition two or three times a season, a tree may flower once a year—while the plants of the sea grow and reproduce rapidly. Here in the morning and gone by the afternoon, a single marine plant can produce, and lose to predation, one million progeny in just a few weeks.

Magnified hundreds, thousands of times, the intricate, delicate structures of individual marine plants are revealed, their separate identities distinguished. Seen with scanning electron microscopes, they appear as clusters of tiny pineapple rings, chains of elegant pillboxes, solitary winged shells. In the sea, these floating plants, the plankton, are invisible. Each is a single cell, embellished for a life afloat. A minute drop of oil or wax embedded within and an array of spines or horns protruding without keep the cell adrift, prolonging its life at the surface, retarding its eventual, inevitable descent. The plants sink slowly, gently spiraling down like falling autumn leaves, or beating their tiny whiplike hairs incessantly against the current.[2]

These floating meadows were long thought to form the base of marine food webs, but it turns out that other plants, far smaller, permeate the ocean. A stream of seawater, bathed in the light of a laser, fluoresces in the presence of life, bringing into focus a previously unrecognized single cell of chlorophyll. The more familiar phytoplankton are tiny, each plant the width of a single strand of hair, but these mere specks are one to two hundred times narrower. *Prochlorococcus,* modern descendant of ancient earth's first photosynthesizing organisms, is the most abundant and ubiquitous plant in the sea. Yet, tiny as it is, we have come to realize that other organisms, even smaller, even more numerous, dwell in the ocean. A bucket of seawater may contain 150 million cells of *Prochlorococcus,* along with one billion scavenging bacteria and ten billion viruses. This vast but minute world, overlooked by us until recently, constitutes a full 99 percent of the sea's microorganisms. Perhaps now we have fully outlined the sea's food web, but then again, perhaps not. Another layer may still lie outside our view, teeming with unknown, unimagined life. The limit to greater understanding seems to rest not in the sea, but in our perception.[3]

We barely know Atlantic, this most studied of oceans. Each exploration opens new paths of inquiry, broadens our sense of the possible. Approximately 900 feet (275 meters) down, on seafloor off the Bahamas, botanists find a large tract of purple algae. The light seems too faint to support photosynthesis, but there the algae are, thriving. Across Atlantic, on seafloor off the Namibian coast, lie strings of giant white bacteria, the largest ever seen in the ocean. They rest in a green ooze made from plant remains drifting to the bottom. The water smells of toxic hydrogen sulfide released by decaying plants, but the bacteria flourish. Making energy from sulfide, they shine like pearls. The necessary nutrient, nitrogen, falls from the surface, but only sporadically. When it rains down, the bacteria collect it, storing enough to last for forty or fifty days, growing fat on the reserves. Their large storage capacity, 98 percent of each cell, gives these bacteria their girth. "Sulfur pearls of Namibia," they are one hundred times larger than any other known oceanic bacteria. Their presence surprises, expands for us the dimensions of life in the sea. In the oceanic realm of the minute, we may have yet to realize how small the smallest, how large the largest.[4]

Here, at the heart of the marine food web, issues of identity are also confusing, as distinctions between plant and animal, producer and consumer, blur. Tiny photosynthesizing "plants" dine on bacteria. Single-celled cone-shaped "animals" browse the meadows of the sea, using the chloroplasts of plants they consume to make carbon in the presence of sunlight. There are even tiny bacteria floating at the surface that convert light into energy. The sea abounds with organisms crossing between the kingdoms of plant and animal, and is richer for it. One such chimera blossoms into some of Atlantic's most productive floating meadows, but only after ingesting another alga and appropriating its chloroplasts. Once the theft is complete, the cell grows and divides, again and again, filling the sea with chlorophyll. As the chloroplasts age or become diluted, the chimera, a ciliate, simply steals more. The plant is pink, the animal without color. Each is invisible to the naked eye, but when one becomes the other, the sea bursts into a red bloom.[5]

The pastures of the sea are generous, breathing oxygen into the water, capturing sunlight and transforming it into carbohydrate, the bread of life for oceanic animals. Tiny animals graze this plenty, turning carbohydrate into protein. Some animals live out their entire lives afloat; others spend their infancy at the surface, then settle on the bottom. One of the sea's most abundant and important protein producers passes almost its entire life grazing in the sunlit waters of the surface. No larger than a grain of rice, the copepod—vegetarian, chief grazer in the pastures of the sea, key link to the fishy carnivores—may be the most multitudinous animal in the world. Net the floating animals of the sea and 70 to 90 percent by

weight will be copepods. "Red feed," or "cayenne," fishermen call them, for the touch of scarlet on their transparent, shrimplike bodies. Wherever there are herring or mackerel, in the North Sea, the Gulf of Maine, or Georges Bank, there are copepods.

Though small, copepods are fast. If need be, they can burst away from a predator at two hundred body lengths per second. For their size, they travel great distances through the water. Each morning as the sun rises and the sky fills with light, they drop into colder, deeper waters, 300 to 1,000 feet (90 to 300 meters) down. Each evening as the light fades, they return to the surface. Scientists cannot fully explain this mysterious journey, which is equivalent to a human hiking fifty miles (eighty kilometers) every day. Perhaps copepods descend to flee predators feeding at the surface, or to escape damaging ultraviolet light streaming from the high sun. Perhaps the energy they conserve in deeper, cooler water exceeds the energy they expend in travel. Since they cannot stem tide or current, vertical migration may carry them to fresh food, to floating crops of yet ungrazed plants. We see the pattern of this tiny steadfast swimmer but, unaccustomed to such exertion ourselves, fail to grasp its meaning.

Copepods thrive in Atlantic's temperate waters where the sea swarms with plants. Other grazers dwell in less congested waters. These animals, too, are tiny. Six of the tiny grazer *Oikopleura* could comfortably fill the space needed to write its name. Tailor-made for waters less fertile and more dilute, *Oikopleura* lives in a walnut-sized mucus house equipped with a screen door. The animal filters water through the mesh; nothing too large or overwhelming gets in. When the filter clogs, *Oikopleura* jettisons the entire house, within minutes manufacturing a completely new one. Scientists sampling the minute marine world with nets often fail to retrieve the fragile house intact; crowding in the net crushes the delicate structure.

Scientists who actually get wet, who dive into Atlantic to observe animals in their own environs, discover a wealth of delicate, gelatinous creatures humans have never before seen: shimmering comb jellies; long, ribbony chains of transparent animals; wide, boxy jellyfish. So little is known of the planet's watery realm. One biologist floats amidst Atlantic's jellyfish, another in a lab analyzes a fluorescing stream of seawater, and another tows a video camera from a boat: each finds a largesse that will require reorganization and expansion of entire taxonomies, a task so enormous it will take years to complete. Each revelation points not to the depth of our knowledge, but to the vastness of our ignorance.[6]

In the first days of her journey, the *Cramer* heads south, plowing through a green sea. The waters are blooming, a final burst of life in the waning summer sun. Less fertile waters are more transparent; this sea is thick. Measuring water clarity, indicator of ocean fertility, we lower a heavy white disk the size of a dinner plate into the water until it disappears. While cloud cover, the angle of the sun, and the visual acuity of the observer render this measurement less than 100 percent accurate, an approximation suffices. The disk vanishes at 11½ feet (3½ meters). Later, when we repeat the test in the clear waters of the Gulf Stream, it disappears at 125 feet (38 meters). Here the water is murky, alive with billions of tiny plants, but this explosion of growth will not last. The light is fading and the turbulence, which makes us all so seasick, signals the onset of winter.

Sea meadows swell and thin according to a tempo and rhythm governed by earth's tilt before the sun, and by the supply of salt. Washed off mountains and borne in on rivers, released from hot springs on the seafloor, and returned in the decay of dead animals, salt fertilizes the sea. Not much is needed. Every bucket of seawater contains only a few spoonfuls. Sodium and chloride, common table salt, are sea salt's most common elements, but the few grains of nitrogen, phosphorus, silica, and iron are critical, turning meager waters into lush meadows.

Populations of drifting plants surge and die away, soar and then diminish, timed to the coming and going of light, and to fluctuation in the supply of salty nutrients. Onshore in New England, redwing blackbirds and pussy willows herald the arrival of spring. Offshore, steely gray seas cloud over, giving way to deep murky green as the first rays of spring sun arrive and the sea's floating meadows burst into bloom. Satellites mark the onset of Atlantic's season of renewal, tracking the advancing line of chlorophyll. It appears in March, in waters between Spain and the Carolinas, and glides north, arriving in coastal Iceland by June.

This spring sea teems with life, but its texture is patchy. Chlorophyll does not ride on every swell. Satellites trail spring blooms larger than Great Britain, but to observers at the sea surface, even pastures this large may be difficult to locate. The wandering grasses, blooming and dying first in one place, then in another, are flighty and elusive. Where they grow, they grow lavishly, though to us watching from a ship's deck, even a green sea seems thin. Sonar and lasers scanning the ocean home in on fine details, locate concentrations of floating plants thick enough to support vast populations of grazing animals.

As summer comes on, surface waters warm. The tumult dies down. In the dog days of July, when the sea is like glass, plant growth slows. Sunlight abounds, but spring blooms have depleted the salty fertilizer. A lens of warm barren water rests on rich stores in the cold below. The next surge of growth awaits the fall, when temperatures drop. The cooled surface, blending with nutrient-rich water below, is replenished, reprovisioned. The bloom continues until the light fades. Throughout the winter, hurricanes and nor'easters churn the sea; turbulence mixes surface and bottom, plowing the waters, readying them for spring, when, with the return of light, populations of plants will explode once again.

There are exceptions. Where tides continually stir the water, in parts of the North Sea and the English Channel, the meadows of the sea bloom all summer, as long as the sun is high. Currents rushing through coastal waters continuously nourish the surface. Off northwest Africa, where winds push the Canary Current away from the shore, cold, rich waters well up from the deep to fertilize large swaths of floating meadows, producing a bountiful feast.[7]

Aboard the *Cramer,* a tiny vessel sailing a seemingly boundless sea, we sample a bare fraction of this abundance. We tow a small-mesh net alongside the ship, haul it out, and hose it down with a high-speed spray, forcing the catch into the collecting or "cod" end. Then we empty the net and scrape the sample across a sieve with a spoon. Jellyfish and other gelatinous creatures are reduced to pulp; only the tough survive. We bring what's left into the lab, pickle the larger animals, and divide the rest. Hunched over microscopes, fending off the hot and clammy feelings that precede seasickness, we cruise the petri dishes, counting and sorting copepods, fish larvae, amphipods, and any other animals we might find.

Our lab work documents the sea's yield, establishing an extensive baseline of information that should prove useful as the water warms or currents change their course. And yet our catalogue, extensive as it is, and our animals, their lives stilled in jars of preservative, barely hint at the fullness of the sea. A butterfly net fails to capture the delicacy of a spider's web, or a sieve the complexity of an anthill. An opossum flattened by a speeding car reveals opossum biology and anatomy, but little woodland ecology. Here in this lab, we deal primarily with road kill. We begin to establish context, measuring the sea's phosphate and chlorophyll levels, but it is just a beginning.

Earth's oceans are living waters. The beauty and wholeness of the sea lie in the rhythms of life of her inhabitants, their interconnectedness, the comings and goings of one tuned and timed to the health and fecundity of another. Much hangs on the richness of floating plants. Populations of grazing copepods rise and fall, explode and die, in exquisite balance with the bounty of floating grasses. In the Arctic, one generation of copepods is born each year, in the long light of the brief polar summer. In the North Sea and the Gulf of Maine, where the sun lingers and where currents and tides mix the water, keeping meadows green, copepods reproduce as many as three times a year. Their numbers swell and subside as the meadows bloom and die. They even feed at a pace finely tuned to the measure of their food; when grasses grow plentifully, copepods gorge, taking large portions, grabbing larger plants with their mouths. When food is scarce, they feed more slowly, filtering small bits from the water with their feathery oarlike feet, never decimating the sparse grasses. The copepod, with a brain neither large nor sophisticated, somehow lives within its means.

Like tiny grazing copepods, larger fish, too, move to the rhythm of life in the sea's floating meadows. Plaice, fish similar to flounder, live on the bottom of the North Sea. In autumn, they leave their feeding grounds near Dogger Bank and migrate south for the winter, to feed and spawn at the edge of the English Channel. Migrating plaice ease the journey by riding the ebbing and flowing tide. What calls them twice each day to rise from the muddy bottom into the mid-water to catch the tide? Although no human has plumbed the mystery of this call, the voyage serves plaice well. It ensures their survival, for the young are helpless drifters, unable to swim to food or navigate to their summer nurseries inshore. Spawned near the English Channel, they are positioned to catch coastal currents. The water courses north, pulling the larvae along, and spring seas bloom, enveloping them with a surfeit of food. They feed on plants in their early days, and on copepods as they mature. By summer, the rushing current delivers young plaice to their nurseries in bays and estuaries along the coast of Holland.

Timing is all. If spring tarries, and the water stays cold, and the eggs hatch before the meadows bloom, the larvae starve. If the wind disappears and the current weakens and fails to bring the young fish north, they die. We humans have long ceased to be such prisoners of weather and the seasons. For many of us, winter turning to spring and then to summer marks the passage of time, painting beautiful scenery for us to enjoy. Winter no longer curtails or challenges our well-being; we can feed our children anytime, shelter them in any season. While we survive by altering and enhancing our environment, plaice survive by synchronizing and dovetailing with theirs. Eventually, we may come to question some of our choices.[8]

We have only begun to comprehend the intricacies and dependencies within marine food webs, the cycles of life and death, loss and renewal, which turn the sea green, endow her waters with life. For years we believed that most carbon made in the meadows of the sea was eaten by grazers, who were then consumed by fish. Any leakage from the system, whether animal wastes or decaying plants, was considered inconsequential. A house abandoned by *Oikopleura,* a tatter of *Prochlorococcus,* a few specks of pollen and soot blown off the land, a fecal pellet from a copepod: these are the remains. They drift slowly toward the seafloor, a cloud of wayfarers gently, continuously exiting from the sunlit surface. Most never reach bottom. Bacteria find them first. They colonize and consume the sinking flakes of "marine snow," releasing nutrients into the water to sustain new generations of plants and grazers. The cycle has run continuously, silently, and, until recently, imperceptibly. When marine scientists could not reconcile the sea's apparently stable level of dissolved carbon with its continuous production, when they could not explain the disappearance of substantial amounts of carbon, it was because they underestimated what they could not see. Fully half the carbon produced by the sea's floating grasses is routed down through the bacterial loop in the marine food web.

Other overlooked but critical participants in this loop are viruses, billions of them. They prevent bacteria from proliferating by invading and reproducing inside them, their viral progeny bursting the host cells apart. Their numbers rise and fall within minutes, within hours, keeping pace with swelling and shrinking bacterial populations. Viruses are a diverse lot. Some prefer bacterial hosts. Others attack the sea's drifting plants, thinning the meadows, reducing photosynthesis. Numerous enough to end a spring bloom, viruses help distribute carbon through the food web. Taking herbage away from grazers, supplying detritus to bacteria, they tilt the flow of energy and nutrients toward the sea's smaller inhabitants.

The rhythm of life and death in a blooming meadow helps determine whether bacteria or fish will dine. In the eastern Mediterranean, bacteria reign. Competition is intense for the few nutrients in these impoverished waters, and bacteria, though small, compete successfully. Scavenging limited supplies of iron with specialized compounds excreted into the water and then reabsorbed, they thrive where others languish. In the eastern end of the sea, they capture 85 percent of the carbon manufactured by phytoplankton, leaving little to support grazers and fish. At the western end, the wealth is shared; bacteria receive 55 percent. In the chilly water off Newfoundland, the energy balance tips toward big fish. Here *Oikopleura*'s cast-off house is inhabited primarily by those making energy from sunlight. Scientists have suggested that the plants thrive while bacteria, enervated by the cold, grow more slowly. When the frigid waters warm, the sea has already bloomed, leaving meadows of plenty for herbivorous grazers and a feast for their predators.

In this single fluid medium of the sea, all that is living and all that is dying, all that is gathered to nourish life and all that is cast off, mingle and merge and flow together. Viruses, bacteria, plant-consuming animals, grazing copepods, fish: their fates are intertwined in ways we can only begin to explain. At this point, we cannot fully grasp the complexities and subtleties of oceanic relationships. We have yet to understand, for example, how only five resources—light, nitrogen, phosphorus, iron, and silica—support such splendor, such richness of species. We have yet to understand how so little nurtures so much, although models of resource competition show multiple species coexisting on meager provisions, one following another to dominance, with no ostensible order or design. Whatever the pattern, repetition of labor gives strength to marine food webs. The work is often the same; drifting grasses make carbon from sunlight, bacteria break it down and recycle it. Their differences, their predilection for one or another nutrient, for strong or weak light, for cold or warm water, enable production to go forward wherever the sun shines. Unity of purpose coupled with difference in style gives the sea strength and resilience and has enabled her floating meadows to nourish life for millions and millions of years.[9]

Life drifting at the surface of the sea responds not only to the mixing of nutrients and the coursing of currents but also to the quality of light. Each spring, as sunlight returns to the icy stratosphere, chemical pollutants carve a hole in earth's ozone layer, the planet's shield from the sun's harmful ultraviolet light. Each spring, refrigerants and fire-fighting chemicals tear open the hole; carbon dioxide emissions from motor vehicles and power plants postpone its repair. More than half the ozone at polar latitudes disappears for a few months each year. In September and October, at the height of the spring bloom, a hole the size of North America opens over Antarctica. Ultraviolet light streams in, inhibiting photosynthesis, thinning the meadows of the sea. Young, immature plants are the most sensitive. Exposed to ultraviolet light, their rate of photosynthesis drops 65 percent, their growth rate 17 percent, and their ability to reproduce successfully 50 percent. Each spring, wherever the ozone hole opens, over seas off Antarctica, Greenland, or northern Europe, diminished plant productivity has the potential to ripple through the food web, to touch populations of grazers and the penguins, seals, crabs, and fish who prey upon them. How small the world has become, that chemicals lofted high into the atmosphere affect the well-being of drifting sea plants, that the health of the distant stratosphere is felt so keenly by the denizens of the sea.[10]

That tiny floating plants power entire food webs is a wonder in and of itself, but the influence of plankton reaches well beyond the sea. Drifting plants receive light from the atmosphere and give back clouds. In ways science is just beginning to articulate, microscopic sea meadows shape earth's climate, absorbing sunlight, cooling the ocean, raising wind and rain. The concentration of plants in the sea may strengthen Indian monsoons and brighten the clouds in the sky. Each plant harbors a small amount of sulfur, protecting it from the sea's drying salt. When plants die or are eaten, the sulfur is shed into the water. Viruses and grazing copepods affect how many plants die, how much sulfur is released.

Once sulfur seeps into the water, bacteria consume it. The little they leave behind diffuses into the air, seeding clouds, absorbing and scattering light, cooling earth. The more wisps of sulfur, the whiter the cloud; the whiter the cloud, the more sun reflected back into space. A high cloud, a floating plant: these are the individual, each imbued with its own singularity. Yet they are joined. Sea and sky seem worlds apart, but the design of a cloud hanging over the sea originates in the death of minute plants floating in the water below. We may never learn how much the interplay guides earth's climate. Just as we discover the dance, we learn there is another partner. From the sea, from ships crossing northern waters, comes another source of sulfur, equal or greater to the contribution of drifting plants.[11]

The relationship between sea and sky is close-knit. Millions of years ago, earth's first plants—mats of algae resting in a hot, briny sea—created the planet's life-giving atmosphere. Removing massive amounts of carbon dioxide from the air during photosynthesis, they breathed out oxygen, the breath of life for the animal kingdom. When they died, bits and pieces of their remains drifted slowly to the seafloor. In some places, the shreds of carbon accumulated. As the sediments of the sea slowly turned into layers of rock, the carbon was transformed, squeezed into thick reservoirs of oil.

Today, factories, power plants, and automobiles burn that oil, spewing carbon dioxide into the air each year, quickly returning what lush meadows of floating plants took out and buried in the sea over eons. The atmosphere has begun to feel the effect. From the end of the last ice age until the dawning of the industrial era, the amount of carbon dioxide in the atmosphere remained stable. Since then it has increased 30 percent, and by 2100 it will most likely triple, with damaging consequences for earth's climate, sea level, health, and agricultural productivity. The difficult, but arguably more sustainable, solution would be to reduce carbon consumption, to redirect oil and gas subsidies to cleaner technologies. Another, more facile response calls on the sea.

The floating meadows are rich. Though each plant is minute, the yield is prodigious, 45 billion tons of carbon each year. Some is returned to the atmosphere as grazing animals and their predators breathe, but approximately one third falls to the seafloor, in the remains of dead plants and animals. Entrepreneurs hope to coax the sea to soak up a little more by fertilizing where waters are low in nutrients. They'd like to do so along the edge of the Gulf Stream, in the Gulf of Mexico, the Pacific, and in the seas off Antarctica, wherever nutrients are in short supply. Although blue, barren seawater turns a lush dark green when iron is added, dramatically increasing phytoplankton productivity and carbon dioxide absorption, the long-term effects of enriching the sea are far from clear. Scientists have sown the sea with iron, and seen its meadows bloom, but these brief experiments, poised in the present, focus only on the local, on the particular. The oceanic food web is finely tuned, delicately and intricately balanced; our understanding of its rhythms and its nuances is coarse.

Nobody knows, perhaps nobody can know, whether infusing the sea with nutrients over an extended period will significantly lower carbon dioxide levels in the atmosphere. Most carbon produced in luxuriant sea pastures leaks back into the air instead of descending to the depths. Large animals, flocking to the new source of food, may crop it back, exhaling carbon dioxide as they feed. In the Greenland Sea, almost all carbon dioxide drawn in during spring and summer blooms returns to the air by the onset of winter.

When waters are rich in iron, some plants are less able to export carbon into the deep. In a rusty sea, golden diatoms bloom. In nutrient-poor waters, their thick shells fall quickly from the surface, carrying carbon down to the seafloor. Strengthened with iron supplements, they grow light and slim. Bacteria consume the slowly drifting shells, recyling carbon back into the food web. In a thin sea, diatoms, scarce on the surface, pile up on the seafloor, while in well-nourished waters, where they are more abundant at the surface, they dissolve en route to the bottom.

Those who would fertilize the sea hope to increase the productivity of marine food webs and replace depleted fisheries, but fertilizing the sea is nothing like sowing a field. In the open ocean, human-induced nutrient enrichment has not yet been found to correlate with the size of fish populations. Marine food webs are simply too complex. When diatoms bloom in the ocean, for example, copepods come to feed, but then their eggs fail to hatch. In a diatom bloom, between 12 and 24 percent of copepod eggs are viable, while in other meadows 90 percent of copepod eggs hatch. In diatom meadows, chemicals produced by the plants impair the reproductive success of copepods.

Coercing the sea into an everlasting spring may have unintended repercussions. Coral, unsuited for nutrient-rich waters, will die. Forcing a bloom in one place may leave waters barren in another. Just as plankton blooms off the Iberian Peninsula deplete water flowing into the Mediterranean, nitrogen and phosphorus taken up in tropical blooms may deprive temperate waters of essential nutrients. Furthermore, a meadow can grow too green. Bottom currents carry fresh oxygen in from high latitudes, but seafloor buried in decomposing carbon may deplete it, depriving bottom dwellers of fresh air. Saturating coastal waters with nutrients touches off toxic algal blooms; saturating the open ocean may do the same.

Fertilizing the sea to reduce greenhouse gases may increase them instead. Decaying plants may release methane, a highly potent greenhouse gas. Scientists use sulfur hexafluoride to trace the movement of iron in the sea. Once released, it dissipates into the air and accumulates, with thousands of times the warming power of carbon dioxide.

The risks of iron fertilization are high, and the benefits uncertain. Though floating plants consume prodigious amounts of carbon dioxide, the sea's capacity to absorb it falls far short of what is needed to stem global warming. In a recent experiment off Antarctica, barren waters turned green, but scientists did not observe any export of carbon into deep waters. We have already experimented with earth's atmosphere, adding carbon dioxide and chlorine. Adding massive amounts of iron to the sea might have consequences equally severe and surprising.[12]

Greenhouse gases continue to accumulate in the atmosphere, and the planet warms. Have her seas begun to feel the heat? Earth's oceans are vast, but oceanographers are beginning to observe temperature changes. In the next decade satellites and sound waves will generate more data on heat stored throughout entire oceans. In the Tropics, surface waters are warming quickly, 1 degree Fahrenheit (0.5 degree Celsius) per decade. Heat has already begun to penetrate below Atlantic's surface, barely but noticeably. Across the width of Atlantic, along the route Columbus sailed to reach America, in waters between 2,300 and 8,200 feet (700 and 2,500 meters) deep, the sea has warmed 0.6 degree Fahrenheit (0.32 degree Celsius) in the last fifty years. In the Mediterranean, warmth has permeated the depths. Over the last forty years, water temperatures in the northwest corner of the sea, 6,600 to 8,500 feet (2,000 to 2,600 meters) down, have climbed 0.2 degree Fahrenheit (0.13 degree Celsius). For Atlantic as a whole, the net effect has been a warming, ever so slight, within the last fifty years.

Human activity insulates the atmosphere, warming earth. The sea responds in kind, its waters begin to warm. Absorbing a portion of our atmospheric carbon dioxide emissions, the sea serves as a buffer against planetary warming, keeping earth cooler than it might be otherwise. Warm waters, though, hold less carbon dioxide than cold, and as sea temperatures continue to rise, the ocean becomes less able to mitigate the effects of a thickening atmosphere. In addition, a warm sea separates into layers; nutrients in the cold water below mix less easily with warm water at the surface. Floating meadows, less fertilized, grow more sparsely and produce less carbon to export into deep water. And the grasses that do bloom draw less carbon dioxide out of the air. As melting sea ice and increased rainfall layer waters off Antarctica, diatoms coming to dominate the plankton community remove half as much carbon dioxide as plants blossoming in well-mixed waters. If oceans warm, and meadows thin and then thin some more, rendering the entire sea as unproductive as today's most impoverished waters, then the deep sea's vast reservoirs of carbon dioxide would be released into the air, tripling atmospheric levels, greatly accelerating greenhouse warming, creating an earth we fear to imagine.

Life abounds throughout the ocean—bacteria thrive in deep-sea hot springs and in the rock hundreds of feet below the sea floor—but by and large, tiny drifters in sunlit waters weave the web of marine life. A solitary whale, a densely packed school of herring, a tangle of seaweed and crabs: this bounty is sustained by lush, watery meadows of microscopic plants. Scanty meadows cannot feed swarms of smaller grazers. When temperatures in the North Sea periodically rise, the number of copepods declines as much as 20 percent. Over the last forty years, one 50,000 square mile (130,000 square kilometer) patch of the Pacific has warmed, its temperature climbing between 3 and 5 degrees Fahrenheit (2 and 3 degrees Celsius). The warm water floated at the surface, barely mixing with colder, nutrient-rich water below. Stripped of nourishing nitrates, surface waters could not produce a rich growth of plants. Scarcity at the center of the web reverberated throughout. Populations of grazing herbivores plummeted 80 percent, and the birds who feed upon them starved. Not so long ago, sooty shearwaters skimmed over these waves, but now they have all but vanished, their numbers down 90 percent.[13]

Will the web unravel? Will we rip it beyond repair? It matters a great deal if gases we cannot taste, touch, see, or smell build up in the sky, causing the meadows of the sea to fade away, seabirds to disappear, and fisheries to decline, for our own lives may be intimately, critically connected to all other life. In no way am I more reminded of this than when I look at a small conch shell I have kept since I was a child. The shell, nothing extraordinary, contains for me a distillation of blissful days I spent on the beach when I was young. Years later, I recall those happy times, holding the shell to my ear, listening. I believe I hear the rush of surf hitting the shore, the crash of breaking waves. It is a universal fantasy, shared by every child ignorant of the facts of science, but it is also the truth. In the voice of seashells, in the echo of blood rushing through our veins, the waters of life, the sea, are singing.

CHAPTER 2

A Diminished Thing

THE *Corwith Cramer* is a sailing research vessel; each day, the science watch samples the sea, measuring its temperature, salinity, chlorophyll and oxygen content, and recording the life snared in the nets. The task is daunting. We glimpse but can in no complete way define the immensity surrounding us; our science watch cannot encompass the breadth or depth of this sea. While lack of sophisticated technology limits our knowledge and experience of the ocean, it is not the only obstacle. The passage of time and the path of human evolution, the shield and shelter of material comfort, and the tendency toward self-absorption have increasingly isolated us from intimacy with the oceanic world.

The first amphibians crossed the great divide between land and sea 370 million years ago. Perhaps they were lost, or hungry, or fleeing an eager predator. Whatever the reason, they dragged themselves ashore and gulped a few breaths of air, thus beginning new directions, new lines of ancestry in the animal kingdom. Though descended from those ancient marine animals, we *Homo sapiens* evolved for a dry life. Ensconced in our terrestrial niche, we cannot return to the sea, even for a short time, with any ease.

On this passage we who would study Atlantic are relegated to an observation post at its surface, sending our probes, nets, and buckets into the water's depths but staying on deck ourselves. We peer from a distance into a place vast, mighty, and obscure, perched on the edge of a world we no longer call home. Virtually omnipotent on dry land, we walk less confidently here, take less for granted in this place where we can slip so easily, so suddenly, from safety into danger, from security to vulnerability. There is much to remind us of our own frailty. The red survival suits stored on deck epitomize our limitations, compensating only feebly for our loss of watery adaptations. For practice, we squeeze awkwardly into the damp, close thirty-pound body suits. At the captain's order, one student jumps overboard. She becomes a splash of red, possessed by the swell, with only a few inches of rubber separating her from Atlantic's icy, numbing waters. The insulation staves off hypothermia for approximately sixteen hours. Should more time elapse, her heart rate will fall, her circulation slow, and her body cool to the temperature of the sea. Without the suit she would succumb within minutes. In the water she is transformed: competence becomes submission, and ego, power, presumptions, and control dissipate.

My shipmate floats in the water, helpless and uncomfortable. While whales swim thousands of miles padded with blubber two feet (0.6 meter) thick, she can barely maneuver, and tires with only the slightest movement of her heavy arms and legs. She flails in the waves, a fish out of water. A descendant of oceanic immigrants returned to her ancestral homeland, she is a stranger, cut off from her distant past.

How much better suited to watery environs, how much more at home, are the fish. The opaque surface hides the water below, but mackerel are known to dwell in this place. Possibly a school swims beneath the boat, wending its way, like us, into deeper waters. When autumn finally fades into winter, mackerel will rest at the bottom, but now, while food is still plentiful, they may surface to feed, leaving a trail of bioluminescent plankton in their wake. The cold, so dangerous to us, barely threatens them; their body temperature runs with the sea.

Their world is wide open, like the sky, empty of obstacle, devoid of architecture. Land's rocky hills and winding valleys, thick forests and grassy prairies shelter their inhabitants, but the open sea offers only darkness as refuge. Swimmers, unable to hide from predators, must outpace or outmaneuver them. Their streamlined bodies minimize resistance and drag, and their tails are elegantly designed for quick turns, sudden stops, or swift propulsion. Look at a mackerel—its pointed nose, smooth velvety scales, long slender body tapering to a powerful forked tail—and see mobility and speed. Submarines are clumsy by comparison. The simple flippers of a scuba diver may still represent man's best imitation of a fish's agility.

The sea tests our ability to breathe. We gasp and splutter for air, while gills enable fish to draw the gift of oxygen from liquid. A fish opens its mouth and oxygen-laden seawater rushes over the arch of flat, layered filaments, each supporting many tiny threads. It diffuses across these thin membranes into oxygen-depleted blood. Gill filaments, squeezed into a tiny space, supply ample surface for the transfusion. Mackerel gills laid flat cover an area ten times the size of the whole fish. With such an efficient design, fish absorb as much as 85 percent of the sea's dissolved oxygen; we use less than a quarter of the air we breathe.

Unable to breathe the sea's oxygen, we bring our own, packaged in cylinders and compressed to match the sea's pressure. The ocean is heavy. The atmosphere sits lightly at the surface, a gentle, unnoticeable pressure of 14.7 pounds per square inch (1 atmosphere). Thirty-three feet (ten meters) down the pressure doubles, then doubles again and again, reaching over 1,000 pounds per square inch (70 atmospheres) half a mile down. Human lungs collapse under the weight, compelling divers to breathe compressed air. Delicate mixes, carefully balanced combinations of oxygen and helium, enable industrial divers in the North Sea to work hundreds of feet down at the base of oil rigs. Other divers defy the pressures of the sea, wearing metal suits built to maintain surface pressure in deep water, eliminating the need for compressed air and lengthy decompression. This kind of equipment has not yet come into general

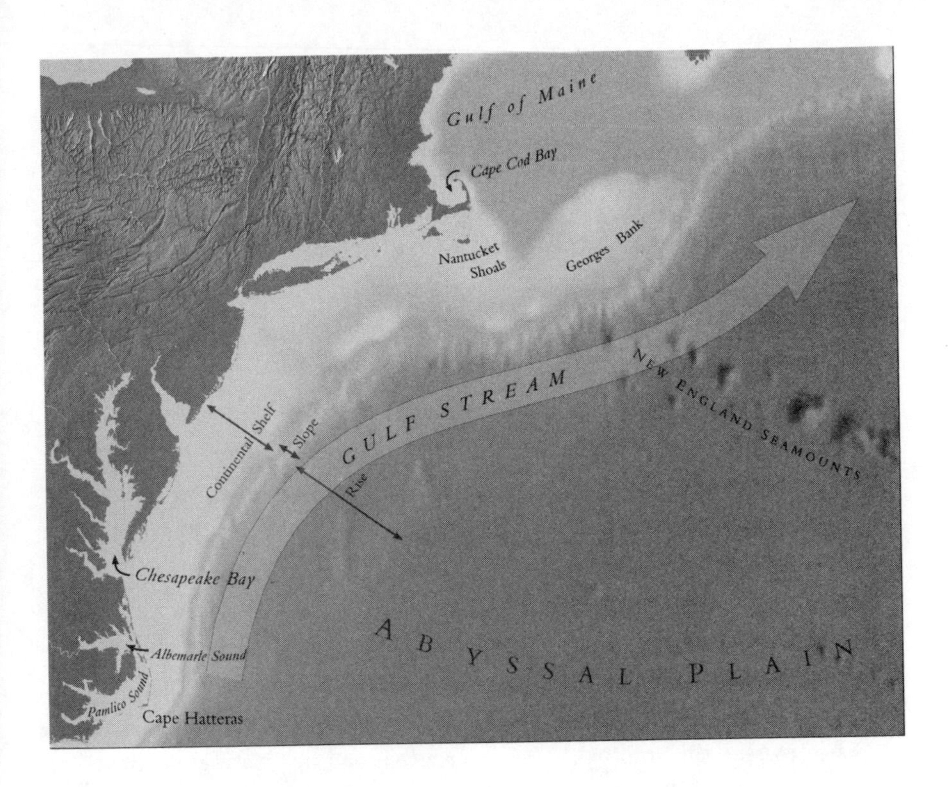

The Gulf of Maine and the northwestern continental slope and rise

use, leaving most divers to breathe nitrogen, with all its risks and limitations—the intoxicating but incapacitating euphoria experienced at depths below 150 or 200 feet (45 or 60 meters), or the painful and sometimes paralyzing "bends," induced by nitrogen bubbling into the blood during a too-rushed ascent. Earth's atmosphere transports poorly into the sea. Until science can replicate the gill, man's descent to the sea's depths will be arduous, and brief.[1]

We struggle to gain entrance to the oceanic realm. Working the deck watch hour after hour, day after day, night after night, we constantly face our limitations, keenly aware that only our ship and sailing skills shield us from a world where we cannot survive. We seem inept compared with the exquisitely adapted life surrounding us, and irrelevant and inconsequential amidst the vastness of the sea. Land, long out of sight, has fallen away, opening a vast expanse of space and time. Time moves slowly here, and in its slack we feel its reach into the past.

Early in the passage, we cross Nantucket Shoals and the Gulf of Maine, places carved out over thousands of years by the confluence of currents and the topography of the bottom. Nantucket Shoals lies on the continental shelf, a piece of North America extending beneath the sea. Its sand and rock, though deep in the water now, belong to land, not sea. Eighteen thousand years ago, when ice covered much of Europe and North America, this shelf was a coastal plain reaching eighty miles to the east. The continents edge into the ocean all around Atlantic, widening as much as 250 miles (400 kilometers) off Europe and North America and disappearing nearer the equator, accounting for 28 percent of the seafloor.

I can discern nothing of the seafloor from the deck, but a depth recorder in the science lab bounces sound waves off the bottom and plots topography. The line is straight, the bottom flat. Research vessels crisscrossing the shelf with sonar have mapped in great detail these relicts of land still prominent beneath the waves. Ancient stream beds cut the French shelf in the Bay of Biscay and off the coast of Brittany; river valleys, deltas, and submerged barrier islands lie hidden in the waters off northwest Africa. There seawater flows over stilled dunes. Once shaped and reshaped by wind rushing over sand, they were cemented into stone when beaches drowned.

In northern Europe, New England, and Canada, glaciers left their mark on continental shelves. Just north of Nantucket Shoals, the seafloor is hilly and coarse where, as the glaciers receded, melting ice left behind sand and gravel, and rocks and boulders the size of houses. Rivers cut through heaps of glacial rubble, sculpting deep valleys and high banks. Newfoundland's Grand Banks and Stellwagen and Georges Banks off New England, cut by glacial rivers, would become ideal habitats for fish when the water returned. The shoals are broad: Dogger Bank in the North Sea is roughly as large as Holland. Massachusetts, Connecticut, Rhode Island, and New Hampshire fit comfortably on Georges. When

New England slept under a sheet of ice, Georges Bank was a cape extending into the sea, and Nantucket and Martha's Vineyard were prominent hills rising from the debris.

As the ice retreated and melted, the sea rose and flooded Georges Bank. By 11,500 years ago, incoming waters had turned it into an island. The air warmed. Across Atlantic, willows and birches flourished on Dogger Bank, in forests inhabited by deer, bear, and people. On Georges, juniper, pine, and oak grew in succession, leaving buried peat, twigs, and pollen to record the climate's turn from tundra to temperate. Mastodons, moose, tapir, and giant sloth once roamed this land now concealed beneath the sea. Their teeth and bones sank into the peat only to reappear, thousands of years later, in the nets of fishing trawlers.

Glaciers chiseled the exterior of Georges, but its core is much older. Beneath the surface sand and gravel lie layers of rock ten miles deep, set down when Atlantic first opened, when the great continent melding Europe to the Americas and Africa cracked and rivers washed sediment down into the newly forming sea. Remnants of an ancient reef underpin Georges, built 160 million years ago when the sea was young and shallow and its shore faced the warm equator.

Six thousand years ago, water crested Georges Bank. The island remains submerged, but barely, giving rise to tales heard in New England fishing ports of fishermen out on Georges playing softball at low tide. The shallows are treacherous. Over the bank, amidst the ripping tide, the cold Labrador Current meets the warm Gulf Stream. Swirling waves stir the bottom, bringing up nutrients to sustain some of earth's most fertile waters. Georges and Stellwagen and Jeffreys Ledge sit in the Gulf of Maine, in waters once rich in fish. The *Cramer,* crossing Nantucket Shoals, passes through.[2]

Strong winds and high seas mix the water, bringing on the season's last bloom, a final burst of life before summer light fades. Copepods graze this abundance, then drop into deeper waters to wait out the winter, living off stored food reserves. A tow carried out by the science watch yields a generous cup of gelatinous green mush, the microscopic plants and animals gleaned from Atlantic's fecund meadows. Later, when the *Cramer* reaches the clear and less productive waters of the Sargasso, the same tow surrenders only a few teaspoons of life.

It takes a lot of greenery to feed a fish. In Atlantic's rich waters, between 24 and 35 percent of the floating meadows go to nourish fish. Approximately five thousand pounds of plants feed five hundred pounds of copepods. Five hundred pounds of copepods feed fifty pounds of mackerel; fifty pounds of mackerel feed five pounds of cod. At each level, only about 10 percent of the energy supplied turns into flesh. Animals use the other 90 percent to stay alive: to breathe, to swim, to eat, to reproduce. The green sea reflects the food web's interior layer. Large fish—cod, salmon, or striped bass—represent the outer layers, the last in a line of feeders.

Every so often, it is possible to view the ascending layers simultaneously. One of my neighbors, a young man who spends every possible moment in the nearby creeks and bay, saw this complexity one hot July afternoon a few years ago, on his birthday, when he went fishing with his father. The ocean's grasses were blooming that day; the sea was murky. Though myriads of scarlet-flecked copepods had been born and had grown to nourish the young mackerel schooling at the surface, the fishing was singularly uneventful. After many fruitless hours, father and son finally gave up and turned their boat home. Just outside the breakwater, at the entrance to the harbor, they came into a boil, a chop of bloodthirsty bluefish in the midst of a feeding frenzy. The young man grabbed a small lure, a mackerel, and cast his line one more time. He hooked a live mackerel, and then a bluefish swallowed the mackerel. Within a few minutes he had reeled in his birthday dinner. Mackerel to mackerel to bluefish: the only link not immediately visible in that moment was the copepod, but they were there, packed in the stomachs of the mackerel.[3]

The *Cramer* sails through a sea drenched in green, but in this sea, which once nourished some of the world's most productive fisheries, not a fish is in sight. Waters that should be swarming are empty. Huge populations of haddock, cod, flounder, halibut, and redfish all once flourished on Georges Bank and in the Gulf of Maine. Now they are gone, vanquished. It's lonely out here, in this sea of ghosts, and disconcerting. Sailing a small boat on a wide sea evokes deep and powerful feelings of insignificance, but these feelings, which permeate so completely, impress so profoundly, are an illusion, a conceit dangerously dissonant with reality. We who are so ill at ease on the sea have decimated entire fisheries, one after another. Now many fisheries are vestiges, shadows of their former selves.

Cod was the last to go. No one ever thought that anyone or anything could imperil cod, this common fish with a jutting upper lip and square tail. When John Cabot sailed into the waters of the New World to claim North America for England, cod were so dense that sailors scooped them from the sea in buckets. Old textbooks boast that behind every rock in the Gulf of Maine waits a cod. Always, there was cod. In Maine's tiny island fishing villages, it was a staple; baked with bread crumbs, broiled with a little butter, cooked in a rich milky chowder, or, when inspiration failed or there were no fresh fish, salted. Its sheer abundance gave Cape Cod its name. Today a replica of a cod still hangs in the Massachusetts State Legislature, symbolizing what was the mainstay of the New England fishing economy for three hundred years.

Cod survived and thrived in the cold coastal waters of the continental shelf, where food is varied and plentiful. They eat voraciously, without discrimination, feeding on sea worms, algae, herring, mackerel, capelin, even their own young. They prey upon squid, crabs, shrimp, sea clams, mussels, and cockles, ingesting them whole, including the shells. They have even sampled wild duck, old boots, and jewelry. Their numbers have dwindled: traits these fish developed over hundreds and thousands of years to ensure their survival, to create a home in the sea, have been consciously and deliberately exploited by man. The watery versatility of fish, so effective against marine enemies, fails when man is predator.

In the sea, where only darkness shelters, coloring protects. To a shark or dogfish gazing down in the water, the gray, black, or dark green upper bodies of cod fade indistinguishably into the green sea. To an enemy hungrily peering up from below, the white bellies of cod, tinged with the brown of the bottom, blend with light streaming down from the surface. Today's fisherman, however, hunts not by sight but by sound. Sonar cuts through the shield of color, seeking substance, rendering the delicate, protective hues of a cod useless.

In November, cod gather on Nantucket Shoals to spawn. Female cod are fecund, each producing three or four million eggs each year. Most die, but the surviving larvae drift in the surface waters, nourished during their first days by tiny bits of yolk from the eggs, and then, when the yolk is depleted, by a rich sea stew of tiny floating plants and animals. After two or three months, when young cod have grown to one or two inches in length and can swim, they descend to the seafloor. When they reach maturity, in two or three years, they spawn together in large schools, a trait that results in a greater proportion of fertilized eggs, and a larger number of progeny.

I imagine schooling cod might be as beautiful as little fish gathering at the edges of a beach. There, in small schools of fifty or one hundred fish, all roughly the same size and age, they sweep through the water in unison, like a flock of swallows in flight. They turn together abruptly, simultaneously, gracefully. Many fish make up a school, but no one individual stands out. The movement of each fish is synchronized, finely tuned to the others. There are many fish, but only one motion. Approach the school and it dissipates, instantly, exploding as fish disperse in every direction, or splitting as fish cascade around the intrusion like a fountain. All this occurs in less than one second, with never a crash or collision. What human endeavor demonstrates a comparable grace? Bomber pilots move as precisely, but the nature of their mission diminishes the beauty of their flight. Perhaps the ballet compares, yet the fluidity of the dancers is practiced, the choreography orchestrated beforehand, while the marine school dances more spontaneously, cued by messages coming through the water.

Schooling fish swim to music we cannot hear, using their vision and the lateral line, a series of gelatinous canals running the length of each fish, to sense the rhythms of the sea. On a cod, the lateral line is pale white, running from head to tail. Pores in the line pick up vibrations in the water column. As fish are alerted to changes in water pressure and flow, the motion of the school adjusts accordingly.

Vision and sensitive lateral lines keep schooling fish together, providing protection in the open sea. Though at the top of the marine food web, cod are not without their predators. Shark and dogfish hunger for cod, and older, larger cod indiscriminately feed on younger ones. The large numbers in a school conceal individuals; a lone fish is more at risk than one hidden among many. The school's large size confuses, even frightens, predators, and the chaos of a dispersing school distracts.

Schooling fish, ably surviving the dangers of the open sea, meet death at the hands of fishermen. Cod spawning off the coast of Newfoundland migrate in dense schools along a "highway" of relatively warm 35 degree Fahrenheit (2 degree Celsius) water. Each school contains millions of fish, packed into several square miles. The cod move inshore in the spring and swim straight into the nets of waiting draggers. Modern fishing trawlers, equipped with depth and temperature sensors, locate ribbons of warm water frequented by migrating fish. Sonar identifies and targets schooling fish, and computerized satellite navigation transforms a blank sea into a detailed grid of latitude and longitude, enabling fishermen to return to the same spots again and again. Fish don't stand a chance. Years ago, fishing was a more uncertain venture. When Gloucester's most talented sea captains steamed past the Ten Pound Island Lighthouse, the paint factory, and the breakwater, heading out to Georges or other destinations in the Gulf of Maine, they traded those fixed landmarks for the subtle signs of wind and wave. Luck, intuition, and seasons of experience guided them to fishing grounds unmarked and unsigned. A captain lowered his net not knowing what the waters held; he hoped and prayed his instincts were correct.[4]

Those days are long gone. Modern technology has transformed commercial fishing from an art into a precise science, becoming our eyes and ears into the opaque sea. With its advent, the contest between man and fish has ended. When Ahab went after Moby Dick, he could not avail himself of spotter planes, cellular phones, mile-long drift nets, satellite navigation, and sonar. If he had, Melville would have had no story to tell. Today's fishermen are engaged not in a harvest, or even a hunt, but in a slaughter.

The ocean is not without violence. Her dwellers do not coexist peacefully, harmoniously. When bluefish are running, the sea bleeds and carnage abounds. From time to time, a hot summer day finds thousands of dead fish lining the outer beaches of Cape Cod, their metallic bodies glistening in the sun, the death row extending for miles. It is not a pretty sight. Hungry bluefish chasing schools of silversides into the shallows provoked this debacle, reddening the water as they chomped into their meal. The silversides, caught between bloodthirsty predators and arid beach, leapt to their death on the sand. This time, on this beach, the silversides were outmaneuvered, but at other times in other waters, they thrive and bluefish find their meals elsewhere. When the contest is between man and fish, man almost always emerges victorious.

Given the enormity of the ocean realm, a relatively small number of large animals dwell there. The swimmers, at the top of the marine food web, constitute only one tenth of 1 percent (by weight) of all the sea's inhabitants. By contrast, the top of the terrestrial food web totters dangerously. Six billion people crowd onto earth: no other carnivorous population comes close to equaling our weight and mass. Dominance exacts a high price. To support such large numbers, humans devour the planet's natural resources. Attempting to satisfy seemingly insatiable needs and desires, we assume the sea's infinite bounty. In relentless pursuit of food and profit, we have tested the sea's limits. Most of Atlantic's fisheries are declining, with the number of fish too low to allow populations to sustain and replenish themselves.

The sea is exhausted. By the time the Canadian government closed the Grand Banks cod fishery in 1992, vast schools of fish, once containing hundreds of millions of cod, had disappeared. They were fished out, reduced to a meager population one hundredth of its former size. The number of spawners had dropped 97 percent. I saw remnants of this once great fishery in a summer seasonal fishing village on Newfoundland's bleak, rocky coast. The freshly salted fish, drying on nets in the sun, were small, barely recognizable. Not one was ten inches (twenty-five centimeters) long. Few were old enough to spawn. Large catches of immature fish sounded the death knell for a fishery that had supported the Newfoundland economy and represented a way of life for islanders for generations. Eight years after the closure, thousands of Canadian fishermen are still unemployed and the Canadian government has spent $3 billion supporting fishermen and processors waiting for cod stocks to recover. They haven't. The Grand Banks are still closed, and recent stock assessments show no signs of growth.[5]

Canadian waters are not unique. The Grand Banks fishery collapsed; others are on the verge. Sixty-five million people live near the edges of the North Sea, and fishermen from at least seven European nations compete there for the haddock, cod, and sole whose abundance will soon be no more than a memory. In the North Sea, forty out of sixty stocks are fished at unsustainable levels. Each year, fishermen remove 60 percent of the cod, depleting the spawning population. The health of fish populations depends on how many fish are spawned each year and how many survive their first year. The year 1996 could have been a good one for cod. High numbers reached their first birthday, but so many were caught and discarded that few lived to spawn. In 1997 and 1998, the number of one-year-olds was the lowest in thirty years. Today, only 4 percent of one-year-old North Sea cod live to reproduce. If catches are not reduced by between 40 and 60 percent, the North Sea cod fishery will collapse.[6]

In New England, once great fisheries have dwindled away. Children growing up in Gloucester hear the story of Howard Blackburn, a stalwart halibut fisherman who, after becoming separated from his schooner in a storm, rowed his dory for five days before reaching shore. Gloucester children know the man but not the fish. Halibut, once the largest fish to dwell on the seafloor off New England, were fished out 150 years ago. Redfish, too, are described in the past tense. Fifty years ago, they were found everywhere New England fishermen might lower a net. Now they are gone. More recently, populations of cod, haddock, and yellowtail flounder have plummeted to record lows. In the 1990s, fishermen took more than half the cod in the Gulf of Maine each year, bringing the fishery to the edge of collapse. In 1993 and 1994, there were fewer cod in the Gulf of Maine than ever before, but for the next four years fishermen exceeded the target catches by 100 percent.

After more than twenty years of management, the government still considers more than 70 percent of New England's groundfish, its high-value cod, haddock, and flounder, overfished. The law requires depleted stocks to be rebuilt within the next few years, but rules adopted in New England stand less than a 14 percent chance of restoring Georges Bank cod, less than a 17 percent chance of restoring Georges Bank haddock, and less than a 1 percent chance of restoring Gulf of Maine cod. At this rate, recovery is a long way off. A recent study of collapsed fisheries found that after fifteen years, most cod and haddock populations were still in a state of collapse.[7]

Year after year, fisheries management, on both sides of Atlantic, follows the same tired script. Time after time, science advisers recommend catch levels required to rebuild and sustain fish stocks, and time after time, fisheries managers, undercut by politicians and fishermen, dilute the recommendations. At the negotiations, which are often heated and always contentious, the fish are absent, their voices silent. Americans romanticize the idea of fiercely independent, rugged fishermen, testing their mettle before stormy seas, and ignore the uncharismatic cod. Public outcry on behalf of a fish no one is likely to recognize unless it is filleted and cooked has been minimal. Without a strong mandate, regulators cannot do their jobs.

Fisheries science is not exact. It cannot predict with complete certainty what level of fishing will rebuild a stock, sustain it, or drive it to collapse. The size of a fish population, the number of fish born, and the number still alive at age one fluctuates from year to year, often dramatically. Scientists cannot know exactly how many fish are removed each year, how many fish are landed, and how many are thrown away. They cannot yet factor in fully the effects of changes in water temperature and currents that occur over the years as the sea shifts under the wind. The fluid medium of the sea is complex; it may be that the watery world of fish can never be understood in its entirety. And yet regulations governing catch levels, mesh sizes, and allowed fishing days leave little room for error, little margin of safety. Regulators who reject the recommendations of their science advisers empty the sea, shrinking vast populations of fish down to the edge of commercial extinction.

European fisheries are now governed by precautionary principles calling for scientists to include a margin of error in their calculations. However, European Union fisheries ministers have rejected a number of these recommendations as "excessively cautious." When cod spawn for the first time, 13 percent of the eggs hatch. When they spawn for a second time, more than half the eggs hatch. By the third time, almost 100 percent of the eggs hatch. The bigger the mother, and the more experience she has spawning, the greater the proportion of eggs that hatch and larvae that survive. If scientists were asked to protect fish rather than maximize catches, to restore populations to their historic levels of abundance, they might recommend that fish be given the chance to spawn at least once, maybe twice, before they are killed. As wild fish head toward the fate of the buffalo and the passenger pigeon, this doesn't seem excessively cautious.[8]

We are still moved by the myth of inexhaustibility. As the great North Atlantic fisheries disappear, fishermen cast their nets in more distant waters. In Europe, 25 percent of the catch comes from abroad. Developing nations, eager to write off their debt and infuse their economies with foreign currency, lease fishing rights in their territorial waters. Distant water fleets plunder foreign waters with the same reckless abandon they use at home, repeating the same mistake of applying too much effort to a limited resource. The windfall is short-lived. The Argentinian hake fishery, depleted until it became commercially extinct, is now closed indefinitely. Spain is quickly fishing out the rich sardine fishery of Morocco. The Canary Current courses by northwest Africa, creating a fertile sea, one of the most productive in the world. Fish is Mauritania's and Senegal's most valuable export: 80 percent of the catch netted in their waters is taken by foreign fleets and sold in international markets at prices too high for locals to afford. In countries where fish is a primary source of animal protein, this practice is likely to result in food shortages in the years to come. Unable to monitor their fisheries, these countries lose both income and fish. Underreporting of the catch, illegal fishing, and failure to land the catch in local ports result in substantial losses of revenue, up to 50 percent. At the same time, catch levels are so high that fish populations are now declining.[9]

According to statistics kept by the United Nations Food and Agricultural Organization, most Atlantic fisheries are worked at or beyond their levels of sustainability. The litany of fisheries disappearing from Atlantic is long and growing. Cod, haddock, plaice, sole, and redfish off the coast of Europe have collapsed or are on their way to collapse. Georges and Grand Banks cod, haddock, red hake, and flounder are fished out. Off the coast of western Africa, octopus, cuttlefish, squid, mackerel, and hake are disappearing. As one by one the fisheries collapse, the circle begins to close, delineating the edges and limits of Atlantic's fecundity.

These are desperate times. Fishing captains flaunt international law, sailing under "flags of convenience," registering their boats in countries unable or uninterested in enforcing the law. Spanish boats sail under the flags of Belize, and captains from Liberia, Panama, and Iceland register their boats in Cyprus. When fishermen violate regulations or exceed quotas, it's difficult to assign blame or seek remedy. Out on the high seas in international waters, disputes are even more difficult to resolve. On the Reykjanes Ridge, south of Iceland, Russian, Japanese, and Icelandic trawlers, some of which are sailing under flags of convenience, squabble over dwindling stocks of redfish. Farther south, at the edges of the continental shelf off the tip of Antarctica, boats registered in Argentina, Chile,

Belize, and Korea, tempted by lucrative markets, plunder the slow-growing, long-lived tooth fish, taking twice the allotted quota. The Nose and Tail of Newfoundland's Grand Banks jut out beyond the two-hundred-mile limit. Although Canada closed the Grand Banks fishery, the Nose and Tail lie beyond its jurisdiction. Fish know no political boundaries. European trawlers wait at the edge of the Grand Banks, showing little concern for Canada's belated efforts to rebuild its fishery. In desperation, the Canadians seized one Spanish boat. Inspection of the vessel produced an illegal small-mesh net, two sets of logbooks, and $2 million worth of juvenile halibut.[10]

The pillage of Atlantic extends through her coastal waters and into her depths, reaching species that until the last forty years have by and large lived undisturbed. Repeating the same mistakes over and over again, blind to the wreckage before us, we move on, searching the sea for additional wealth to exploit, turning to so-called underutilized species, fish whose appearance or taste has not traditionally appealed. The monkfish is mostly mouth, with a tail. American fishermen first found markets for the tails, and then the belly flaps, and then the livers. Catches soared as liver prices quadrupled. Today, scientific surveys of the population reveal sharp declines in abundance and decreases in fish size. As this short-lived fishery ends, fishermen and regulators argue over how much catch levels need to fall. Spiny dogfish experienced the same fate. Once considered a trash fish, good only for discard, creative marketing produced buyers. Dogfish grow slowly, reaching maturity when they are twelve years old. They bear live young, in small numbers. It only took a few years to wipe out this fishery, and the recovery plan, like so many others, has been weakened by political wrangling.

Now fishermen are moving on again, this time to deeper waters, on ventures financed by government grants. Combing the deep canyons and continental slope off New England, the pickings so far have been slim. Perhaps that is fortunate. Fifteen years ago, Canadian fishermen wiped out the roundnose grenadier, a fish dwelling in deep waters between Baffin Island and the Grand Banks. The stock has yet to recover. In

Europe, deep-sea fisheries are managed inconsistently. While European governments give their fleets financial incentives to fish for orange roughy, blue ling, and tusk living in deep water, their science advisers recommend cutting catches of these fish by 30 percent.[11]

Current fishing practices penetrate deep within oceanic food webs. Fishermen, having skimmed off the top predators at the web's outer layers, now take from the interior, reweaving the web in ways that hinder the recovery of hurting stocks. In the North Sea, European fishermen fished out cod, haddock, plaice, and saithe and then targeted their prey, the smaller mackerel, herring, capelin, pout, sand eels, and sprat. The introduction of another major predator reverberates throughout the food web. Pout are an important layer in oceanic food webs: they eat and are eaten. Taking large quantities of pout deprives cod, haddock, and saithe of food. Pout prey on krill, who prey on copepods, who are eaten by cod and haddock. As pout declines, krill increases and copepods decline, further depriving larger fish of food. Linkages among dwellers of the sea are complex, and often unrevealed, but in the outer layers of the food web, the effects are noticeable. When overfishing collapsed herring and capelin stocks in the Barents Sea, cod failed to thrive. If fishing further and further toward the base of marine food webs continues at the present rate, it won't be that long before sand eel, sprat, and pout fisheries also collapse, leaving a sea of plankton.

Oblivious to the complex, intricate relationships among animals living in the sea, we disrupt delicate balances of predator and prey, all for fish meal and fish oil. Fishermen compete at the top of the food web with fish and seabirds. The latter seek sustenance; the former are selling to pig, poultry, and aquaculture farms, and to companies making margarine and cookies. The generous sea cannot accommodate this demand. Puffins and guillemots living in Norway's Lofoten Islands failed to breed when humans took out the herring. When capelin stocks collapsed, hungry seals and dead seabirds crowded Norway's coast. When the sand eel fishery off the Shetland Islands crashed, thousands of terns, guillemots, kittiwakes, and fulmars starved. Migrating seabirds rely on these fatty, high-energy fish for strength and nourishment during long hours of flight.

Breeding birds rely on the fish to feed their young. Puffins and kittiwakes live on Scotland's Isle of May, east of the Firth of Forth. While the Isle of May itself is a protected national nature reserve, the health of young birds depends on a supply of sand eels targeted by fishermen in waters just offshore on a fishing ground known as the Wee Bankie. Kittiwakes are already experiencing breeding failures; the population is dropping, and two pairs of birds are no longer able to raise even one young chick between them. European fisheries ministers now plan to close the Wee Bankie to sand eel fishing during the kittiwake breeding season. Catch levels had already dropped below quota when the decision was made, indicating that sand eel populations may already be crashing. Hopefully, the seasonal closure will save both birds and fish.

The intricate patterns of marine food webs are less than fully understood. Humpback whales eat capelin and herring; white-beaked dolphins, minke whales, and gray seals eat sand eels. If seabirds starve when sand eels are depleted, whales and seals may suffer also. The Canadian government, after impatiently waiting four years for Grand Banks cod stocks to return, subsidized a seal hunt, justifying the slaughter on the grounds that hungry seals hurt the recovery of cod. Scientists piecing together the workings of marine ecosystems disagree. While seals eat cod, they also consume illex squid—another, and perhaps more important, predator of cod. Overfishing has already unbalanced the marine ecosystem; additional human intrusion may only upset it further.[12]

Fishing is a wasteful business. While mixed species of fish swim together—herring with sprat, whiting and haddock with pout—the gear is nonselective. Fishermen targeting whiting often take young haddock in their nets. Fish that are undersized, unmarketable, or in excess of target catches are tossed overboard, where, once again, their marine adaptations fail them. Most fish live suspended in the sea, a world virtually boundless in every direction, while we humans walk a thin line at the meeting place between land and sky. Rooted to the ground, our bones and skeleton keep us upright on this narrow plane. To control buoyancy in water, where there is no edge, we attach cables, tethers, battery-operated thrusters and lead weights to diving suits and submersibles. A fish carries its own natural flotation device, a small pouch of oxygen located just below its backbone. As air in the swim bladder compresses under the weight of the sea, it collapses. As a fish swims deeper, its swim bladder shrinks. If a fish descends sixty-six feet (twenty meters), its swim bladder contracts by one third. To compensate, additional oxygen diffuses into the pouch from a *rete mirabile,* or "wonderful net," a dense web of minute blood vessels that absorb and concentrate oxygen from the fish's bloodstream, then route it to the swim bladder. Cod are buoyed by their swim bladders. When they rise to the surface to spawn or chase schools of herring or mackerel, the pressure decreases and air in the swim bladder expands, then exits, to be slowly reabsorbed into the bloodstream, a process that can take days. Just as it can be lethal for divers to ascend too quickly, so it can be for fish. Cod and pollock dragged up from the bottom in fishing nets arrive on deck dead or dying, their internal organs damaged by the too rapid ascent.

Every dead fish, whether landed or discarded, is a loss to the spawning population. In 1991, otter trawls on Georges Bank threw back 43 percent of their catch. On the other side of the Atlantic, dab and whiting fishermen throw away more fish than they bring to port. Half the North Sea whiting catch consists of immature cod and haddock. In 1999, when New England fishermen had fished almost all the cod targeted for that season, managers lowered the landing limit to thirty pounds (fourteen kilograms) per trip, with the result that, in a fishery on the edge of collapse, fishermen threw thousands of pounds of cod overboard.

Over the years, the lack of selective fishing gear and the tremendous size of the bycatch has, without anyone noticing, brought one large fish to the edge of extinction. Barn-door skates—wide, flat, and thin—undulate through the water as if on wings. Bottom dwellers, they once grew to a length of five feet. They are slow-growing, reaching maturity when they are ten or fifteen years old, and they produce only a few young each year. The eggs incubate for almost two years in leathery cases lying on the seafloor. Trawls and dredges easily crush both egg case and baby fish. Fifty years ago, 600,000 skates skimmed over the bottom of St. Pierre Bank off the coast of Newfoundland. By the 1960s, the population had dropped to 200,000, and then to 500 by the 1970s. Then they disappeared. In the last twenty years, not one has surfaced in a trawl net, although a small population still lives farther south, on Browns and Georges Banks. The barn-door skate, once common throughout the Gulf of Maine, is now a rarity. Fishermen, uninterested in fish with no commercial value, never noticed its disappearance. Nobody deliberately fished out the barn-door skate. It was a casualty of other efforts, other aims.

The losses mount up so slowly, incrementally, it's hard to notice, let alone appreciate, their meaning. It used to be that every time my family and I walked the beach, we'd find the tideline littered with the empty egg cases of skates. Deep, dark green, almost black in color, the rectangular pouches curled into hooks at the corners. "Mermaid's purses," they were called, and they looked and felt tough enough to protect the egg within from anything. Every once in a while the fish itself would wash ashore, not the barn-door but the smaller, common skate. Both egg case and fish, with its blunt nose, close eyes, and rippling flat body, spoke of the strange and unfamiliar world of the sea. They were so abundant, though, we took their presence for granted, overlooking them as we watched for other, uncommon treasures washed in with the tide. And now they are gone. On our last trip to the beach, we walked the entire wrack line and found not a one.[13]

Fishermen combing the sea leave no stone unturned. Their gear cuts through the seabed, overturning stones and boulders, uprooting communities of animals, disturbing the bottom more than currents, more than underwater storms. Boats drag heavy trawling gear and dredges across the bottom of the North Sea, in some places as often as four hundred times in a year. Nantucket Shoals is more heavily trafficked than the North Sea. The *Cramer*'s depth finder doesn't reveal every detail on the bottom, but if it did, we might see the tracks of trawlers, their imprint held in the sand and mud for weeks, sometimes months. Snow crabs, basket stars, sea urchins, sand dollars, brittle stars, and soft coral all live on the bottom. Their numbers decline significantly when they are dislodged from their homes and damaged by fishing gear. Losses are high, whether the bottom is mud, sand, or gravel. Fishermen describe smoothed-out hills, scoured beds of mussels and coral, hardened burrows once full of brittle stars and mussels and sea squirts, now smashed and empty. Where cobbles and boulders were once overgrown with organisms, now they are bare. Scientists have likened the effect to clear-cutting.

Damage caused by fishing gear forces the pace of life on the seafloor to accelerate. Tube-dwelling anemones living in the muddy bottom reproduce infrequently over their fifty-year life spans. Sea pens and brittle stars live for decades if left undisturbed, and over time amphipods turn their tube dwellings into condominiums. Each time the trawls and dredges come through, the community must begin building anew. There isn't time to progress very far. Only the shorter-lived, more quickly reproducing species can establish themselves between the passes of fishing

gear. In any particular place, fishermen pick up tube-dwelling anemones in their trawls only once.

Flattening the seafloor eliminates nurseries for cod, taking away their food and shelter. Cod larvae settle all over Georges Bank, but the survivors are those that land on the gravelly, rocky three-dimensional bottom, where they can hide. Where the bottom is undisturbed, bushy plantlike animals grow over the rock and communities of animals move in, providing food and shelter for growing fish. When the bottom is dredged or trawled, the animals are crushed, or buried, or exposed to predators. They disappear, and scavenging hermit crabs move in.

A thin gravel pavement, roughly 1,000 square miles (2,600 square kilometers), lines the northern edges of Georges Bank. It is critical habitat for three species. Juvenile cod and haddock find sanctuary and nourishment in the gravel, sea scallops unable to survive burial in shifting and swirling sands live there, and herring spawn in the stones where tidal currents are strong. When scallop dredges and groundfish trawls regularly gouged the gravel pavement, its communities declined in abundance and diversity. When the Georges Bank cod fishery collapsed in 1994, regulators closed a portion of the bank. Within months communities began to recover, as sponges, bryozoans, crabs, sea urchins, and scallops reestablished residence.[14]

The law has not been strong enough to protect and restore Atlantic's ailing fisheries, has not been strong enough to preserve a valuable resource for future generations. In discussions over catch levels, mesh sizes, and effort levels, the economic interests of fishermen frequently prevail, resulting in a situation where, by the end of the 1980s, New England fishermen would have had to kill 50 percent fewer fish just to maintain fish stocks at their diminished levels. As fisheries crash, the industry is left with an oversized fleet. Europe's is too large by 40 percent. Although fisheries are collapsing, governments on both sides of Atlantic continue to support the industry with major subsidies in the form of low-cost loans, loan guarantees, gear improvements, tax credits for fuel, vessel buybacks, and grants to exploit new fisheries. In Europe, this subsidy is approximately 14,000 euros ($12,600) per boat per year. Since 1994, the fishing industry in New England has received $50 million in subsidies.

Some scientists now suggest that establishing marine protected areas could help restore depleted fisheries and maintain the richness of life in the sea. Barely one third of 1 percent of United States territorial waters are designated marine sanctuaries. Few sanctuaries exist, and those that do hardly provide refuge: despite the designation, fishing is prohibited in only one, a tiny section of the Florida Keys National Marine Sanctuary. To protect seabed communities and spawning and nursery areas from damage by fishing gear, to restore depleted fisheries, and to ensure against the risks and errors of fisheries-management practices, scientists recommend that 29 percent of New England's seabed be designated as a marine protected area. The designation includes traditional fishing grounds, nurseries, and spawning areas in the Great South Channel, Georges and Stellwagen Banks, and Jeffreys Ledge.

Sanctuaries can help restore depleted populations, but only if they are large enough. To be effective, sanctuaries may need to encompass as much as 50 percent of the remaining animals. In addition, regulators must restrict fishing outside the reserves, to prevent fishermen from merely redirecting their effort there, and they must have the means to prohibit poaching within the reserves. The only hope for the barn-door skate may be for fisheries managers to list it as an endangered species, to declare it overfished, and to promulgate a rebuilding plan that would prohibit trawling and scallop dredging in areas where barn-doors still live, and where they once thrived. In all likelihood, fishermen would vigorously resist such a plan.[15]

In November 2000, the United States National Marine Fisheries Service and Fish and Wildlife Service listed Maine's wild salmon as an endangered species in need of protection. The agencies had considered the listing five years earlier, but salmon farmers, logging companies, blueberry growers, and the politicians who support them objected. The government agencies withdrew their proposal, settling for a protection plan developed by the state of Maine. It didn't work: wild salmon populations continued their precipitous decline. When, in 1999, federal agencies again stepped in, proposing to add Maine's North Atlantic salmon population to the endangered species list, Maine's governor and one of its senators objected. When salmon were officially placed on the list, they objected again, but in the summer of 2000, only 127 fish were estimated to spawn in eight of Maine's rivers. Time is running out for Maine's wild salmon. Decimated by overfishing and damage to their habitat, they are now in danger of being edged out by farmed fish. The struggle over Maine's salmon, once focused on issues of habitat destruction and pollution, now also turns on whether, as we cultivate more and more of the sea, we care to leave room for its wild inhabitants.

New England's coastal waters once teemed with salmon, a fish exquisitely adapted to make the perilous journey between salt water, where it lives, and fresh water, where it spawns. The dual residency demands accommodation, and salmon adjust particularly well. Not all marine creatures are so blessed. Sea cucumbers, for example, tolerate only a narrow range of salinity. Outside the range, osmosis kills them: too much salt, and they lose fluid and shrivel. In fresh water, they swell to the bursting point. In the salty ocean, salmon avoid dehydration and loss of fluid to the surrounding water; they drink and drink, replenishing fluids lost through osmosis and expelling salt through their gills. In rivers where they come to spawn, salmon drink less. Fresh water still diffuses in, but their kidneys compensate, working continuously to prevent the bloating so fatal to sea cucumbers.

Year after year Atlantic salmon, following the scent of home, traveled from salt water to fresh, from open ocean to the clear-running streams of their birth. The particular blend of soil, plants, and animals gives each stream its own unique odor. Holding the memory of that smell, fish feeding hundreds of miles away, in waters off Labrador and Greenland, almost always find their way home. When white settlers arrived in eastern North America, nearly every passable river and stream north of the Hudson was clogged with salmon. It's hard to imagine that salmon were once considered a poor man's food, that indentured servants fed fresh salmon every day were considered ill treated. As the number of settlers grew, salmon became less able to complete their long journey home. The new residents built cities on the rivers, on the Housatonic, the Merrimac, the Kennebec, the Androscoggin, and one by one dammed them. Salmon, unable to jump the dams, were shut out of their spawning grounds.

Few are left. The number returning to spawn in the rivers of eastern North America once ranged between two and a half million and five million. By the 1970s, it had plummeted to 800,000. It keeps falling: 300,000 in 1991, 80,000 in 1998. In 1998, only 9,500 brood salmon (large females two or three years old) returned to New Brunswick's large salmon river, the Miramichi. Since 1987, fewer than 150 wild fish return each year to spawn in the rivers of Maine. I saw spawning salmon one summer jumping the falls on a cold stream in the woods of eastern Quebec. As the water hurtled over the rocks, the large, lean, muscular fish flipped, twisted, strained against the current. Some fought the water, pushing up under the falls. Others leapt. They landed, exhausted, in quiet pools at the top where silent men, patiently waiting on rocks at the river's edge, netted them. Though thousands of years of evolution superbly prepared salmon to travel hundreds of miles to their spawning grounds, they cannot meet these newer challenges.[16]

The risk is not only from fishermen. Upland from the spawning streams are blueberry producers and forestry companies whose activities alter water quality. A female salmon builds her nest in the gravel on the bottom of a stream bed, swishing her tail over the stones, creating a large bowl-shaped depression to hold her eggs. The survival of her progeny depends on an abundance of clean, cold, well-oxygenated water. The young are not doing well. Maine blueberry growers, whose production is expected to double by 2005, compete with salmon for water, extracting it from streams to irrigate their bushes. Growers apply more than a dozen herbicides and insecticides; some find their way into streams where salmon are spawning. With what effect, no one is certain. Endocrine-disrupting insecticides used in the forest industry do hurt young salmon. When foresters sprayed to prevent infestations of spruce budworm, fewer salmon growing in affected streams returned to spawn. As salmon ready for their time in salt water, hormones prepare their bodies to get rid of salt. The insecticide seems to block this critical adaptation, rendering exposed salmon unable to survive at sea.[17]

As wild fish disappear, people turn the sea to pasturage, farming it for food, further accelerating the decline of wild stocks. Wild Atlantic salmon may be the first to go. Salmon are farmed in Norway and the United Kingdom, Canada and the United States, and they far outnumber wild fish. In Maine alone, salmon farmers produce 30 million each year. They are raised like chickens, in close confinement, thousands of them squeezed in underwater pens the size of swimming pools. In a year when approximately 150,000 wild fish swam the rivers of Canada, 15 million farmed salmon were grown in the Bay of Fundy. A salmon farm is a big, messy operation. Huge quantities of uneaten food and waste accumulate beneath the pens. Salmon farms are major polluters in Norway; their nitrogen discharges approximate the amount of untreated sewage produced by 3.9 million people. Along the Norwegian fjords, where the sea cannot absorb this quantity of waste, water quality has noticeably declined.

It is impossible to keep wild and domesticated populations apart. Animals and waves routinely tear aquaculture pens, releasing thousands of salmon into the sea. In Norway, one million escape from aquaculture farms each year. By now, somewhere between 25 and 40 percent of salmon in the North Atlantic come from farms. The populations have already begun to mingle, and where they have, the prospects are not good for wild fish. Farmed salmon, susceptible to disease, require generous doses of antibiotics. Susceptible to parasites, they require treatment with potent pesticides. The residues accumulate in the water. Ivermectin, used in Scotland on salmon infected with sea lice, killed other marine animals who are a source of food for wild fish. Viruses and bacterial infections afflicting farmed salmon require the slaughter of thousands of farmed fish to prevent further contagion. It may only be a matter of time before fatal diseases spread into the wild population. Sea lice already have. In Norway, sea lice infestation of wild salmon is ten times higher in areas near salmon farms. In Scotland, lice originating on salmon farms has already caused the collapse of the sea trout population.

Wild fish are sleek and muscular. Fat and pudgy farmed fish are not particularly well adapted to a life of hard swimming, but some escaped juveniles do go out to sea and return to spawn, competing successfully with wild fish for food and nesting space. In one Canadian river where three hatcheries operate in a watershed, the number of wild salmon returning to spawn has dropped by 90 percent. Interbreeding between populations does not bode well for wild fish. In the rivers of Ireland, fewer progeny of farmed fish live to make the journey downstream to the sea, but those that do, replace their wild counterparts en route. Maine salmon farmers import European stock; wild hybrids will further stress North American populations.

The coup de grâce for wild salmon is still to come. Hundreds of transgenic salmon swim in the penned waters off Prince Edward Island. They grew from eggs injected with two genes, a growth hormone from Pacific salmon and a promoter gene from ocean pout. The promoter gene keeps the growth hormone going all year long, producing market-sized fish twice as quickly. Scientists analyzing the effect of this "Trojan gene" suggest that when transgenic fish escape their pens, wild salmon may disappear. Genetically modified salmon may mate more successfully than their smaller counterparts in the wild, but fewer will survive to maturity. Sixty transgenic fish, introduced into an unaltered population of 60,000, can eliminate that population in forty generations. Millions of genetically modified salmon eggs have already been sold, pending government approval. If that happens, the days of wild salmon are surely numbered.

The burgeoning salmon industry threatens not only wild salmon but other fish as well. Salmon are carnivorous, hungry fish. In the wild, they feed on herring and capelin. Farm-raised salmon also eat from the sea. Their rich diet—45 percent fish meal and 25 percent fish oil—is made from herring, sand eel, pout, and menhaden. Into every pound of salmon goes almost three pounds of these smaller fish. Already there are not enough small fish in the North Sea to support Europe's salmon farms, sending salmon farmers farther afield to import their feed from waters off South America. Almost one third of wild fish taken from the ocean is made into feed for pigs, chickens, and farmed fish. Aquaculture's share—only 10 percent in 1988, 33 percent in 1997—is rapidly rising. Ultimately, the sea cannot sustain these high levels of extraction.[18]

The future of wild fish hangs in abeyance. Since the Fisheries Conservation and Management Act banished foreign vessels from U.S. territorial waters, fisheries law has become stronger, requiring managers to protect habitat as well as fish, and to restore depleted stocks. In Europe, policies encourage the use of precautionary principles. Yet at the moment of decision, where law becomes practice, ideas of sustainability and replenishment become subservient to the immediate economic and social needs of fishermen. Language describing the health of fisheries embodies the more narrow vision, the emphasis instead on single species rather than whole communities, on taking rather than protecting and nurturing. It seems to assume that fish exist for one purpose—our consumption. We measure not their health, but our need. No better verb describes man's relation to fish than "exploit," which the American Heritage Dictionary defines as "to employ to the greatest possible advantage," or "to make use of selfishly or unethically." Fisheries, according to scientists, are "underexploited," "fully exploited," or "overexploited." It is a peculiar state of affairs to call a fishery not yet ransacked "underexploited," as if there were some task unfinished.

Fisheries managers occasionally speak of catches as harvests, or as production, as if wild fish were agricultural crops. This language deceives. Fishermen do not farm the sea, sowing and reaping and replenishing its waters. Taking 60 percent of the cod and flounder population year after year is not crop management but plunder. Regulators categorizing different fisheries engage in other odd contortions of language. Fisheries brought to the edge of commercial extinction are often described as "depleted." Despite the severe language, the designation has not brought about immediate and powerful remedial action. Consequently, other phrases have appeared, such as "seriously depleted" and even "dramatically depleted." These are surely redundancies, a sad testament to the desperate and unsuccessful efforts of scientists and managers to rouse a public deep in denial.[19]

We look at the sea with an eye toward its utility. This view—clinical, dispassionate—has diminished the ocean's bounty and dulled us to the rhythms and richness of its life. Time and the earth were once endless, and we have collapsed the immensity of both, causing vast communities that flourished for millennia to wither away. Ravaging one species after another, raiding our own and each other's waters, we now witness the limits of the sea's generosity. We have found her edges, measured her bounds.

Out on the water, we reenact a practice first applied to dry land. In *Changes in the Land: Indians, Colonists, and the Ecology of New England,* William Cronon vividly recounts how, as the subsistence economy of Native Americans gave way to the international market economy of the colonists, the resources of the land began to disappear, transformed from means of survival into commodities sold for profit in worldwide markets. Within a short time, hunters obliterated the beaver, wolf, fox, wild turkey, and moose. Merchants cut down forests to create fields and to supply wood for fencing, fuel, lumber, and ships' masts. Their practice of culling the most marketable species and burning the rest bears an uncomfortable resemblance to fishermen tossing unmarketable species back into the sea.

The urge to dominate the earth runs deep. Genesis, a chapter in what is perhaps the most familiar and widely read book in Western civilization, establishes the positions of man and nature in a hierarchy:

> *And God said, "Let us make a human in our image, by our own likeness, to hold sway over the fish of the sea and the fowl of the heavens and the cattle and the wild beasts and all the crawling things that crawl upon the earth."*

The notion is repeated as God blesses His human creations and commands them to take their place at the top.

> *And God blessed them, and God said to them, "Be fruitful and multiply and fill the earth and conquer it, and hold sway over the fish of the sea and the fowl of the heavens and every beast that crawls upon the earth."*

Some translations substitute "rule" or "have dominion over" for Robert Alter's "hold sway," but Alter believes the Hebrew language suggests "an absolute or even fierce exercise of mastery," an idea made explicit after the Flood, when God says to Noah and his sons:

> *And the dread and fear of you shall be on all the beasts of the field and all the fowl of the heavens, in all that crawls on the ground and in all the fish of the sea. In your hand they are given.*

Genesis suggests not only a hierarchy but an antagonism between man and nature when God banishes Adam and Eve from the lush and fertile garden to a place where the soil is "cursed," and when, in cursing the snake, He establishes "enmity" between man and nature. The ideas expressed in Genesis, passed from generation to generation over two thousand years and considered sacred to millions of people, play in our unconscious, permeate our attitudes toward earth. While they are by no means the only biblical injunctions concerning the earth, they are among the most powerful, the most memorable, and, sadly, the most prophetic.

The legacies of Genesis continue on in Western rational, analytical ways of thinking and describing. When, at the end of the Middle Ages, the hierarchy established by the divine right of kings gave way to new hierarchies of reason and science, *cogito ergo sum* unleashed the dichotomies of body/soul, secular/sacred, earthly/heavenly, elevating man and submitting nature into his service. These distinctions between body and soul, matter and spirit, man and nature still persist, contributing to our continuing perception of ourselves as separate and superior to the natural world, and further justifying a wanton attitude toward earth's natural resources.

Today, our technological superiority, our quest for material wealth, our inattention to the intricacies of marine food webs, and our solipsistic worldview continue to imperil the sea. Today, our culture, with its emphasis on the self, on autonomy, and on independence, asks how we as individuals can be served, rather than how we as communities can care for each other. The self, which extends to the edges of our skin and which is separate from the rest of the world, dominates. When we strive for self-sufficiency and consider dependence a weakness, it may be difficult to acknowledge our need for earth's resources and change our patterns of consumption accordingly. It is easy to blame fishermen for the devastation of Atlantic's fisheries, but their actions are firmly rooted in modern values.

Fishermen are no different from all of us on the *Cramer,* dipping into our fresh water supply as if the tank were bottomless, resisting the captain's repeated requests to ration. Every day, someone has the job of presenting the water update, giving us figures on how much we have consumed and how much we have left. The discussion that follows is similar in some ways to fisheries management meetings I have attended. When it appears we're depleting our water supply, we question the validity of the calculations, and when that fails, we find someone else to blame. Fishermen who take too many fish from the sea are no different from all of us at home, filling wetlands and depleting oil and gas reserves.

There are voices from psychology, religion, economics, science, and philosophy urging us to see through another lens, to mend the rifts between body and soul, man and nature, to live within a broader ecological perspective. These voices speak from the edges of our institutions, but their audience is growing, perhaps because somewhere the message rings true. We devastate the sea, but at the same time its lure is strong, not unlike the pull of the wilderness that once brought pioneers across North America. Today, more than half the world's population lives within forty miles of the coast. We yearn to keep in sight a distant horizon, to see beyond all we have tamed and conquered, to know there is still something out there greater than us, beyond us, something pristine, something we can look upon with awe as did the settlers who first arrived on these shores, or the pioneers who inched their way across the vast expanse of prairie and plain.

Who hasn't felt calm and peace beneath an open sky, beside a wide sea? Sailing through the Gulf of Maine, I feel the blessing of a wide horizon, the confinement of closed space. Below deck, there is little relief from tight, claustrophobic quarters, where I feel irritable and disoriented. The exception is my cabin. Two of us share a room not much larger than a walk-in closet, but it doesn't seem small, for a porthole opens out onto the spacious sea. Porthole and deck, even in heavy weather, offer a greater context, an expansive view that dissolves the edges of cramped space.

For centuries people have turned to the sea for material and spiritual nourishment, for its once endless abundance, for inspiration, for solace. Babylonian legends tell of the powerful king Gilgamesh, grieving over the death of his friend, desperately searching the ocean for a seaweed containing the gift of eternal life. Walt Whitman gathers and merges himself with Atlantic in "As I Ebb'd with the Ocean of Life." Ishmael, homeless wanderer in *Moby-Dick*, finds that when he is depressed, when there is "a damp, drizzly November in [his] soul," he is inevitably drawn to the sea. There is magic in water, he says, and almost all men are led to it.

The Fog Warning, painted by Winslow Homer almost 150 years ago, portrays a relationship between a lone man and a vast sea which still lingers in the modern imagination. In the painting, a fisherman quietly, steadily rows a dory heavy with halibut back toward a distant schooner. The waves are high, the sea dark and menacing. The fog rolls in, the bank of low-lying clouds threatening to engulf the schooner long before the lone man in the boat can safely return. The uplifted face of the fisherman is calm, almost curious, as he accepts his fate and yields to the sea.

Ironically, by the time Homer painted *The Fog Warning,* that fisherman, along with his contemporaries, had already depleted New England's waters of halibut. Today, the sea continues to yield to the desires of man, and we continue to cling to the idea evoked in Homer's painting of ourselves as small and insignificant on boundless, eternal waters. As individuals, we cannot plumb the depths of Atlantic, and so we think her bottomless. We cannot see her edges, and so we think her boundless. The sea appears so vast and, by comparison, each fisherman so small that we have trouble understanding how the catch of one fisherman affects the health of a fishery, how the catches of all fishermen empty a sea. Overlooking the harbor in Gloucester stands a large, weathered statue of a fisherman in oilskins and sou'wester, standing at the helm of his boat, looking out to sea. Standing by the Man at the Wheel, following his gaze out into the harbor, to the fishing boats wending their way home, he and they seem more real than charts of fish population statistics.[20]

The charts are just as real. The view from the deck of a boat may be wide, expansive, but it seduces, misleads. The sea is contracting, the window closing. If we empty Atlantic of her abundance, make barren her waters, replace her wild fisheries with farms, we will be alone, with nothing to stand in awe before and no place to turn to when we, like Ishmael, are in need. Together, our impulse to dominate the ocean and our need to believe in something larger than ourselves blind us to the denigration we have caused.

We are poor stewards of the watery realm. Scientific and technological advances allow us to see the sea more clearly, to dive deeper and stay longer. They provide exciting passports to a frontier still largely unexplored, a frontier where we may ultimately find refuge. At the same time, we have done considerable damage from our perch at the surface. We seem not quite worthy of the knowledge and skill we are acquiring, but perhaps, as we continue to open the dark sea, to draw vivid connections between the work of fishermen and the health of fisheries, and to understand the delicate balances within marine ecosystems, we may return to what the heart has long known, and suppressed.

Some 370 million years ago, the first animals crawled out of the sea, abandoning the watery realm for dry earth. It was a journey made possible by tiny grasses of the sea breathing oxygen into the atmosphere over millions and millions of years, building up a protective layer of ozone, shielding the emergent life from the harmful ultraviolet rays of the sun. The offspring of early land dwellers incubated in a watery egg, remnant of the sea encapsulated in a hard shell.

Today's earth barely resembles that ancient time, and human beings share little with ancient amphibians, but one strand of evolution persists. Fully of dry land, we still carry the sea within us. When we reproduce, sperm joins egg in a watery environment resembling that ancient sea. Gill slits, vestige of our watery beginnings, emerge during the first weeks of life in every human embryo and then disappear, fusing into the lower jaw and larynx. Throughout a woman's pregnancy, the fetus grows into a child floating in its own sea of amniotic fluid. At birth, each of us, leaving the nurturing waters of the womb, repeats our ancestors' journey from the sea.

Cast up on land, we are cut off from the sea that first nurtured us, yet we still retain our watery heritage. We are held together by strong bones and tough skin, but inside we are mostly salt water. The salt of the sea rushes through our veins, courses through our blood, and we taste it in our sweat and tears. It is the immediate physical embodiment of a union formed long ago, in distant moments when we and the sea were one. We have relegated those moments to the dim recesses of memory, but that time is imprinted on our psyche, indelibly ours. We have lost touch with it, buried it under the accretions of our culture. Somewhere within us it sleeps. Someday we may hear its call, and wake.

CHAPTER 3

Along the Slope

IN the first days of our passage, the *Cramer* crosses the continental shelf, the fertile, shallow province of Atlantic at the edge of dry land. The crossing is rough. The sky is overcast, the seas heavy, the weather rainy and cold. Storms disperse their energy across long distances of open ocean: swells seven feet (two meters) high roll in under our ship, one after another, from a gale raging on the other side of Atlantic, off the coast of Africa. We feel the effects. The words nausea and nautical share the Greek root *naus*, meaning "ship," with obvious reason. Almost all of us are pale and sick. While we make it through our watches, few of us have the appetite or equilibrium to sit at a meal.

For us at the surface, the transition from land to sea is difficult. Below, on the seafloor, another transition takes place, as continent gives way to sea. That passage is quick, and almost unnoticeable. For me, watching from the surface, only the depth finder measures the change. Some 100 miles (160 kilometers) out of Woods Hole, the water deepens as the bottom drops away. The sea fills and fills. One day Atlantic is 600 feet (180 meters) deep. Twenty-four hours later, a full mile (1,600 meters) separates us from the bottom. We are leaving the last traces of land behind, crossing the far outposts of the continent, entering the realm of the deep sea. The continental slope, the edge of continents, one of earth's most well-defined boundaries, is imperceptible from the sea surface, but profiles built from sound waves show the rim of North America gently giving way, slipping softly into deep water. Not every passage is this mild. Off the coast of Florida, currents eroded this shelf and slope built of cemented seashells, and left a steep cliff plunging abruptly down to the seafloor.[1]

Here and in Europe, where dry land once disappeared beneath the march of glaciers, and off the coasts of Africa and South America, where sediment-laden rivers still run to the sea, jagged canyons and gullies scar the slope. Long before oceanographers could map the steep canyons, men felt their presence. A canyon lies unseen on the seafloor, but waves above it break gently, turned by the deep water toward the canyon sides. Where the calm, deep waters of canyons approach the shore, men built fishing communities and seaports, in Dakar, at the Trou Sans Fond ("hole without a bottom") in Abidjan on the Ivory Coast, in Marseilles, in Lisbon.

Many canyons of Atlantic descend from the mouths of great rivers, from the Congo, the Hudson, the Rhône, the Amazon, the Senegal. Silt and sand, rocks and boulders, brought in by rivers or currents, left by departing glaciers, pile up on the seafloor near the canyon heads. The mounds grow steep and collapse, either under their own weight or when they are disturbed by earthquakes. Collapsing piles of sediment set off underwater avalanches, dense currents of water and rock hurtling at high speeds down the slope, sweeping through the canyons, scouring them clean.

When glaciers receded from North America, meltwater flowed into Atlantic through the St. Lawrence River. Debris from the retreating glacier piled up south of Newfoundland near the Grand Banks. In November 1929, an earthquake there set off an avalanche and created a swift current of debris and water rushing down the continental slope at

35 miles (55 kilometers) per hour, snapping telegraph cables as it advanced. The last cable, 500 miles (800 kilometers) away from the first, broke thirteen hours after the earthquake began. The current left a broad signature, scattering sediment over 40,000 square miles (100,000 square kilometers) of seafloor, but nobody would have known of its existence if the cable hadn't broken. Underwater avalanches can move large volumes of sand, mud, and silt. On the seafloor of the Mediterranean, between France and Algeria, lies a broad plain a little larger than Portugal. The top 26 to 32 feet (8 to 10 meters) of this plain were built from sediment carried off the slope in an avalanche twenty-two thousand years ago.

Mountains of debris are moved along the seafloor, and by and large we are oblivious to their motion. We are deaf and blind to rocky slides from the slope, to high-speed currents pushing through weighty water, but from time to time we are alerted to their presence. Cables in the Congo Canyon break about once every two years, whenever the river is particularly heavy with sediment. Colombian workers fixing broken cable from a submarine canyon off the Magdalena River have hauled clumps of marsh grass up from water 5,000 feet (1,500 meters) deep. Turbulent currents gouging out the continental slope had dragged grass from the shallows at the mouth of the river, where it grows, into the deep water. A clump of grass far from its home, volumes of river sediment scattered across the slope, a deep canyon etched into the seafloor: these are signs of the large and powerful forces that carve the shelf and slope at the edge of continents.

We see so little of the sea. The forces at the edges of continents have been in motion for thousands, possibly millions of years, but how often they carry cascading piles of sediment down to the deep sea, we don't really know. We have never observed turbidity currents, as they are called, directly. One close encounter followed a 1997 earthquake on the coast of Venezuela. Scientists measuring water clarity in a seafloor basin 55 miles (90 kilometers) away felt the pulse of the current triggered by the upheaval. In the weeks that followed, the bottom of the basin clouded over as particles of sand and silt loosened by the earthquake swept across the seafloor in a speeding current. By September, when the water finally cleared, the current had moved in 100,000 tons of sediment.[2]

Not only do such large motions across the slope escape us, smaller ones do as well. The muddy, seemingly monotonous continental slope may be home to the greatest diversity of life on earth. Long considered barren, the continental slope may be as rich in species as rain forests or coral reefs. Scientists have sampled little of the slope, but where they have, they have found an unimagined opulence. The wealth is in worms. Living amidst the particles of mud, in the tiny interstices between grains of silt, are hundreds of species of tiny worms. More than half the worms found in a few cores from the slope off the United States, between New England and South Carolina, were previously unidentified. Across Atlantic, the slope off Scotland shelters a world of worms too small to be trapped by anything less fine than filter paper. The samples contain other animals, tiny mollusks and crustaceans, never before noticed by men. Each new sample yields new riches, leading scientists to think that the sea contains anywhere between ten and one hundred million species of living organisms. The ones we have seen, and listed, may constitute a mere 1 percent of what the sea actually holds. Our catalogue of the sea is mostly blank.

Species represent fine delineations, small twigs on life's taxonomic tree. Twigs stem from branches, branches from boughs, boughs from trunks, as species join to genera, families, orders, and classes, in increasingly inclusive classifications. The phyla constitute the broadest categories, the greatest distinctions within the animal kingdom. When one considers these higher levels of taxonomy, the sea emerges as a friendly host. Almost all earth's phyla are found in the oceans, many exclusively.[3]

Gazing out into the opaque water, I have no sense of the riches lining the seafloor, and I have no way of sensing a swift underwater current racing along the bottom, but as we sail out farther and farther out over the slope, we can sense the water changing. The sea itself grows warmer and saltier. One night, standing at the bow watching for freighters, I see a cluster of twinkling lights off in the distance, a small village lighting the blackness. The lights come from fishing boats seeking the tuna and swordfish feeding in the outer reaches of the slope waters. Not far from our boat, tuna are schooling in the waves. They have plenty to eat. Drifting plants and small fish are concentrated here, pushed against a watery edge in the sea, the liquid wall of the Gulf Stream.

Part 2

THE OPEN SEA

CHAPTER 4

The Rush of Water

SOME ships crossing into the Gulf Stream leave behind a broken wake marking their passage, but the *Cramer,* being a sailing vessel and not a tanker, cuts through the waves with barely a trace. Squalls often hover at the edge of the current, but on this night the clouds have rolled away, opening a clear and starry sky. Our announcement is hot air rising off warm, inky blue water. The science tow offers further evidence, capturing travelers in the current, transparent eel larvae, each one a quarter of an inch long, and by-the-wind sailors, small, delicate jellyfish buoyed on the surface by hidden air-filled floats. Many sea animals, like North Sea, Grand Banks, and Georges Bank cod, are confined to specific geographic areas defined by water temperature. These animals tend to stay near home. By-the-wind sailors roam the world. Catching the prevailing wind in tiny translucent sails set to blow them offshore, they move with the Gulf Stream, riding the current up the east coast of the United States and across Atlantic to England and Ireland. We will leave the Gulf Stream after only a few days and head south, barely tasting its power, but by-the-wind sailors go the distance, heeding the call of currents whose reach extends throughout Atlantic.[1]

The ship runs before a fresh, gusty breeze, plowing through spray and foaming whitecaps. The waves break turquoise. The sea, viewed from the porthole, below the water line, turns a deep and clear indigo. By evening the air feels hot and moist. The water temperature rises rapidly, to 80 degrees Fahrenheit (27 degrees Celsius). The sea radiates heat, melting away the winter chill. The taffrail log, a mechanical device towed from the stern to measure distance traveled, clocks our speed at four knots, but when we calculate our position, the actual speed is seven knots. Wind pushes the ship along, but the current carries it as well. The Gulf Stream, blue river of Atlantic, doubles our speed.

The sea at the opening of Genesis belongs to the void, to an unformed earth. Only after God divides the waters, lifting some up to the heavens and gathering the rest into an ocean, can dry land make an appearance. God keeps this sea, this uncontrolled remnant of disorder preceding Creation, at bay, outside the tamed, cultivated paradise of Eden. Many generations later, He cites pushing away the encroaching sea, along with laying the foundations of earth and commanding the sun to rise, as signs of His greatness. Thundering to Job from the whirlwind, God speaks of stopping the threatening, violent, and chaotic sea, forcing it back, away from man:

> *Or who shut up the sea with doors,*
> *When it broke forth, and issued out of the womb;*
> *When I made the cloud the garment thereof,*
> *And thick darkness a swaddlingband for it,*
> *And prescribed for it My decree,*
> *And set bars and doors,*
> *And said, "Thus far shalt thou come, but no further;*
> *And here shall thy proud waves be stayed?"*[2]

Man and sea are of separate realms in this accounting. Between the two there is no connection, no meeting, no joining. This ancient view still permeates our thinking, but it could not be further from the truth.

Advances in physical oceanography suggest that the ocean is not unbridled chaos, removed and set apart but, rather, that its waters flow around us in unending circles, whole and indivisible. The Gulf Stream is one arc of the circle.

The swiftest rivers of the continents pale beside this mighty current of Atlantic. Together, all rivers washing into the sea from dry land do not begin to equal its surge. Coursing up from Cape Hatteras, the Gulf Stream delivers forty billion gallons of water each second to the Grand Banks of Newfoundland. Carrying one thousand times more water than the Amazon, the Gulf Stream is only one of many currents crossing Atlantic and sounding her depths. Throughout the sea, along the dark bottom, in the sunlit surface, in the fluid midwater, run the currents, streams of water of differing weights, temperatures, and saltiness. Their wanderings and workings are a mystery, slowly unfolding. Saturated with oxygen, rich in silica, or flecked with radioactive tracers, each carries its own unique signature, travels its own path.[3]

Even the most able and knowledgeable physical oceanographers readily admit that Atlantic's circulation, though more studied than that of other seas, is far from fully understood or defined. The sea defies easy description. Much of the year high winds and rough waves preclude research ships from taking her pulse. Science has charted portions of the circulation; it has yet to map the whole. New, sophisticated technologies are beginning to fill the gaps, bringing the substance of the sea into sight. Floats designed to withstand smashing waves, corrosive salt, and heavy pressure drift for months with an individual current, tracking its speed, following its path. Heavy-duty ships equipped with powerful ice-breakers penetrate the Arctic ice pack, one source of the ocean's cold, deep water. Distant satellites hundreds of miles above the water, measuring the height of the sea's surface with radio waves, identify the motion of currents. Taking in a broad swath of water within seconds, mapping the entire surface weekly, satellite images are instantaneous, capturing the reach of the sea at a glance.

As the whole of the sea comes into view, the weave of currents is revealed. Winding across Atlantic's surface, diving into her depths, sliding across her seafloor, they flow one into another. None is solitary, none is independent. Cold water sinking beneath Antarctic ice surfaces again, years later, off the coast of Scotland. Fresh water seeping into the Arctic through the Bering Strait shapes the strength of the Gulf Stream, thousands of miles away. The pulse of one current determines the vitality of another, and all respond to the touch of winds wheeling through the atmosphere and the delicate balance of salt in the water. The relationships are subtle, only partially described.

As soon as man took to the sea, he felt the pull of currents. Norse sailors found their way to Iceland, Greenland, and eventually Newfoundland following the path of water circling polar seas. Columbus sailed from the Canary Islands toward the Caribbean pushed by the trade winds, skirting the edges of a steady current flowing from Africa to South America. Ponce de León, anchoring off Florida to search for the elusive fountain of youth, saw one of his ships swept out of sight by a swift stream flowing north, the same stream that carried Hernán Cortés's men home to Spain after they razed Tenochtitlán.

These currents, noted and identified in early logs and rudders of sea captains, form an arc. Warm equatorial waters stream across Atlantic from Africa, bump against South America, and turn north, pouring through the necklace of Caribbean islands, seeking passage between Cuba and Haiti, the Virgin Islands and Anguilla. The water piles up in the Caribbean Sea and shoots through the Florida Straits, hugging the slope of the seafloor, a narrow current racing north at a speed of 50 to 150 miles (80 to 240 kilometers) each day. At Cape Hatteras, wider, deeper, carrying five times more water, it gushes into the open sea and rushes to the Grand Banks, then splits into slow, diffuse ribbons, continually circling back on itself, drifting on toward Europe. The North and South Equatorial Currents, the Florida Current, the Gulf Stream, the North Atlantic Drift: each is part of the others. It is the same water, known in different places by different names.

Following the path of strong, swift surface currents, explorers, con-

Atlantic currents

quistadors, and colonial merchants opened a green continent to European conquest. Traders pressed the currents into their service, saving time and money, while those who ignored the flow of water paid a price. British mail boats made slow crossing to the colonies, prolonging their passage by as much as two weeks as their captains, stubbornly ignoring the advice of New England whalers, foolishly attempted to stem the Gulf Stream.

The Stream is an old, well-traveled route, ferrying passengers across Atlantic long before men and boats appeared. High in the forest canopy of central America, well away from the shore, seed pods from large tropical vines grow heavy and drop into creeks and rivers, washing into the sea. Hard, tough, impervious to salt and water, the seeds, dark woody sea hearts and pale banded sea beans, ride the currents month after month, for thousands of miles. These talismen of good fortune and safe passage endure, eventually washing up on the beaches of Norway, the sands of Ireland and Scotland, even the edges of the Azores, where they are gathered by beach gleaners. The safe arrival of a small seed after a long, rough, watery transit still inspires awe. To hold such a treasure is to touch a foreign shore. Columbus walked the beaches of the Azores; some believe his desire to find the source of the beautiful sea heart steered him east across unknown waters. Frail upon the sea ourselves, and still wary of its waters, we cherish exotic drift seeds, our tie to faraway places.

Drifters still ride the currents, their lots cast with rushing water. In 1980, a fifteen-year-old boy and his geography class, visiting the beach at Plum Island, Massachusetts, threw bottled messages past the breakers out into the sea. Twelve years passed. The boy grew up, finished school, went to work, married, and moved into his own home, forgetting his class and the lone missive bouncing in the waves. The current wound its way across Atlantic and beached the bottle on an island in the English Channel, 3,000 miles (4,800 kilometers) away. Though the reach of the sea is great, its caress joins distant shores. Like the boy standing on the beach at Plum Island, I too am touched by the magic of currents, by the possibility of rushing water ferrying a tiny object across a vast ocean.

While at sea, I prepare a bottle myself, writing a message, and sealing the cap with hot wax. I confess this is not my first missive; I suspect it will not be the last. I heave my bottle overboard and watch it bob in the waves. It quickly disappears from sight, but I believe it will find safe passage; the only question is where on the great circle it will come ashore. Like those who toss their message bottles beyond the waves and those who seek the sea heart hidden in a clutch of flotsam at the tideline, I too believe that all waters are joined.[4]

Other passengers, less visible, less easily observed, drift with the current across Atlantic. In the summer of 1993, when the Mississippi River overflowed its banks, record-breaking floodwaters poured into ocean currents moving through the Gulf of Mexico. By September a pool of fresh river water had rounded the tip of Florida, appearing off the coast of Miami. Silty, green, surrounded by the clear blue of the Florida Current, it contained tiny floating animals and fish larvae far removed from their homes at the river mouth. The pool of Mississippi River water, fourteen miles (twenty-two kilometers) wide, barely one third the width of the current, floated to Cape Lookout, North Carolina, and headed into the Gulf Stream. Scientists thought it might reach New England by late October.

The *Cramer* crossed the current then, but none of us noticed any oddities in the sea's salinity; we were taking our measurements along a very thin line. None of us saw the pool of green water. As wide as it was, and as expansive as we thought our view to the horizon to be, our paths never crossed. Our track was a narrow line in this vast sea, the pool of river water a mere puddle. From our small boat, a piece of the Mississippi, floating beyond the reach of our sampling bottles, beyond our sight line, didn't exist. I learned of its presence in the Gulf Stream only after I returned home. Perhaps, unknown to us, ship and river water shared passage in the Gulf Stream, the Mississippi water continuing on to Europe when we turned south. Or perhaps it disappeared, diluted by the sea. No one knows where its journey ended.

This shred of Mississippi River water may have left a mark before it disappeared. Fresh, buoyant, it blanketed the warmer, saltier sea, preventing its heat from warming the air. As the pool drifted toward Europe, the sea breeze might have blown a little colder, chilling the English air, giving winter an unaccustomed rawness. Other pools of water riding the currents have affected our weather. Forty years ago, a large, unusually warm patch of water floated up from the Caribbean. It deflected the moisture-filled wind blowing off the sea, pushing away coastal Norway's heavy fall rains and triggering a drought. Only now are scientists beginning to identify pools of fresh, salty, warm, or cool water carried on the currents and to understand their meaning and implications for weather and climate. The sea sings in our midst, but the notes are confusing: we have yet to make out the full melody.[5]

The most precious commodity carried by currents arcing across the Atlantic, more valuable than the gold or spices, timber or rum traded by colonial merchants, is heat. From 90 million miles (140 million kilometers) away, the sun radiates heat toward earth, but earth receives this warmth in uneven portions, absorbing more at the equator, where the days are long, where land and sea face squarely into the sun, concentrating its hot rays. Less penetrates the Arctic and Antarctic, where sunlight hits earth more obliquely, and where, each winter, earth's tilt shrouds the days in cold and dark. The imbalance would make earth's climate severe, but for the accommodating influences of wind and wave.

The Tropics unburden their heat to the air, which, warmed by the sun and laden with moisture evaporated from the sea, grows light and rises, then drenches the earth with rain. The equatorial climate is hot and wet. On land, abundant rains water the Amazon forest. Out on the open ocean, in the Doldrums, rising air produces only the gentlest of breezes; towering thunderheads fill the sky. Torrential downpours return water to the sea, throw heat to the sky. The air, empty of moisture, drifts away from the equator, to 30° north and south, then sinks back to earth, dry. Beneath this blue sky and hot sun bake the deserts of Africa, the Sahara and the Sahel. On the water, in this windless region known as the Horse Latitudes, colonial sailors were often becalmed. Desperate to catch any light fickle breeze, they tossed their cargo of live horses overboard to lighten the load.

The rising and falling air gives birth to the wind. As warm, moist air rises at the equator, cool air blows in from the Horse Latitudes to replace it. The strong, steady trades bring tidings of Africa, lifting sand from storms on the Sahara and, within one or two weeks, blowing it across Atlantic to Barbados. Over the course of a year, millions of tons of red Saharan dust come in on the trade winds to the Caribbean, leaving a trail along the seafloor and a touch of haze in the air of the Virgin Islands. Wind carries dust and pushes water, driving broad, gentle equatorial currents across Atlantic.

Air and sea are partners, redistributing earth's heat, carrying it from the equator to the Poles. The play of the wind guides the motion of currents; heat from the ocean powers the wind. Spawned by cool air sinking at the Horse Latitudes, strong prevailing westerlies blow toward higher latitudes, soaking up moisture and heat from the sea. Rains release heat, in tempestuous storms off the Cape of Good Hope and in raging gales off the Grand Banks. The heat is palpable. As measured by the *Cramer,* water temperatures in the Gulf of Maine are a cold 59 degrees Fahrenheit (15 degrees Celsius), while Gulf Stream waters peak at a comfortable 81 degrees Fahrenheit (27 degrees Celsius). Heat from this sea, carried east by water and wind, gives Europe a pleasant climate. Each winter, below freezing temperatures plunge Greenland and Baffin Island into bitter cold, icing both land and water. Across Atlantic, the warm current brings heat to the coast of Norway, keeping the sea from freezing. Winter temperatures in the Lofoten Islands, high in the Arctic Circle, are a full 45 degrees Fahrenheit (25 degrees Celsius) warmer than elsewhere along the same latitude. In the summer, armadas of drifting icebergs float through the sea off Labrador, while in Ireland and Scotland palm trees and tropical gardens flourish in warm breezes coming off the Gulf Stream.[6]

Water rushes north in the Gulf Stream but then returns. Cooled, it flows south, drifting past Spain and northwest Africa in the broad, slow-moving Canary Current. Prodded by trade winds, it veers offshore, south of the Canary Islands, off Western Sahara and Mauritania. Deeper, colder water, rich in nutrients, in phosphorus and nitrogen and silicate, rises to take its place, seeding a prolific fishery. When the wind slackens, the current slows, the fertile water stays deep, and the fishery disappears. Eventually, the Canary Current swings west, becoming an equatorial flow. The current slides back across Atlantic, back into the Caribbean, completing its passage around the sea.

The sea gyre turns, its waters one. Occasionally a drifter completes the circle, its track pointing to the integrity of the ocean waters. In the late 1800s, Atlantic was awash in derelict ships, wooden-hulled schooners ravaged by sea storms and abandoned. Dismasted and waterlogged, hundreds of ships drifted for months on end, captive to the leaning of wind and wave, a hazard to navigation. Pilot charts published monthly by the Navy recorded the ships' positions. Physical oceanographers, analyzing the charts one hundred years later, traced the wanderings of the *Fannie E. Wolston,* a ship adrift for three years. The derelict schooner rode the Gulf Stream to the Grand Banks, floated south toward the Canary Islands and west across Atlantic, rejoined the Florida Current at the Bahamas, and then finally closed the loop, returning to waters where she first went down.

The path of ships like the *Wolston,* adrift in Atlantic's currents, is convoluted; flowing currents make many detours and digressions as they circle the sea. Some ribbons of current keep to the central gyre, repeatedly making the large circle around Atlantic. Others follow smaller, tighter loops, flowing from Newfoundland to the Azores and back again. Still others flow north, warming Norway and Iceland, then slipping around Greenland and over to Labrador, rejoining the Gulf Stream in foggy seas off the Grand Banks. A particle of water may take ten or twelve years to make this circle. The water leaves but always returns.

When Timothy Folger, a Nantucket whaler, first mapped the Gulf Stream for his cousin Benjamin Franklin, he drew a river in the sea. In the sea, though, both the river and its bank are liquid; the position of neither is firm. Drift buoys and satellite images reveal that from month to month, week to week, even day to day, the rivers of the sea are mutable, often shifting course dramatically. Take, for example, the waters feeding the Florida Current. Half consists of water recirculating from the Gulf

Stream, returning across Atlantic in equatorial currents. A fickle current flowing north along coastal South America supplies the rest. For six months each year, between January and June, this latter current rushes steadily north into the Caribbean, but between July and December it turns east instead, toward Africa, leaving behind a few splotchy remnants, two or three eddies spinning north. It pushes one thousand miles out to sea before turning back, taking a long and circuitous route to the Caribbean. When the year begins anew, it returns to the shorter path, its peregrinations ended until the winds shift and the seasons change.[7]

The Gulf Stream wanders as well, looping along its course, doubling back, then taking shortcuts, as winding rivers cut new channels. There the similarity to terrestrial rivers ends, for what remains are not stationary oxbow lakes but swirling eddies. Atlantic is dappled with eddies, as many as one thousand at a time, spun from currents rushing along the coast of South America, from deep water pouring out of the Mediterranean through the Strait of Gibraltar, and from the Gulf Stream. The Gulf Stream sheds its eddies between Cape Hatteras and the Grand Banks, where the current widens and slows, meanders and shifts course, one week veering toward the continental shelf, the next wandering farther out to sea.

Gulf Stream eddies marble the cold slope water, filling 40 percent of that realm with splashes of warm water pinched from the Sargasso Sea. Each eddy or ring houses a core of warm water surrounded by a filament of the Gulf Stream. The warmer, lighter water rides a foot and a half (one-half meter) above the surrounding sea, giving the ocean a bumpy texture. Each ring is a self-contained aquarium: tropical and subtropical fish, Portuguese men-of-war, and loggerhead turtles all dwell there. Tropical terns and frigate birds fly overhead, following their food. From time to time, eddies carry fish from the Caribbean into New England, to the shores of Martha's Vineyard, to the coves of Rhode Island. Early one fall, expatriates from the tropics floated into the bay near our home. The water was filled with regulars, schools of herring, and stripers coming in to feed, but swimming slowly amidst them, lethargic in the cold, were fish hailing from Florida: jack, snapper, sheepshead.

Eddies breaking off the Stream to the south whisk cold green slope water into the blue Sargasso. These colder rings teem with life, making up in abundance what they lack in diversity. In the early days, weeks, and even months of an eddy's trajectory, copepods, marine snails, and tiny fish thrive in its rich nutritious stew. Copious greenery gives the Sargasso, a seeming desert compared with the fertile shelf and slope waters to the north, a brighter bloom. Each ring, about the size of New Brunswick, is a microcosm, a detached piece of a temperate world drifting south into a subtropical climate.

Yet life brought down from the north in an eddy cannot survive the Sargasso heat. Under the hot sun, surface waters in the ring eventually warm. The nutrients become depleted and the plants die. What remains of the cold-water core moves deeper, sinking between one and three feet each day. The surviving animals follow the descending cool water, moving farther and farther away from an already meager supply of food. Their growth rates slow, the females cease to produce eggs. Eventually they starve. Smaller, less fertile, less abundant drifting plants and animals, comfortable in the warmth of the Sargasso, move in.

Rings of the Gulf Stream with cold cores of slope water can spin through the Sargasso for as long as three years before they spin back to fuse with the current. The Stream itself courses on, sometimes weaker, sometimes stronger, carrying passengers like *Velella velella* around Atlantic's gyre and bringing heat to northern waters. Not all Atlantic's currents, however, ply the surface.[8]

Surface currents, heeding the wind, are shallow, only a few hundred feet deep, a mere 2 percent of the sea's volume. Beneath them, great masses of deep water slip across the seafloor. No one can describe this deep circulation with complete confidence. The measurements of cold currents moving along the seafloor are sparse, the data insufficient. For a long time we believed that deep currents were a world unto themselves, separate and isolated from the flow of water above. Now we know otherwise. Dark, deep currents sliding along the bottom are born of surface waters sinking in icy latitudes. Sunlit surface currents are refreshed by cold water rising from the depths. Each derives from, and shapes, the other. What has always been implicit is now within our technical grasp: water, a liquid, is indivisible.

Not all warm water traveling up from the Caribbean in the Gulf Stream is destined to stay at the surface, kissed by sun and air. A portion of the current's northern limb, water flowing to the seas of Greenland, Norway, and Labrador, will have but a brief sojourn at the surface, under the light. In these Arctic waters, the current sinks to the deep-ocean floor and begins a long journey south, lasting hundreds of years and covering thousands of miles. It glides along the sea bottom, spreading out, filling earth's seas. Eighty percent of ocean water originates in currents sinking in polar latitudes.

Two critical ingredients, cold and salt, give life to bottom currents. A cold, salty sea is a heavy, sinking sea. In the northern hemisphere, only salty Atlantic can spawn deep currents. While fresh water pours into Atlantic from earth's fullest rivers—the Amazon, the Congo, the Mississippi, and the St. Lawrence—far more water evaporates than returns in rivers or rain. Moisture-laden air evaporated from Atlantic and blown by trade winds sweeps unimpeded across Central America, delivering fresh water to the Pacific. The same winds blow dry, dusty air into Atlantic from Africa. Each year, in waters just west of the Cape Verde Islands, evaporation exceeds precipitation by 55 inches (1.4 meters). The imbalance leaves Atlantic with a salty residue carried north with the currents.

Another crucial pinch of salt flows into Atlantic from the Mediterranean, where the hot, baking sun gives this sea its brine. When the dry mistral blows down from the Alps, salty Mediterranean water cools and sinks, then spills through the narrow gateway at the Strait of Gibraltar, sending a salty plume three thousand feet deep into Atlantic. The plume, bearing its signature of salt, spreads south along the coast of Africa, west toward the Bahamas, and north along the Iberian Peninsula toward Ireland and the Norwegian Sea. Each eddy spun off the plume carries two billion tons of salt, essential to the formation of deep water. The final grains come in on bottom currents born in the cold waters of Antarctica, on the ice shelves of the Weddell Sea. Throughout the long journey, the current loses its icy touch, dispersing into the warmer, shallower water, but the salt marker remains, its traces appearing in the coastal waters off Great Britain and in waters as distant as the Labrador Sea.[9]

Each dash of salt renders Atlantic a little saltier, a little heavier. In the cold, salt-laden waters of the Norwegian and Greenland Seas, the deep-water circulation in the northern hemisphere begins. When temperatures fall in winter, the ice pack edges south from the Arctic and east from Greenland, out into the sea. When bitter winds blow off that glacial continent, needles of ice form in the open waves and coalesce into soupy sheets of "grease ice," bending and yielding to the swell. In the chill, the ice thickens and breaks into round floes, or "pancakes." The Greenland Sea turns to a slurry of ice, freezing, melting, and refreezing on the pulse of blustery storms.

As the sea freezes, most of the salt remains in the water. What little is frozen is flushed out by rain, leaving freshwater ice floating on dense brine. The brine sinks. Each day, a mass of water the size of Los Angeles descends 10,000 feet (3,000 meters) from the surface. Arctic bottom water is dense, the most dense in earth's seas. Much of it rests on the seafloor, but some leaves this polar sea to become Atlantic's deep bottom current. It flows in through a channel south of the Faeroe Islands, and it spills over notches in a volcanic ridge connecting Iceland to the Faeroes. From time to time, it rises up over the sill, lifted by pressure changes induced by passing storms. The young current bends to the west, where it swells, joined by more cold water cascading in over a sill in the Denmark Strait between Iceland and Greenland.

The countenance of the deep lies below the sea's inscrutable, opaque surface, its contours but dimly perceived. Water spilling over the Denmark Strait, unseen by dwellers at the surface, unknown to all but a handful of scientists, may be one of earth's great wonders. Angel Falls, plunging a little more than half a mile from a mesa in the rain forest of Venezuela, is supposedly the world's longest continuous waterfall. Before a hydroelectric dam submerged it, Guairá Falls, on the Brazil–Paraguay border, carried the most water. The cascade of deep water at the Denmark Strait eclipses both these terrestrial waterfalls, reducing them to mere trickles. Over 100 miles (160 kilometers) wide, the Denmark Strait cataract descends more than 2 miles (3 kilometers). As it flows, it pulls in the surrounding water, increasing its strength by half, moving 176 million cubic feet (5 million cubic meters) of water per second. No scientist has ever seen this immense cataract; its presence is inferred, from temperature sensors giving shape and dimension to the water mass, from current meters tough enough to withstand the turbulence.

Water pours in from the Denmark Strait feeding the great deep, cold current. Moving west, it grows, engorged with spillover from the Labrador Sea. The current has a name, the Deep Northern Boundary Current, and then, as it turns south along the edge of North America, the Deep Western Boundary Current, but these sleepy titles understate its weight and substance. Pulling water in from its edges, this deep current swells to four times its original size, carrying enough newly made cold water south to balance warm water flowing north on the surface. The volume is massive, twenty to forty times greater than that brought in by all freshwater rivers draining into Atlantic.[10]

In the icy waters of the Poles, where surface water slips into the deep, the sea breathes. Surface water, bathed in the atmosphere, saturated with oxygen, and weighted with salt, ventilates the deep sea, bringing oxygen to the depths, to deep dwellers in the far reaches of Atlantic. The deep-diving water also carries with it synthetic chemicals that trail the bottom currents, revealing their paths through earth's ocean. In the early 1960s, radioactive fallout from hydrogen bomb tests conducted by the United States and the Soviet Union rained into the Norwegian and Greenland Seas, releasing a spike of tritium, an isotope of heavy hydrogen that would track deep water born in polar seas. By 1972, tritium-splashed water sinking in Atlantic had traveled south as far as Cape Hatteras. Ten years later, it had reached waters off Abaco Island, in the Bahamas. Water samples taken in 1981 along a line between Scotland

and Greenland showed tritium at all depths; twenty years after the last explosion, remnants of the atmospheric atomic-bomb tests were still falling from the sky, still sinking in polar seas.

Atmospheric nuclear testing has ceased, but man continues to make his mark on sky and water. Chlorofluorocarbons (CFCs) from aerosol sprays and refrigerants have drifted high into the atmosphere and then returned to the surface of the sea, sinking in polar waters to join the deep current moving south. Today marine geochemists track deep water, revealing that it, like the currents at the surface, has frayed edges that seep into the surrounding sea. Off Cape Hatteras, where the lip of the continent plunges to the seafloor, the deep current passes beneath the warm Gulf Stream. Their meeting is more than two ships passing in a dark sea, two roads crossing at an intersection. When the currents collide, they bend and yield. Emerging from the meeting, they are altered. CFCs trace the engagement. The trail of CFCs shows the deep current splitting, the shallow half shearing off and moving offshore, its lot cast with the Gulf Stream. The deeper water bends with the Stream, feels its sway, but resists and continues south. Each current is distinct, yet at this junction off Cape Hatteras, the two intertwine, and each bears a portion of the other when they divide.

The pattern repeats itself throughout Atlantic. An armada of floats, equipped with temperature and salinity sensors, follow the deep water. They show individual currents dividing and merging, wandering and returning, giving and receiving from surrounding waters. Ribbons of bottom water continuing south spin off the deep current, making their own circulations and recirculations, abandoning the main flow, then turning back. Filaments of deep water drift hundreds of miles out to sea, looping around mountainous ridges and over the seafloor plains, rejoining the current only to depart again. Currents swell, infused with icy bottom water flowing from the Antarctic, and thin, as water warmed by heat from the surface rises.

Scientists know little about the gentle upward diffusion of deep water. In one experiment, a chemical released in waters 700 miles (1,100 kilometers) west of the Canary Islands rose 60 feet (18 meters) within six months, 100 feet (30 meters) within one year. Questions of where, how often, and why the deep sea is softly stirred still remain. Perhaps internal waves, breaking over hills on the seafloor, lift the deep water. Or perhaps the distant moon, tugging the tides at the surface, also calls to currents in Atlantic's depths.[11]

Eventually, deep currents surface. North Atlantic deep water flows thousands of miles from the polar seas of its birth to the chilly waters off Antarctica. There its time in the darkness comes to an end. The deep water wells up, joining thick currents circling Antarctica. Warmed, diluted by melting ice and snow, it divides and begins a new journey. One strand keeps to Atlantic, sinking back into the Weddell Sea, then pushing back north along the bottom. Other strands wind through the rest of earth's seas before returning to Atlantic hundreds of years later. Along the way, the waters divide and loop, eventually returning, riding the Gulf Stream to northern seas where they will plunge to the bottom and begin the cycle anew.

Atlantic's deep currents are guided by the topography of the seafloor. Bottom waters originating off Antarctica come a long way, following a circuitous path. To the east, an imposing ridge of ancient and extinct volcanoes blocks their passage. These relicts from Atlantic's birth, from the time, millions of years ago, when Africa and South America split apart, deflect water around the tip of Africa, into the Indian Ocean. To the west, deep cuts in the seafloor carry the current through the ridge to the basins beyond. The large mass of water slides slowly north, filling almost one quarter of Atlantic. Once past the equator, it crosses oceanic plains and then slips east, cutting through tall volcanic mountains via a deep gash in the seafloor.

At the same time Atlantic's deep currents give form to the bottom. They carved away the continental slope off Florida, turning a gentle incline into a steep cliff. Where the current moves slowly, its touch is lighter. Photographs show deep dwellers, feathery sea pens, leaning into the current as the water flows by. Ripples mark the sand. Time passes, thousands and millions of years, and the current adds its name to the topography. Water pouring out from polar seas slows, and suspended silt and mud and sand settle out into huge drifts of sediment, each hundreds of miles long, tens of miles wide. Parallel rows of drifts fashioned by deep currents line the seafloor between Great Britain and Iceland; the Feni, Hatton, Gardar, Bjorn and Erik Drifts, fed by deep water, grow slowly, 4 inches (10 centimeters) every thousand years.

Though the mean speed of bottom currents is slow, less than an inch each second, the water is moody. Deep-water currents are like the wind, blustery and turbulent one day, calm and tranquil another. The water often moves slowly, creeping gently south for days, sometimes weeks, at a time, but just as the atmosphere produces high winds and heavy rains, so violent storms churn up the seafloor. Then the currents race along the bottom, 13 to 20 feet (4 to 6 meters) each second, the deep-sea equivalent of a terrestrial gale. On calm days, the bottom water is clear; during a storm, cameras resting on the seafloor are blinded by blizzards of mud and silt. The storm water is thick with suspended mud and clay; one ton of sediment hurtles downstream each minute, gouging and scraping the seafloor. Whatever is scoured from the bottom will journey long and far. Currents have swept pieces of red clay and shale, washed into the sea from the rock of Nova Scotia and New Brunswick, as far south as the Bahamas.[12]

Drifters ride Atlantic's currents along the reach of earth. The genes of tiny unicellular animals floating in icy polar water confirm the idea that Atlantic's currents are linked together, that her circle is unbroken. Identical species of foraminifera live in both Arctic and Antarctic seas. DNA sequencing suggests that sometime in the last two hundred thousand years, their populations began to mingle. Somehow, despite the great distances, despite the barriers of equatorial warm water, these cold-water species exchange genes. How they are able to meet, what route they follow to find each other, is still a mystery, but Atlantic's currents, delivering water from one polar sea to another, are the likely conduit. Where the wind slows, cold currents well up at the surface, off the coasts of Namibia and northwest Africa. Perhaps foraminifera use these cold, nutrient-rich pools as stepping-stones between the Poles, as a cold corridor for their long journey. Perhaps, when they reach the intolerable heat of the tropics, they somehow drop into cooler waters moving below the surface and continue on. Even if their path eludes us, their genes speak of a destination reached.

What Atlantic's currents receive, they take away, but eventually the passengers arrive at distant shores. Bipolar foraminifera demonstrate this, and those who comb a beach for exotic sea hearts feel it as well. At the same time, we still hold to ancient biblical ideas of a sea apart. Heeding the words of Micah, "And Thou wilt cast all their sins into the depths of the sea," we set other seafarers afloat to join by-the-wind sailors drifting at the surface, and sand and silt pushed by deep-sea storms. We launch our castoffs, not as benign as these, hoping they will vanish over the horizon, disappear in the vastness. Atlantic's currents travel slowly, over long distances, but already we have begun to see the unanticipated consequences of our inattentiveness.

Earth is riddled with persistent pollutants. PCBs, though they have been banned in many countries, still leach into rivers and streams, still wash into the sea. Many countries continue to allow the application of toxic pesticides, and even where they have been banned, their residues remain. Farmers spray their fields, companies throw away transformers, but the DDT, chlordane, dieldrin, toxaphene, and PCBs linger. They are thrown away, but they do not disappear into the air or melt in the water. Volatile and poisonous, they persist, riding water and wind, circling earth. Flowing with currents, blowing with wind, they travel toward polar regions. Years, sometimes decades later, they arrive. Small portions, absorbed by the tiny plants of the sea, dissolved in the fat of marine animals, become more and more concentrated as they wind through marine food webs.

Persistent organic pollutants cause cancer and neurological impairment, inhibit successful reproduction, and suppress the body's ability to fight disease. In the Arctic, they have worked their way through the food chain and accumulated in top predators, in birds, seals, polar bears, whales, fish, and people. In this cold and remote place, seemingly far removed from industrial, agricultural, and household pollution, glaucus gulls, fish-eating cormorants, and puffins harbor levels of PCBs high enough to endanger the health of their eggs. Toxaphene concentrations in Greenland halibut are almost high enough to affect bone development and fertility. Arctic belugas and narwhals nursing their calves pass on their burden of PCBs and DDT. Polar bears in Svalbard have high levels of PCBs; some have given birth to hermaphroditic cubs.

Inuit peoples living high in the Arctic, on frozen ground above the tree line, depend on fishing and hunting for sustenance. They take halibut and whales from the sea, birds and caribou from the land. From their diets of fish, sea mammals, and seabird eggs, Inuit women living in Greenland and northern Canada receive both nourishment and high levels of pollutants. Though they live far from agricultural fields and industrial complexes, their blood and breast milk is contaminated with PCBs and pesticides at levels high enough to put them and their children at risk.

At the edge of the Irish Sea, in Cumbria, England, sits Sellafield, a nuclear fuel reprocessing plant where spent uranium fuel rods, used to generate electricity in nuclear power plants all over the world, are chemically treated for reuse. Radioactive wastes released from Sellafield and two other reprocessing plants, in Dounreay, Scotland, and Cap de la Hague, France, move with the currents around Scotland and along the coast of Norway. Five or six years later, they arrive in the waters of the Arctic, where they will circulate for several hundred years, along with radioactive fallout from the accident at Chernobyl, which still flows in from the Baltic. Radioactive wastes enter the food web through animals feeding at the bottom of the sea, and end up in caribou and reindeer. There is no safe dose. While peak discharges from Europe's reprocessing plants have declined, they have not ceased. Levels of one radionuclide in Irish sea lobsters are forty times what they were in 1993. Members of the OSPAR Commission for the Protection of the Marine Environment in the North-East Atlantic have agreed to halt the discharges, but France and the United Kingdom have yet to implement the policy.

Unseen passengers in atmospheric winds and oceanic currents, chlorinated pesticides, industrial by-products, and radioactive wastes travel thousands of miles away from where they were manufactured, used, or discarded. The persistence of toxic chemicals and their concentration and magnification as they move through the food chain give lie to the notion that what is cast away disappears. Toxic chemicals do not cease to exist simply because the sea is vast, because flowing water and moving wind disperse and dilute them to undetectable levels.

Drift buoys, satellite images, chemical tracers, tropical seeds, messages sealed in bottles: all point to the wholeness, the unity of the sea, the connectedness and interdependence of disparate waters. A drop of water swept off a beach in Sitka, Alaska, may eventually appear off the coast of New England, seawater swept through the Anegada Passage off Cuba may touch the Grand Banks of Newfoundland, the sands of Scotland. Chemical tracers, identifying the sea's tangle of currents, tracking their paths and destinations, light the darkness of the sea, show the weave of water, the liquid path to the distant reaches of earth.

Currents carry water away, but bring it back. What is cast away returns, if not now, then later, if not here, then on a distant shore. The supply of water is not endless. What comes in on a current, and then flows out to some faraway place, is not "new" water. Earth has little new water. Only a tiny fraction of juvenile water seeping in from hot springs on the seafloor is new. The rest, which circles earth's oceans, bringing heat from the equator to warm the planet's extremities, which sinks to the bottom of polar seas, bringing oxygen into the deep, is old, as old as earth itself. It evaporates into the atmosphere, blows with the wind, rains onto the land, washes into the sea, over and over again. Water flowing through earth's seas is four billion years old. A mere two hundred years of industrial and agricultural growth threaten that water, the only water we have, the only water we will ever have.[13]

CHAPTER 5

Climate and Atlantic

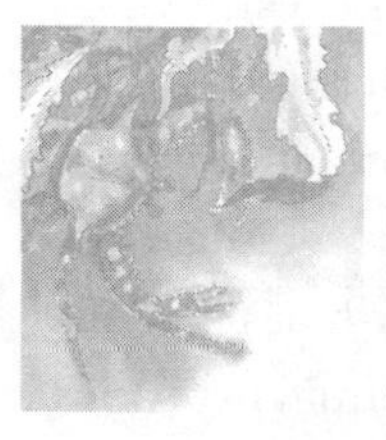

THIRTY-FIVE hundred years before the birth of Christ, an ancient civilization flourished in the fertile valley between the Tigris and Euphrates Rivers. Its people were successful traders, productive farmers, and highly skilled artisans. They built magnificent cities and temples, created the world's first script and carved it into stone and clay tablets, and harnessed the river floodwaters to irrigate fields of barley and wheat. A little more than one thousand years later, beset by internal strife, threatened with invasion by desert nomads, and weakened by a failing agricultural economy, these once robust Mesopotamian cities collapsed.

Sumerian agricultural practices help explain the sudden decline of one of the world's earliest civilizations. Tree cutting, dam building, and river diversion destroyed soil fertility. Fields, silted and salted by extensive irrigation systems, could no longer produce rich harvests of wheat. A long and severe drought added a final stress, forcing people to abandon their cities and farms. When the drought began, rains high in the Taurus Mountains of eastern Turkey slackened. Less water drained into tributaries feeding the Tigris and Euphrates. These rivers, whose floodwaters nourished the fertile crescent, the cradle of civilization, subsided. Windblown sand fell instead of rain, and an ancient civilization came to a close. Falling lake levels in Turkey, at the headwaters of the Tigris and Euphrates, and in the Dead Sea implicate drought as a possible cause for the demise of this ancient civilization. The seafloor in the Gulf of Oman, downwind of the rivers, offers further evidence of aridity at the time the cities were abandoned. Dust, stirred up dry winds, blew from the Tigris–Euphrates Valley out to sea, where it settled on the bottom. High levels of eolian dust in the Gulf of Oman hold the memory of an ancient time that left a sparse record of its end.

Water continues to be a precious commodity in the Middle East. The Tigris and Euphrates still rise and fall in tune with rainfall in mountain highlands, rising in years when rain is abundant, falling in times of drought. In the dry periods, the waters of the Euphrates River subside as much as 40 percent. A scarcity of water has provoked political tensions. Turkey, controlling the headwaters, has turned off the water supply to downstream Iraq and Syria. Iraq and Syria, arguing about the effect of a Syrian dam on river flow, have called in their armies to help settle the dispute.[1]

At the southern edge of the sand seas of the Sahara lies a vast semiarid plain, the Sahel. On this steppe reaching across Africa from Atlantic to the Red Sea, clumps of short grasses and clusters of thorny acacia and mimosa rise from the bare sand. Some rain falls here, more than in the Sahara, but less than in the grassy savanna and lush forest further south. In the Sahel, this place of conversion where barren desert gives way to green savanna, nomadic herdsmen roam, seeking water for their cattle, goats, sheep, and camels. At the plain's southern edge, where the growing season is longer, sedentary farmers cultivate crops of sorghum and millet. Whether nomads or farmers, dwellers of the Sahel await the rains, but they are unreliable. The entire year's water, 4 to 8 inches (10 to 20 centimeters) in the north and 20 to 24 inches (50 to 60 centimeters) in the south, can fall in a few torrential downpours. The heavy rain washes away seeds and crops and rushes toward the sea. Even after a storm, the soil, only lightly touched by the rain, often remains dry.

Sahel, translated from Arabic, means fringe or shore. Its boundary, like the inconstant line between land and sea, ebbs and flows with the coming and going of rain, advancing and retreating from season to season, year to year. When rain is plentiful, the Sahel recedes. When rain is scarce, it advances, drying out the savanna, cutting shorter the already short growing season, leaving the herds thirsty. Satellite images of the Sahel show its edges keeping pace with the rain, growing and shrinking as much as 190 miles (300 kilometers) over the years, as the rains depart and return. Beginning in 1968, drought has devastated the Sahel, withering crops and killing animals. Thousands of people have died of starvation, and millions more have abandoned their sand-drenched villages for the poverty of urban shantytowns. Rainfall is half what it was in earlier years.

Life in the Sahel was not always so tenuous. In the years between 1200 and 1600, the wealthy city of Timbuktu stood at the crossroads of trans-Saharan trade. Prosperous merchants gathered in this city on the bend of the Niger River to trade salt, copper, dates, and figs from the north for gold, ivory, and slaves from the south. Farmers tilled soils fertilized by the Niger's abundant floodwaters and exported wheat throughout West Africa. Muslim scholars came to study at the city's great mosques. Today, the fabled city of Timbuktu is but a shadow of its former glory. Shifting sands encroach upon the decaying mosques, and the Niger's subsiding waters have cut off access into the town. A scant 5 to 9 inches (12 to 22 centimeters) of rain falls each year. Nearby, sand swallows a village built around an oasis that has dried up.

To the east, Lake Chad is disappearing. Once fed by swift-flowing rivers, once spreading over 10,000 square miles (26,000 square kilometers) the largest lake of the Sahel has shrunk to two ponds half its former size. Fishing villages built at its edge now sit stranded in dry sand and grass, miles away from the receding shore. Outside the villages, the land is bare. Great herds of elephants, zebra, and giraffe once flourished in the Sahel, when rivers were full and grasslands were lush and green. Today, fossilized bones and empty gravel tracks mark the riverbeds where large animals once gathered.

What is the difference between feast and famine in the Tigris–Euphrates Valley, between life and death in the Sahel? In the Sahel and in the Middle East, nonsustainable agricultural practices—excessive woodcutting, overgrazing, and misuse of water—may further degrade land stressed by drought, but drought—and its relief—are caused by rain. And rain is the gift of the sea; its scarcity or abundance is determined hundreds of miles away, in currents circling Atlantic, in warm water rushing north with the Gulf Stream and cold deep water sinking to the bottom of the Labrador Sea. In the life-giving rains that once drenched the Sahel, in the floodwaters of the Tigris and Euphrates Rivers, even in the bitter chill of a London winter, the voice of the sea sings. The signal from the sea is muted, subtle, carrying the memory of wind long since dissipated. Though hard to read and to understand, its influence is immense, shaping and sustaining earth's climate. Atlantic currents surge or thin, water temperature and wind rise or fall, and faraway lands feel the effects. During the Mesopotamian drought, North Atlantic cooled. When her tropical waters warm, desiccation comes to the Sahel. Changes in climate have led to the poverty and wealth of nations and the rise and fall of civilizations, and climate answers the call of distant, enigmatic seas.

The sea stirs, ever so slightly, and the winds and rains reply. A minute sea change resounds and fills the atmosphere. The Labrador Sea chills, barely perceptibly, between 1 and 2 degrees Fahrenheit (.5 and 1 degree Celsius), and the water disappears into the depths, unleashing heat into the air. The westerlies strengthen, infusing northern Europe with heat and moisture from the sea. Abundant snows thicken the glaciers of Scandinavia, while to the south, Alpine ice shrivels under dry skies and the edges of the Mediterranean and the Middle East thirst for rain. Such has been the climate in Europe for the past twenty-five years; such was the climate when the Mesopotamian civilization collapsed.

This gentle song of the sea, dryly known as the North Atlantic Oscillation, swells and subsides, rises and falls, and the climate adjusts. During the 1950s and 1960s, the waters of the Labrador Sea warmed slightly. The seawater floated—too warm to sink and release its heat to the wind. Europe, its radiator turned down, cooled. The westerlies shifted south, causing drought in Scandinavia but bringing needed rain to Morocco, Spain, and the Tigris–Euphrates headwaters. Atlantic still spawned deep, cold currents, but they originated farther east, in the Greenland Sea, and easterly winds conveyed their heat to Labrador.

In the marginal lands of the Sahel, the sea brings forth life, announces death. There, throughout most of the year, a dry, dusty wind, the Harmattan, part of the northeast trades, sweeps across the Sahara and through the Sahel. From the south come the monsoons, moisture-laden winds blowing off the Atlantic. Where they meet, the warm, moist air rises, thunderheads fill the sky, and it rains. This point of convergence, near the equator, migrates with the sun, moving south in the winter and north in the summer, bringing rain when the sun is highest. When the rains come, lines of storms known as Lignes des Grains cross the Sahel, bearing the awaited water.

Moving north across the Sahel, between lush savanna and desert, annual rainfall declines precipitously, decreasing approximately one quarter of an inch (six millimeters) each mile (1.6 kilometers). Should the moist winds tarry or fade, depriving the Sahel of even these sparse rains, drought soon follows. For the last thirty years, monsoon rains have forsaken the Sahel, lingering 150 miles (240 kilometers) to the south, out over the sea. Some scientists attribute their absence to the destruction of coastal rain forests in Nigeria, Ghana, and the Ivory Coast. Dense forest, thick with ebony, mahogany, and palm trees and drenched in water, once returned its moisture to the wind, but 90 percent of West African rain forests have been felled, and falling rain washes back into the sea. Wholesale disappearance of the West African rain forest confuses our understanding of climate, blurring distinctions between the mark of man and the rhythms of nature, the former perhaps prolonging the effects of the latter. With the cutting of rain forest, drought may persist in the Sahel, but lack of rain has desiccated this land before, when the rain forest still stood.

The disposition of tropical seas gives rise to rains watering the Sahel. Ships rerouted through South Atlantic when the 1968 Arab–Israeli War closed the Suez Canal unknowingly produced the evidence. In their logs of sea-surface temperature, as well as in data oceanographers have collected more recently, a pattern emerges. Minimal warming of tropical Atlantic, amounting to only a degree or two, accompanied by minimal cooling in waters immediately to the north, foretells death in the Sahel. When equatorial waters warm, the monsoons pause and rains fail the Sahel. When currents carry their heat a little farther north, toward the Tropic of Capricorn, monsoon winds and rains follow, and dry lands drink.[2]

The sea holds the memory of fleeting winds and rain. Imprinted on the sea surface, where air and water touch, it is carried into deep waters and spread with the motion of currents. Eventually, it graces the shore. Once, between 1908 and 1914, and then again in the late 1960s, when westerly winds tracked south and blew warm air and moisture over the Mediterranean, strong northerly winds engorged northern Atlantic with cold, fresh polar water. The saltiness of the sea declined. In both cases, the appearance of freshwater pools coincided with drought in the Sahel. The more recent, and more studied, pool was large, hundreds of miles wide. For fifteen years it drifted through Atlantic's icy waters, following currents across the Labrador Sea to Newfoundland, then circling back across the ocean, along the coasts of Ireland, England, and Scandinavia, up into the Arctic Circle and back to Greenland. Ripples from this pool affected the very depths of the sea. Cooler by one or two degrees, fresher by 1.4 percent, this Great Salinity Anomaly, as it was so unimaginatively called, slid over the northern waters of Atlantic, cooling sea and land.

More than three thousand years ago, an Egyptian pharaoh dreamed of a cycle of feast and famine, of lean years and fat years. Today, science can describe, if not explain, the cycle intimated in ancient literature. Computers scanning the temperature logs of ships crisscrossing Atlantic have now identified large parcels of surface water, hundreds and sometimes thousands of miles wide, some warm, some cold, trailing the currents, each winding through the sea in about twenty years. They carry the signature of yesterday's winds, and they freshen today's. They summon the rains and beckon to the deep sea. Atmosphere and ocean, sea surface and depths, are wedded. The path of prevailing westerlies, the transit of warm- and cold-water parcels through Atlantic, the birth of deep water in the Greenland and Labrador Seas synchronize in twenty- to thirty-year cycles, accounting for warming in Europe, aridity in the Middle East.

The patterns of rainfall in Europe and the Tigris–Euphrates Valley follow the cadence of the North Atlantic Oscillation. It is Atlantic's El Niño. When westerly winds blow warm, moist air over northern Europe, circulation in the Labrador Sea intensifies. Surface water throws heat to the wind, then runs cold and deep, penetrating the bottom current, spreading to the Grand Banks and on to the Bahamas. At the same time, the Greenland Sea quiets. The overturning water subsides, conveying less to the depths. The surface water warms, the sea ice retreats, and the deep current diminishes. Between 1993 and 1995, churning Labrador Sea water reached depths never before witnessed by science, while in the Greenland Sea a great deep-water pump, the Odden ice, disappeared.

During winter storms, polar winds chill and freeze the Greenland Sea, and salty brine sinks to fuel the deep current. The ice breaks apart in high waves and turbulence, and melts when storms die, only to form again in the next gale, injecting more brine into the deep. Warm temperatures in the Greenland Sea during the 1980s and 1990s kept winter sea ice at bay, preventing extensive formations of Odden ice, shutting off the deep-water pump. We cannot see water disappear into the depths of the Greenland Sea, but tags of tritium fallout mark its descent, measure its speed. When deep-water production slowed in the Greenland Sea, it declined by 80 percent. When the pump is running, surface water sinks 10,500 feet (3,200 meters). In the 1990s, it barely reached 3,000 feet (900 meters). In the winter of 1996–97, Odden ice suddenly reappeared in the Greenland Sea, but then it receded, ending the millennium without a further appearance.

The sea murmurs and the land responds. Europe feels the shifting wind, the changing sea temperatures. When the Odden ice returned to the Greenland Sea, a cold, harsh winter returned to Europe. In 1997, for the first time in over a decade, ice in Dutch canals thickened sufficiently to allow the Netherlands to hold its historic 125-mile (200-kilometer) skating competition. The rains, subsiding between 50 and 75 percent in northern Europe, shifted south, soothing parched lands in Spain and Africa.

Whalers have long felt these rhythms of Atlantic, noting three hundred years ago that cold in Baffin Bay foretold warmth in Europe. Scientists have begun to articulate how the movement of sea and wind is expressed in the sands of the Sahel, the snows of northern Europe. They know the dance begins with a temperature change at Atlantic's surface and a corresponding shift in the wind, but they cannot predict the end of one sea cycle and the beginning of another. They cannot say when water will turn the Sahel green, or when the Tigris and Euphrates Rivers will swell with rain. Undoubtedly, that knowledge will come, but the wonder lies not in the predictability of the model, but in the idea that life-giving rain falling on land is borne in on the pulse of a distant sea.[3]

The natural rhythms of Atlantic measured in the droughts of the Sahel and the rains of Norway are short. We can hear the whole song in our own lifetimes. Overlying the shorter rhythms are longer melodies, playing out over generations. These voices of the sea also reach land, warming the climate and chilling it, over periods of 1,000 to 1,500 years. The deep sea records these songs. Year after year, century after century, millennium after millennium, the remains of tiny foraminifera accumulate in sand at the bottom of the Sargasso Sea, near Bermuda. The chemical composition of their calcium carbonate shells, measuring temperature at the sea surface, suggests that 1,000 years ago, the Sargasso warmed. It was a small increase, 2 degrees Fahrenheit (1 degree Celsius), but Europe basked in the warmth. Wineries flourished in England as far north as York, Gloucester, and Hereford. Farmers, to the annoyance of

Northumbrian sheepherders, expanded their fields into the Scottish uplands and hills. Mountain passes blocked by ice cleared, opening more routes across the Alps.

To us, living in more temperate climates, North Atlantic currents such as the Irminger and the East Greenland Currents have names but no faces. The pulse of these currents, running along the coasts of Iceland and Greenland, is vital to lives and cultures onshore. Nuances in the strength of a northern current, in the amount of heat it carries, can have dire or fortuitous consequences for land dwellers. Warm water carried north by oceanic currents melted the ice pack off the Icelandic and Greenland coasts, opening these once marginal and desolate lands to Norse settlement. Norse voyagers first reaching that volcanic island between Norway and Greenland in the 860s named it Iceland, for its severe winters and ice-filled fjords, but within a decade the sea warmed, the ice retreated, and the Norse came and stayed. A longer growing season yielded ample grain and grasses to support people and cattle, and the settlements thrived. Even the Arctic ice may have melted in the balmy seas. Cores of sediment drawn from the Arctic ice pack contain fossils of algae that grew in open water one thousand years ago.

The Vikings sailed on through warmed seas to Greenland and established farms on its southern and western coasts. They grew vegetables and hay in the fertile soil and raised livestock, but these colonies lasted only until the sea cooled. This history, too, is recorded in the shells of organisms buried in the Sargasso. Within a few hundred years, the warm Irminger Current, carrying heat from the Gulf Stream, gave way to the chill of the East Greenland Current. These polar waters flowed south from the Arctic, bringing pack ice to choke off the Denmark Strait between Iceland and Greenland.

The ice lingered through the year and clogged the fjords, preventing ships from landing, cutting off supplies and communication from Iceland and Norway. Norse settlements in Greenland declined and ultimately disappeared as the air chilled and the ground froze. The inhabitants starved, unable to feed their cattle on the sparse grasses of fading summers, unable to grow vegetables on the encroaching tundra. Unable or unwilling to trade their thin European clothing for warm animal skins, they froze. The nomadic Inuit, living off the bounty of the sea, following their food across the ice, survived.

Between three thousand and six thousand Norse lived in the two Greenland settlements. By 1540, all had perished. The chilling sea wrought tragedy in Iceland as well. The population, which had grown to seventy-seven thousand during the few hundred years of warmth, plummeted by half as the Little Ice Age set in and the land could no longer support such large numbers of people. A small decline in temperature, 2 degrees Fahrenheit (1 degree Celsius), cut summer short, reducing the yield of the soil. Farmers once cultivated wheat and barley throughout Iceland, but as the air chilled, production ceased, first in the north and then in the south. Livestock—sheep, horses, and cows—starved for lack of grass, and the sea ice, which once came no closer than a day's journey across the water, now encroached upon the shore, shutting in the fishermen. In the south, toward the still open water, the cod disappeared, unable to thrive in the chilly polar waters. Perhaps the Irminger veered south and ceased to temper the cold Arctic currents hugging Greenland and Iceland's shores. Perhaps a pool of polar water cloaked northern seas, slowing the deep circulation. Perhaps, as the deep current slowed, the Gulf Stream became sluggish and delivered less heat. Only now is science beginning to understand how the quiet motions of a sea ravaged a culture.

The drop in sea temperature wreaked havoc in Europe. Farmers descended from the Scottish Highlands, abandoning their fields and farms to sheepherders. Snow crowned Ben Nevis all year. Grapevines withered in England. In Germany the vines survived, but cold weather soured the grapes. For more than a century, vintners sweetened the sour wine with lead oxide, inadvertently poisoning those who imbibed, sometimes fatally. The Thames, Rhône, and Guadalquiver froze in the winter, and pack ice drifted down to England. Glaciers spread down from the mountains. In Chamonix, Outzal, and the Italian frontier, Alpine glaciers engulfed farms, threatened churches, and felled trees. Meltwater streams and bursting glacial dams washed away houses and barns, roads and bridges, leaving arable fields flooded with rubble. Entire villages were destroyed and abandoned. The late springs and cool summers produced massive crop failures and widespread famine. Ireland's potatoes, stored in the ground after the harvest, froze. Mediterranean fruit and olive trees perished in the frosts and biting wind. Disease and starvation were rampant.

Atlantic's northern currents pause, and Europe plunges into a deep cold. The Little Ice Age is only the most recent cooling in a series of cooling cycles, whose sign is etched into the seafloor. The cycles reach back thousands of years, during a time when earth, released from the icy grip of glaciers, supposedly basked in a warm, stable climate. The record of the seafloor asks that we not take benign winds and salubrious seas for granted. Between Newfoundland and Ireland, pieces of volcanic glass and iron-stained silt litter the bottom, at intervals of approximately 1,500 years. They were rafted down from their original sites in Iceland and Greenland on armadas of icebergs released with each pulse of cooling. They, as well as the shells of polar foraminifera layering the seafloor, track the course of cold water moving south. Polar species buried off Cap Blanc, Mauritania, suggests that even tropical seas felt a chill as the Canary Current brought cold water south.[4]

The Little Ice Age profoundly disrupted life in Europe, but its name does not mislead. By ice age standards, the Little Ice Age was little. Europe has experienced much deeper chills, over much longer periods. For thousands of years at a time, massive ice sheets, 10,000 feet (3,000 meters) thick, have buried Great Britain and Germany, forcing plant and animal life on a long retreat south. As science learns to read the deep sea more adeptly, tantalizing links between circulating currents and earth's ice ages emerge, shedding light on perplexing questions of how the sun's energy heats and cools the planet, and how earth's seas move with the dance of stars.

Earth's distance from the burning sun fashions our climate. Under a protective atmosphere, earth's temperature hovers at 59 degrees

Fahrenheit (15 degrees Celsius), but over millions of years it shifts, cooler, then warmer, cooler, then warmer, as the planet glides through a delicately poised orbit. Earth's axis tilts between 22 and 25 degrees every 41,000 years, its orbit stretches to an ellipse and shrinks back to a circle every 100,000 years, and it approaches the sun at the same season every 23,000 years. As earth moves through these cycles, the intensity of sunlight striking the planet waxes and wanes, strengthening and weakening the force of seasons. Skies fill with rain, then turn to dust, the sea's heat-bearing currents swell and thin, and ice sheets advance and retreat. As earth spins through her journey, sunshine lighting the Arctic Circle increases and then diminishes, by as much as 20 percent.[5]

Orbital variations leave their mark on earth's climate. For long stretches between 5,000 and 10,000 years ago, the Sahel and the Sahara were wet and moist. In the hidden caves of Tassili n' Ajjer, a desolate sand-swept plateau in the desert, are thousands of ancient paintings portraying the herds of wild animals that once flourished there, before lush grasses turned to sand and swift rivers emptied, before brimming lakes dried up and marshes drained. In those days, earth leaned more on its axis and passed closer to the sun in her orbit. Perihelion occurred in September, rather than in January, as it does today, increasing the intensity of northern hemisphere summers. The monsoons blew stronger, reached farther north, and heated the sea. Sun, sea, and wind carried more rain, increasing precipitation in northern Africa by at least 25 percent. Bathed in fresh water, the soil turned rich and loamy and the sands grew to grass. The savanna held more moisture and released it to the sky, multiplying the rains. The Sahara shrank by almost one third. As the desert turned green in the rain, giraffe and elephant, hippopotamus and crocodile came to dwell in places that today are as dry as dust.[6]

The relationship between earth's track through the sky and its climate is confusing. The advance and retreat of ice sheets coincide with the wobble of earth's orbit, but not always simply or directly. Layers of seafloor, holding the memory of climate change, call into question accepted truths about exactly how astronomy governs temperature on our planet. Sediment from the slopes of the Bahamas suggests that the heat of northern sunlight may not be the only cause of glacial retreat. The cores record the withdrawal of ice sheets a full 5,000 or 6,000 years before the northern hemisphere would have warmed from a change in orbit. Hidden in the sea are the secrets of our climate, and with each new core we come closer to understanding the weight of water, its prerogatives and persuasions.

Once we believed the ocean immutable, the deep water dark and still. Now shells of tiny animals and chemical pollutants raining from the atmosphere reveal a sea whose waters call to each other from the far ends of the planet, whose message is heard not only on the surface, but in the cold deep water sliding along the bottom. Deep currents heed the call of planetary motions, but also ebb and flow to their own rhythms as well.

The trail of CFCs sinking into the abyss of Antarctica's Weddell Sea has lightened in the last hundred years. Scientists cannot say why the deep circulation has mysteriously slowed. They suggest that perhaps deep currents counterbalance each other, Antarctic deep water thinning while northern currents swell, southern waters warming while northern seas chill. Scientists suspect that during the Little Ice Age, Antarctic deep water surged while deep circulation in Atlantic's northern waters slowed. Today's waning of Antarctic deep water may be the next phase in the 1,500-year sea cycle.[7]

The swings of deep-sea currents help explain the tumultuous end to the last ice age. The planet began to warm, but the glaciers left in fits and starts, the ice retreating and then returning, retreating and then returning. Gradual changes in solar insolation resulting from earth's steady spin though the sky do not account for these sudden and frequent bursts of glaciation. Rather, it is the motions of deep currents that jag the greater rhythm of climate prescribed by earth's orbit. When northern currents swell, the northern hemisphere warms, and when they fade, the ice returns. The seabed records this partnership between ocean and climate, tracking the peregrinations of Atlantic's deep currents thousands of years ago when the ice ages waned.

Microscopic single-celled animals, buried in dark obscurity, illuminate the larger history of ocean circulation and climate change. The single-celled inhabitants of the deep, dwellers on the walls and floors of channels where the deep water flows, are picky, partial to one current over another, even though the water temperatures may differ only slightly. Some species find their niche with the cold current spreading south from the Arctic; others prefer icy water creeping up from Antarctica. Their shells rest in the sediment long after the animals themselves have died, long after the currents have departed.

The Vema Channel, off the coast of southern Brazil, connects two large basins of Atlantic. Cold, heavy water moves through this deep, winding passage no more than twenty feet (six meters) wide. Today, 13,000 feet (4,000 meters) down in the channel, cold Antarctic water flows north. Above it, 9,700 feet (3,000 meters) down, deep water from the northern reaches of Atlantic moves south. Buried in the wall of the channel in this upper layer are shells from dwellers of the colder Antarctic water. They are 18,000 years old. Their presence suggests that when ice crept down onto the northern continents, the current from the north weakened and icy southern waters swelled to take its place.

Other shells, buried on the continental slope off Bermuda and in undersea mountains near the Azores, also chronicle the undulations of deep-sea currents during the ice ages. The relative amounts of cadmium and calcium in seashells reflect the fertility of surrounding waters: the less cadmium, the fewer nutrients in the sea. The northern current, having been at the surface more recently, where blooming meadows of sea grasses have depleted the sea, has fewer nutrients. Where the more impoverished current flows, shells of bottom-dwelling microscopic animals show low cadmium/calcium ratios. When the flow of deep water slows and the sea fills with more nutrient-rich water from the Pacific or from Antarctic seas, the cadmium/calcium ratios rise. Scientists, vaporizing the pinhead-sized shells with powerful currents, reducing them to individual atoms, have measured their cadmium/calcium content and discovered that during the ice ages the cold deep current flowing down from the north shrank by 50 percent.

Other pieces of the seafloor also reflect the larger circulation of water. Iron and manganese precipitate out of the sea, forming tiny particles lining the seabed. The particles contain birth certificates, isotopes of a silvery rare-earth metal that, thousands of years later, faithfully identify the waters from which the particles came, distinguishing the deep water originating in the Arctic from the cold currents of Antarctic seas. Cores of sediment taken from the southeast Atlantic seafloor tell of a weakening of the cold, northern current during the ice ages.

As the deep current grew weak, the Gulf Stream languished as well. Today, Atlantic's blue river fills the North and Norwegian Seas with warm water and warm-water animals, but warm-water species did not always dominate there. Charcoal paintings of penguins line the walls of sea caves along the Mediterranean, testament to the chill brought in from the sea. Polar shells in the seafloor bear witness to a time during the ice ages when cold water crept south and lapped against the shores of Ireland, France, and Spain, where forests gave way to Arctic grasses and tundra and animals scurried south. The Gulf Stream languished all along its course. In the Straits of Florida, where the current flows swiftly today, it grew sluggish, its flow weakening by 35 percent.

The ice ages convulsed to a close. The deep current swelled and subsided, swelled and subsided, and the ice withdrew and came back, withdrew and came back. When the deep current gained strength, the sinking water relinquished heat to the air, and the Gulf Stream, bearing even more heat, coursed in. When the deep current eased, the Gulf Stream slowed. Each time the currents weakened, the land shivered and plunged into a deep and bitter cold. The Tropics, long thought safe from the cool breath of glaciers, shuddered. Coral from waters off Barbados and shells of sea animals who once dwelled on deep underwater plateaus register a drop of 9 degrees Fahrenheit (5 degrees Celsius) throughout the depths of tropical Atlantic. As the water chilled, the land responded. Pollen and dust imprisoned in ice high in the mountains of the Andes record the descent of the tree line, the conversion of tropical rain forest to dry grassland. When Atlantic's circulation slowed, the entire basin felt the chill; no place was immune. A benign climate, which we take so much for granted, is but an impermanent gift.[8]

Living in the present, in the moment, we believe that the latency of spring yields the bounty of autumn, and that barren winter turns to ripe summer in smooth continuous cycles that describe the past, reach into the future. The record of the sea, reaching back for hundreds and thousands of years, points to a different reality. Only a 9 degree Fahrenheit (5 degrees Celsius) difference in temperature separates the harsh ice ages from the balmy climate that nurtured the first agricultural civilizations, and earth flips easily, quickly, from one state to another.

Orphan Knoll, a lone piece of North America stranded at the very edge of the continental shelf, sits in the deep water 300 miles (480 kilometers) north of Newfoundland. Deep-sea coral embedded in its steep rock walls show the deep current languishing in as few as fifty years. Bits and pieces of Atlantic's past—solitary coral living in other parts of Atlantic, ice cores from Greenland, shells of foraminifera once drifting in icy waters—all describe currents whose paths can be quickly rerouted. The motions of deep currents determine whether large portions of earth are hospitable or hostile to human habitation. When the deep current slows, surface waters chill rapidly and the land responds. The climate we take for granted can deteriorate within our lifetime. We walk a thin edge, unseeing.[9]

What urges, impedes the deep currents? Whose voices do deep waters heed? Global truths reside in tiny sea animals. The proportion of heavy to light oxygen atoms in their seashells mirrors that of heavy and light oxygen in the water. As earth cools, heavier oxygen rains out first, into the sea, and lighter oxygen condenses into the ice and snow of high-latitude glaciers. When earth locks into an ice age, sea and shells grow rich in heavy oxygen. When earth warms and ice melts, lighter oxygen floods the sea and materializes in seashells. Changing proportions of heavy and light oxygen in shells record pulses of fresh water pouring into the sea from melting glaciers, pulses timed to the withdrawal of deep currents.

Atlantic's deep currents are exquisitely sensitive to the levels of salt in the water. They can be weakened by heavy rains or water surging from melting glaciers and pack ice. A flood of fresh water dilutes salty Atlantic, enfeebling deep circulation. At the end of the last ice age, when the sun shone more brightly on the northern hemisphere and the warm Gulf Stream awakened, glacial meltwater pooled in southern Canada, forming a great lake reaching through Manitoba and Saskatchewan, North Dakota and Minnesota. Lake Agassiz, blocked by ice to the north and east, drained through the Mississippi River and out into the Gulf of Mexico. As earth warmed, ice to the east of the lake receded, and 11,000 years ago it uncorked the Gulf of St. Lawrence. Lake levels suddenly plunged 120 feet (37 meters), flooding Atlantic's northern regions with fresh, buoyant water. The effect was cataclysmic; Atlantic's cold, deep water current stalled, and earth's warming ceased. With the sea's heat turned off, earth cooled and once again ice draped northern latitudes. Atlantic's deep circulation shut down for 200 years; the cold lasted for 1,000.

The sun continued to shine, and eventually the ice began to melt again. Lake Agassiz still existed in southern Canada; it was larger than the Great Lakes of the United States. When the last of the ice clogging Hudson Bay melted 8,200 years ago, the lake suddenly drained again, leaving behind a few remnants scattered throughout Manitoba. Again the release of vast amounts of fresh water proved catastrophic. The lake poured out into Atlantic, stalling the deep circulation, precipitating another retreat into the cold.[10]

Ice melting in the high glaciers of Atlantic may once again flood the sea with fresh water. The waters of Lake Agassiz are minuscule compared with what is frozen in the glaciers of Greenland and Antarctica. Together these glaciers hold three quarters of earth's fresh water. The Greenland glaciers, long considered a fixed, unchanging part of the landscape, insensitive to changes in temperature, are proving otherwise. Scientists studying cores of Greenland ice 100,000 years old, realize that when earth last experienced a warm period, Greenland's thawing glaciers raised sea level twelve or sixteen feet (four or five meters). They are thawing again. At higher elevations still untouched by summer sun, the ice stands firm, but along 70 percent of the coast it is thinning, in some places by as much as three feet (one meter) each year. No one can yet say why. Higher temperatures can thaw a glacier, and water melting at the surface can percolate down through the ice, lubricating it, smoothing and easing the way to the sea. These conditions explain some melting, but not all. Whatever the cause, Greenland now delivers 1.8 trillion cubic feet (51 cubic kilometers) of melting ice to Atlantic each year, little by little raising sea level. Lands once frozen are now losing substance to the liquid sea, and as the glaciers shrink, their fresh water may dampen Atlantic's deep-water circulation.[11]

Melting glaciers carry fresh water into the sea, but there are other sources as well. One is the Pacific, returning water originally given by Atlantic. Trade winds evaporate water from the tropics and blow it across Central America. The salty residue left behind spawns Atlantic's deep currents, while Atlantic water taken up by wind rains into the Pacific, freshening that sea, lowering the salinity by a seemingly insignificant one fifth of a percent. This difference, too small for us to perceive without sensitive instruments, drives the circulation of an entire ocean. Buoyed by lighter, fresher water, the Pacific rides one foot (0.3 meter) higher than Atlantic. The water runs downhill through the Bering Strait, freezes into the Arctic ice pack, and eventually melts out into Atlantic's northern waters. It has been doing so for thousands of years.

Visible markers have traced the path. In 1879, Arctic pack ice trapped an American vessel, the *Jeanette*. The boat fractured under the press of the ice, and the wreckage drifted across the North Pole, washing up on the shores of southern Greenland five years later. More recently, in 1992, a rough storm swept a large container of bath toys from a cargo ship crossing the Pacific. Thousands of plastic ducks, frogs, turtles, and beavers scattered with wind and wave. Approximately ten thousand rode the currents north and floated through the Bering Strait. Frozen into the drifting pack ice, they are crossing the Arctic and melting out into waters off Greenland, where perhaps they will catch a ride on polar currents flowing south. Eventually, they may show up on the rocky coast of Newfoundland and the beaches of New England, faded from the long journey. Visible passengers crossing the Bering Strait are rare. Most often the currents glide through unseen.

Today, the Bering Strait is a narrow and shallow passage, only 50 miles (80 kilometers) wide and 150 feet (45 meters) deep. During the ice ages, when snow and rain froze into glaciers, lowering sea level, it dried up, creating a land bridge connecting Asia to North America. Some of America's early human settlers, hunter-gatherers from Asia, arrived via this land bridge over 12,000 years ago. During times of warming, when copious rains fall, the Bering Strait grows deep and wide and fills with rushing water. When earth last warmed, between 145,000 and 110,000 years ago, sea level rose fifteen feet (five meters), inundating the Bering Strait with fresh water. The water flowed through the Arctic and into Atlantic, diluting the Norwegian Sea, weakening the deep circulation, giving earth a push into the last ice age. It could do so again.[12]

As the glaciers recede, so does the winter sea ice, further buoying the ocean with fresh water. Each winter, as the northern hemisphere tilts toward darkness, the Arctic ice pack, about 3 million square miles (7.8 million square kilometers), doubles, reaching down through the Canadian Archipelago, around Baffin Island and Greenland, and into Hudson Bay. As winter turns to spring and summer, the ice shrinks back. In recent winters, more and more open sea resists the grasp of ice. Many an explorer has sought and failed to find the fabled Northwest Passage. Iced in for hundreds, perhaps thousands of years, this shortcut from Europe to Asia is opening. In the summer of 1999, surprised Inuit fishermen watched a submarine surface in a watery hole in the ice off Baffin Island. In the summer of 2000, a cruise ship making an annual excursion to the North Pole found that the far end of earth had melted. Gulls wheeled overhead, drifting plants grew where the sun shone through thin ice. Passengers expecting to walk across the North Pole couldn't.

Data gathered by passing satellites indicate that the reach of Arctic ice has declined 3 percent in the last two decades. The ice itself has thinned as well. Multiyear ice, the thick ice that accumulates each season, has declined by 14 percent. Observers on submarines view ice from another perspective, peering up through the dark water underneath. Their data show that in the last three decades Arctic ice volume has declined by 40 percent. Some scientists, analyzing the losses and finding them too large to be accounted for by natural variations in wind, water, and temperature, ascribe the Arctic thaw to global warming. Others predict that if present trends continue, the Arctic will be free of ice during the summers within the next fifty years.[13]

We know so little about the Arctic, this forbidding sea linking two oceans, but her waters are stirring beneath the ice. Today, this distant region is undergoing monumental change, warming to the highest temperatures seen in the last 400 years. Inuit fishermen know; for the first time they are catching coho and sockeye salmon in the Beaufort Sea. Polar bears are stressed by the heat. Melting ice floes in Hudson Bay are forcing them onto the tundra earlier and earlier in the season. Unable to fatten up on the seals they catch during their stay on the floating ice, the bears are losing weight and bearing fewer cubs. Farther south, the International Ice Patrol reported in 1999 that, for the first time in 85 years, Iceberg Alley was free of ice. Icebergs shed off the coast of Greenland drift south on the currents past Labrador. Usually five hundred or so slip into the Grand Banks shipping lanes each year, but now the water is too warm. If the *Titanic* had sailed in 1999, she would have made the crossing. Scientists clocking the speed of low-frequency sound waves traveling through the Arctic confirm the warming.

The warm water comes from Atlantic. A layer of Atlantic water flows in from the Fram Strait, off northwest Greenland. It winds around the Arctic, beneath the sea surface, around basins bounded by ridges of long-extinct volcanoes, and then exits back into Atlantic by the same route.

Recently the water has warmed and shallowed, thinning the sea ice. It has also strengthened, spreading farther into the Arctic, pushing aside water circulating in from the Pacific. Scientists are quite concerned about where the displaced Pacific water is going. Fresh Pacific water escapes into Atlantic along two routes, shooting through the Fram Strait into the Greenland Sea, or winding through islands of the Canadian archipelago into the Labrador Sea. At both sites deep currents are born. At both sites deep water is suppressed by infusions of fresh water.

It is already possible that warming Atlantic water has inhibited the sea's deep currents. Salinities in deep water flowing south of the Faeroe Islands are dropping, more than they have in the last hundred years, and the water is warming. In the winter of 1996–97, water cascading over the Denmark Strait warmed to the highest temperatures ever recorded there, and the bottom layer thinned by 80 percent. We are poised on the edge of an unknown.[14]

Years of research implicate Atlantic's northern deep current as the Achilles' heel of the sea's circulation. When the current falters, dry land shivers. The fresh water that slows deep currents may originate in melting glaciers or sea ice, or water coursing in from the Bering Strait. In a warming world, it may also come with rain. The burning of fossil fuels and the cutting down of rain forest has already warmed earth by 1 degree Fahrenheit (0.6 degree Celsius) in the last one hundred years. Scientists predict that global warming will raise the planet's temperature by another 2.7 to 10.8 degrees Fahrenheit (1.5 to 6.0 degrees Celsius) by 2100. A warm planet receives more rain. A warmer ocean evaporates more water, a warmer atmosphere holds more moisture and releases more rain. We have already begun to feel the watery effects of earth's rising temperature. Within the last hundred years, the number of intense, heavy downpours soaking and flooding areas of the United States has increased by 20 percent.

Yet the rain we see is only a tiny piece of earth's water cycle, which by and large remains veiled. The atmosphere holds very little of earth's water, only .001 percent, but where that water falls determines the balance of salt in the sea, the shape and strength of currents, the climate. Most of earth's water, 97 percent, resides in the sea. Most water entering the atmosphere evaporates from the sea, and most of earth's rain falls there, invisibly, without measure, without mark. Our very survival depends on the transport of water between sea, air, and land, but we know little about earth's water cycle. Evaporation exceeds precipitation in the equatorial Atlantic, but no one can say with certainty by how much. The best estimates differ by an amount equal to thirty times the flow of the Mississippi River. What we can't see in the ocean has monumental consequences on land. If only 1 percent of what we believe to be the rain falling in Atlantic blows west into the Mississippi watershed, the amount of water flowing through that river doubles.

Rocks strewn throughout the Mississippi River basin suggest that these calculations are not merely hypothetical. Geologists piecing together the history of flooding along the Mississippi over the last 7,000 years find that only slight changes in temperature or rainfall, smaller than those occurring under global warming, triggered frequent and catastrophic flooding. Huge boulders three feet (one meter) in diameter have been swept down the river, whose floodplain has drowned in water sixteen feet (five meters) deep. This past may be reawakening in the present. In 1993, the Mississippi River, swollen with torrential rains, rose and breached its banks, flooding out homes and highways, causing millions of dollars of damage. Europe also feels the intensification of earth's water cycle. In 1995, heavy storms, which normally rain out at sea, swept in off Atlantic, drenching England, France, and the Netherlands, filling rivers to overflowing, forcing thousands of people to abandon their homes.[15]

Scientists cannot yet measure rain falling in the ocean, but they know the effects are considerable. Pulses of fresh water, from melting glaciers and sea ice and from rising waters in the Bering Strait, have shut down Atlantic's deep circulation in the past. Those moments from Atlantic's history speak of the potential consequences today, in our time, of disturbing the atmosphere and sea with global warming. In what may prove to be a monumental imposition of man's will upon nature, we have conspired to create the conditions for earth to receive an increased abundance of fresh water. Whether it comes from rain, melting Greenland glaciers, disappearing sea ice, or some combination, there may be enough additional fresh water flowing into Atlantic to weaken and collapse the deep circulation within the next hundred years, and perhaps sooner.

And if the deep circulation of Atlantic is extinguished? Or if the current shallows, and surface waters follow ancient tracks east, to Portugal, instead of north? Water flowing up from the tropics, fueling deep currents emanating from the Greenland and Labrador Seas, releases precious heat to blow through Europe, warming that continent by 9 degrees Fahrenheit (5 degrees Celsius). Inundated with fresh water, Atlantic's northern deep current may weaken, and shallow. Then Europe's heat pump, which has run steadily for 8,000 years, may turn off, plunging mild Dublin into the icy, lonely cold of inhospitable Svalbard, 700 miles (1,100 kilometers) away from the North Pole. Having no collective memory of this past, we cannot imagine its reemergence, despite the record before us. Arctic chills have hit earth suddenly, and then lingered. Climate models suggest that the sea might adjust to greater infusions of carbon dioxide, but only if the increases come slowly, not at today's frenetic pace.[16]

A warming earth may diminish Atlantic's deep circulation. Whether the slowing or shutdown of deep currents could further accelerate warming, scientists cannot fully say. Of the six billion tons of carbon dioxide man expels into the atmosphere each year, half remains there; the rest disappears, dissolved in ocean currents, absorbed by the sea's floating meadows, inhaled by the grasses and trees of dry land. Soaking up carbon dioxide, the sea buffers earth's climate against increased emissions of greenhouse gases. Slowing the deep circulation may weaken this buffer. A portion of the sea's carbon dioxide falls to the depths with currents. More is breathed in by plants, or built into the shells of animals, before descending to the seafloor in their remains. The tiny grasses of the sea that anchor the marine food web bloom profusely in seawater bathed in nutrients. Nutrients drift toward the bottom; winds and rising currents lift them from the depths. When deep currents slacken, nutrients vital to the health of surface dwellers may stay cloistered on the dark bottom. The boundary between layers of surface water and the cold fertile sea below hardens as surface waters warm, and winds fail to penetrate the barrier. If deep-water circulation in North Atlantic waters collapses, the sea's ability to absorb carbon dioxide from the atmosphere may decline by as much as one third or one half.

The sea feels the dimming of deep currents, but science has only begun to grasp the fine workings, only begun to decipher the delicate balances between carbon dioxide drawn down by currents and that imbibed by floating plants. Climate models cannot fully describe how the biology of the sea responds to a warming earth, or how melting glaciers in Greenland, thawing sea ice in the Arctic, and rising seawaters in the Bering Strait combine to influence deep circulation. Our computers are not fast or powerful enough to capture all the complexities that make up a planet imbued with life.[17]

It has been argued that man's contribution to the carbon dioxide content of the atmosphere pales in comparison with the immense volume pulled from the air through photosynthesis and returned through respiration and decomposition. Each year, earth and her atmosphere exchange almost 200 billion tons of carbon dioxide. However, tiny quantities can loom large. Just as barely discernible fluxes in salinity command the currents, tiny additions of carbon dioxide can raise earth's thermostat. Records of climate contained in air bubbles trapped in Antarctic ice show that for the last 160,000 years, earth's carbon dioxide levels and temperature rose and fell in concert. Temperatures ranged over approximately 12 degrees Fahrenheit (6.7 degrees Celsius), a fluctuation far greater than any we have known in the last 10,000 years, far greater than anything we might find comfortable. If present trends continue, man will fill the atmosphere with more carbon dioxide than earth has seen in 50 million years.

Thousands of years of climate change embedded in the fossils of the deep sea suggest that whether earth is buried in an ice age or basking in a brief interlude of warmth, the planet is feverish, heating and cooling rapidly, with deserts turning to grassland and forests to tundra and back again. The rains and waters that have sustained human civilization for so long answer the call of the sea. Small motions of Atlantic influence life and death in the Sahel, civilization and famine in the Tigris–Euphrates Valley. Larger motions open and close Greenland and Iceland to settlement, provide fields of plenty or starvation and disease in Europe. Still larger motions summon hundreds, thousands of years of ice. No matter how poorly or dimly we understand these cycles, we feel their presence.

It is likely that man's arrival on this planet was a response to the rhythms of the sea. *Australopithecus* lived on in the woodlands of Africa for one and a half million years, climbing trees to find leaves, fruit, and seeds to eat during the day and seeking refuge from hungry leopards, hyenas, and dogs at night. Nimble *Australopithecus* babies scampered among the branches at an early age; their tree-climbing mothers could not carry them. Generation after generation of *Australopithecus* survived, untouched by evolution, until, on the other side of Atlantic, the Isthmus of Panama rose from the sea. Before that time, Atlantic and Pacific waters mixed as trade winds blew water across the Caribbean into the Pacific. The rising isthmus blocked water but not wind, freshening the Pacific with rain blowing in on the trades and leaving a salty Atlantic to initiate the circulation of deep, cold currents. Warm water pulled north in the Gulf Stream, sinking in the Labrador and Greenland Seas, released heat and moisture into the air. When earth swayed in her orbit, it snowed and snowed, freezing the Arctic, beginning the ice ages.

As ice descended from the Poles, as the tundra crept into Europe and the dunes of the Sahara swept south, the African woodlands shrank, giving way to grassland, forcing *Australopithecus* down from the trees. Some *Australopithecus* mothers, not needing all their limbs to climb, carried their babies. These babies matured more slowly. They couldn't climb or clamber or fend for themselves as their tree-climbing forebears had, but while they rested in their mothers' arms their brains grew to an unprecedented size. The abrupt change in climate and habitat created an evolutionary opportunity. Slow-moving *Australopithecus* became easy prey for swift meat-eating predators; *Homo* compensated with its large brain. Within a split second of geologic time, in little more than 100,000 years, *Australopithecus* died out. *Homo* survived, filling a niche created by a change in climate wrought by the sea.

The sea brought us here. We have adjusted to its more gentle rhythms. As ocean-born rains deliver or withhold life-giving water, our civilizations and cultures flourish and fade. Now, toying with our atmosphere, we break the rhythms of the sea, nudging the climate, ignorant of whether we can adapt to the new niche we are creating. It may be an arrogant gamble.[18]

CHAPTER 6

Long-distance Swimmers

MUCH of the *Cramer*'s course takes us five hundred, six hundred, sometimes one thousand miles offshore, sailing an unmarked sea. Although we recognize the shallows and thick green of the continental shelf, the inky blue and heat of the Gulf Stream, the weeds and calm of the Sargasso, we can't, without external aids, determine precisely where we are. The sea within our view, three miles (five kilometers) from the deck, three times that from the rigging, is masked; we cannot locate ourselves, cannot read the clues held in the water.

We cannot bear not knowing where we are, where we stand, and so we have bisected and dissected this watery unmarked space, crisscrossed it with lines of longitude and latitude, reduced it, defined it. It took many years before man could locate himself at sea, could find his bearings. In the interim, many a boat returning home to Europe laden with treasure or victorious from naval warfare met an ignominious end foundering upon unanticipated shoals. Mounting losses of ships, soldiers, and booty compelled the British Parliament to offer a king's ransom to the man who would make it possible for a captain to find his bearings. The English clockmaker John Harrison devoted his whole life to assembling a chronometer to keep accurate time at sea, a clock that would keep steady pace with the march of time, undaunted by the roll of the waves, untouched by corrosive sea air, and immune to swings in temperature and pressure. The clock could falter for no more than three seconds a day, for hidden in time is the measure of distance: the difference in time between high noon at sea and high noon at home reveals longitude. An error of only a couple of minutes produces a thirty-mile mistake.

While Harrison toiled over his clocks, astronomers sought the key to longitude in the starry heavens. Astronomers, mathematicians, and map-makers spent their nights recording the positions of the sun, moon, planets, and stars crossing the sky with the onset and departure of darkness. The fruit of their labors appears in pages and pages of fine print in the Nautical Almanac and the sight reduction tables. Equipped with watch, sextant, almanac, and tables, a sailor can find himself at sea, calculating his position on the water by observing the position of stars in the sky and converting celestial time into nautical miles.

On the *Cramer,* we practice the art of celestial navigation, with varying degrees of success. The process demands excruciating attention to detail. At each step, opportunities for error abound. We measure time, locally, to the exact second. We coddle the sextant, cajoling the stars into a perfect alignment that gently touches the fine edge of the horizon, not hanging a mote above nor dipping a speck below. We make multiple corrections, adjusting for idiosyncrasies in the sextant itself, for our own height above the water line, and for the refraction of light entering the atmosphere, which causes stars to appear higher in the sky than they really are. Any sloppiness—noting an inexact time, aligning a star imperfectly on the horizon, measuring imprecisely the angle between horizon and star—comes back to haunt us. Seemingly small errors, compounded by minor mathematical mistakes in the long computations, yield untrustworthy and occasionally bizarre results. One early and arduous calculation places us firmly on shore, in the middle of Texas. As time goes on, we become more attentive, more fastidious, but Barbados is a speck in a great ocean. Two days before we arrive, some of our fixes are still off by 80 or 100 miles (130 or 160 kilometers), errors large enough for us to miss this landfall.

The captain is deeply devoted to the science of mathematically converting the location of stars in the heavens into a precise location on the grid of longitude and latitude. He painstakingly corrects our amateur fixes and compares his own findings with the instantaneous readings spewed out continuously by the computerized satellite navigation system we also carry on board. One flick of a switch, and the Global Positioning System (GPS) spits out latitude and longitude, accurate to within 5 or 10 feet (1.5 or 3 meters). Our captain, wary of the GPS, refuses to surrender his independence to electronics and distant satellites, but the GPS is quickly coming into widespread use, its ease of operation welcomed by many navigators.

Yet to view the world in its terms is to court another kind of ignorance. We gain in absolute precision from electronics but lose in relative understanding. We know exactly where we stand, but no longer see ourselves in relation to our surroundings. The GPS requires of us no accountability, no discipline, no diligence, no need to observe carefully, no need to see ourselves in context. I saw this most clearly a few years ago, when my family and I were hiking in the White Mountains and a lost hiker called the Appalachian Mountain Club for rescue on his cellular phone. Holding his GPS, he identified his exact longitude and latitude but could not name the trail he had been walking on or describe terrain he had traversed or landmarks he had noticed. Perfectly situated on an imaginary grid, he was lost in the real landscape on the ground.

The GPS, housed in its black box like a digital clock, produces a number that sits in splendid isolation on a screen. Our GPS is mounted in the pilothouse, in the heart of the boat, and the number on the screen, that single point of longitude and latitude, centers us in a vast circle of water. As we scan the horizon, our view radiates from this hub. As we move, the center moves with us. Navigating by the stars requires the humbler perspective of seeing oneself in relation. To navigate by the stars is to know, always, that we and our boat can never be the center of anything. The whole process of celestial navigation, laborious and time-consuming, demands this recognition. While the GPS produces its readings instantly, by the time the last celestial calculation is completed and the position "fixed," the ship has long since drifted off.

Each observation and computation makes comparisons, establishes relationships. To begin the task is to place oneself in time, attuned to its passage, for only in those thirty minutes of dawn or twilight is it possible to view the stars and the horizon simultaneously and to observe where they stand, one in relation to the other. Where we float on the broad expanse of water depends on the height of the bright evening stars Vega, Deneb, and Formalhaut, or where Sirius, Aldebaran, and Capella shine in the sky at dawn. To locate each star is to take in the grand sweep of the constellations: Vega in Lyra, the Lyre of Orpheus, Aldebaran in the eye of Taurus the Bull, and white Deneb in the tail of Cygnus the Swan, flying across the band of the Milky Way. Our gaze widens, encompassing stars, constellations, and galaxies, and we become smaller.

Celestial navigation situates a ship in a sea of stars, deriving true position by comparing the altitudes of stars observed by the navigator with computed altitudes of stars from nearby locations. The magic of trigonometry establishes the correspondences, linking the transit of stars in the sky with the passing of time, the passing of time in the heavens with longitude over the water. Ultimately, the calculations produce azimuth and intercept, intimations of direction and distance that place a ship not on a point, but on a great line circling the earth, a line with no beginning and no end. Three stars yield three lines of position, and where they intersect, the boat will be. The GPS carries out its calculations invisibly, revealing only the end result. Celestial navigation exposes the inner workings of the watch. To see them is to know that wherever the boat may be, its location is relative.

Even if one cannot appreciate the beauty of the math—even if one loses oneself in the complexities of sine and cosine—the idea of relativity, of context, still stands in relief. Without the Nautical Almanac and sight reduction tables, we could, in a rough way, track our passage south, observing Polaris sink in the sky as our latitude dropped, or greeting the southern constellations of Cetus the Whale, Eridanus the River, Pisces, and Aquarius, whose stars turn the firmament into an infinite ocean enveloping earth. The GPS spits out numbers; the stars tell us that water is our lifeline. Ever since the dawn of Western civilization, the motions of the morning star Sirius have heralded the coming of rain and the flooding of the Nile, the approaching season of life and growth. The GPS has for us no resonance, no mythic sense of the connection of life and liquid. Although, over time, it will likely prove more precise and less susceptible to error than even the most skilled human navigator, in abandoning the dome of the sky, retreating inside the ship to find our bearings, we stand alone, separate, disconnected from the larger world, and ignorant of the means by which we can measure our place. With no context, we, like the hiker, are lost.

In between the morning and the noon sighting, the noon and evening sighting, we measure our distance and direction traveled by dead reckoning, reading direction off the compass and distance traveled off the log, a rotator towed behind the boat, turned by flowing water. A dial mounted at the stern on the inboard end of the line counts the revolutions and converts them to nautical miles. These readings, oblivious to currents and the press of wind against hull and rigging, are alarmingly inaccurate, unlike those of Christopher Columbus, one of the world's finest and most precise dead-reckoning navigators. Tossing a piece of flotsam overboard, chanting to time its journey from stern to bow, Columbus measured speed and distance, and his log was off by less than 10 percent. He accepted, as did most cartographers and geographers of his time, Ptolemy's low estimates of earth's circumference. He assumed an earth 25 percent smaller than its actual size, and relied on a compass that only partially corrected for discrepancies between true and magnetic north, yet Columbus found landfall where and when he expected to.

Northern navigators, making their way to Iceland, Greenland, and then Newfoundland, sailed by the stars and by the quality of light in the sky. A chilly Arctic ice pack lying beneath a mass of clear clean air reflects, bends, and magnifies light, raising distant lands beyond the horizon into sight. The Arctic mirage, lifting Greenland's glacial mountains into the sky, guided Norse and Irish sailors across uncharted seas, led them across to Iceland, to Greenland, and into the currents that drew some of them to Newfoundland.[1]

While we fumble our way over the sea surface, in poor imitation of Christopher Columbus and Leif Erikson, Atlantic's truly great navigators, animals large and small, roam the waters below. Their dwellings encompass broad swaths of sea, reaching from summer feeding grounds in fertile northern waters to calving and spawning grounds in the tropics. How they navigate through ostensibly unsigned waters, to what degree they are dead-reckoning sailors like Columbus, using magnetic compasses embedded somewhere in their brains or psyches, or how much they are celestial sailors, somehow following the change of light in the sky, we are only beginning to fathom.

In the waters of tropical Atlantic, midway between Brazil and Angola, a lone peak of volcanic rock, no more than five miles (eight kilometers) across, rises from the sea. Part of a vast undersea mountain range, its name is Ascension. A distant outpost of Great Britain, Ascension's inhabitants are few, but its visitors are many. Each December, hundreds of green sea turtles leave their feeding grounds in rich sea grass pastures off the Brazilian coast to converge on this island's remote shores. In the darkness of night, female turtles slowly haul themselves out of the water and up the steep beaches. In places beyond the furthest reaches of the highest tides, each one slowly, awkwardly, laboriously brushes her flippers across the sand and little by little, bit by bit, excavates the large pit, which will be her nest. It is a struggle, for millions of years ago, when freshwater turtles took to the sea, their flexible limbs evolved into rigid paddles more suited for swimming than digging.

As the night wears on, each female turtle lays about one hundred eggs. Then, exhausted, she buries the nest in sand and drags her heavy body down the beach, into the lightness of water. She stays nearby, in the sea, for the next two weeks. She eats little. There is little to feed on in these waters, but that doesn't matter; she has another purpose. She comes ashore again, to dig another hole, to lay another clutch of eggs. At the end of the nesting season, when she has left two or three or maybe four nests in the sand, she leaves Ascension and makes the long journey home.

Seven or eight weeks later, the eggs hatch and the babies, no bigger than silver dollars, dig their way out. The tumbling mass works its way toward the light, those on the bottom tamping down sand scraped from the ceiling by those at the top of the pile. Finally, in the cool of night, they emerge and scramble down the sandy slope toward waters they cannot see. The hasty dash takes only a minute or two. Some turtles don't make it, snatched by birds and crabs before they reach the sea's edge, but on lonely Ascension, predators are few. Those hatchlings that do reach the sea will never forget this beach. In the first hours of their young lives, they are imprinted with a sense of this remote place, which they will never lose. When the female green turtles mature approximately thirty years later, they swim straight back across 1,400 miles (2,200 kilometers) of sea, called to nest in the sands of their birth.[2]

We cannot set a direction at sea without compasses and clocks, satellite readings and celestial almanacs, yet baby sea turtles quickly, easily learn to find their way. Emerging from their nests, they crawl away from the blackened silhouettes of sand dunes toward a distant horizon lit by the night sky. Instinctively, they move, preprogrammed to cope with the unfamiliar, toward waters they have never known. They are inexperienced, but when the first crashing wave lifts the hatchlings off the sand, they know to dive, and ride the undertow away from the beach. Waves passing overhead lift the turtles up, tossing them forward if they move with the water, back if they resist. Their tiny bodies feel the difference, and calling on an ancient wisdom passed down from generation to generation through their genes, they respond, facing the waves and swimming out to deeper water. In these first minutes on the beach, in that frenzied race over sand and into wave, the hatchlings set their direction. Then they hold it, long after the sea's signposts have disappeared, long after the land has slipped from view and the bottom has dropped away, turned too deep to bend the waves toward shore.

The baby turtles swim on in the faceless sea, keeping their bearings with internal compasses set to the inclination of earth's magnetic field. Wherever these compasses reside, in magnetic receptors in their eyes or magnetite particles embedded in their brains, they guide young turtles out into the currents. Turtles born in the sands of north Atlantic waters circle the Sargasso Sea. They live in the currents, enveloped by warm water, for many years. Outside this gyre of heat circling the central Atlantic, the water is dangerously cool. Where the Gulf Stream divides, a cooler strand flows north to Great Britain and Norway, another south toward Portugal. At the split, young sea turtles turn south, toward the warmer, more hospitable water, cued by changes in the inclination of earth's magnetic field, which increases with latitude. The turtles must inherit knowledge of these magnetic signposts in the trackless sea, for they cannot learn by trial and error. Experience is not a kind teacher; errors are costly. The hatchlings perish if swept into cold water.

Even older, hardier turtles who lose their way succumb. In December 1999, over two hundred endangered Kemp's ridley turtles were stranded on the beaches of Cape Cod. Carried north on the Gulf Stream, they lingered in autumn's warmth, feeding in Cape Cod Bay. When the cold snap came, they dove to the bottom to wait it out, but the low temperatures persisted, stunning the turtles when they were forced to the surface to breathe. Immobilized, unable to swim or eat, they washed ashore. Half died. The others spent up to six months in rehabilitation, their body temperatures raised slowly to enable them to return to warmer waters. Occasionally the turtles wash up farther afield. In 1992, two loggerheads washed up on Ireland's Dingle Peninsula. Weak and tired and way out of their range, they were flown home to the Gulf of Mexico.

Sea turtles, itinerant mariners, travel the breadth of the sea over and over again during their lifetimes, and year after year return to the same familiar places, swimming straight for their destinations, knowing precisely where they are going. Year after year, season after season, female green sea turtles make the same 2,300-mile (3,700-kilometer) round trip between Ascension Island and feeding grounds in the shallow waters of the continental shelf off easternmost Brazil. How they find particular nesting beaches, particular waters with abundant pastures of sea grass, eludes us; they seem to perceive the broad reach of water in ways we cannot.

Scientists are coming to believe that sea turtles may navigate with global positioning systems we humans can neither see nor feel. Earth's magnetic field varies in intensity and angle of inclination, and sea turtles are alert and responsive to both. Lines of similar intensity and inclination cross the sea, forming a grid, a map of magnetic latitude and longitude. Baby green turtles seem to learn the magnetic coordinates of Ascension's volcanic sands as they crawl into the sea and head into the waves. Many years later, when the nesting season begins, they leave the rich feeding grounds off Brazil and, guided by their internal compasses, navigate a straight course toward the island, swimming directly and unerringly to this small, remote rock in the middle of nowhere.[3]

Sea turtles have crisscrossed the Atlantic, navigating between home and pasture, for millions of years. The oldest known sea turtle abandoned its terrestrial life for the waters of what is now coastal Brazil more than 100 million years ago, when the Atlantic itself was a young and growing sea. At that time, the turtle had developed around its eyes enormous tear glands, nearly as large as its brain, large enough to cry away the vast quantities of salt it would ingest in its new home. As time passed, this ancient turtle gave rise to other marine turtles, to herbivorous greens, carnivorous loggerheads, and sponge-eating hawksbills. When the land bridge that became the Isthmus of Panama rose above the sea between three and four million years ago, Atlantic and Pacific populations of ridleys became isolated and evolved separately. By the time the Pacific ridley swam around the Cape of Good Hope and met its Atlantic cousin, they had diverged into two species. Between half a million and a million years ago, green turtles of the Atlantic shared a common ancestor, but today the populations have divided and diversified, becoming genetically distinct.[4]

It is the female sea turtle who has brought forth these separate populations. We cannot follow sea turtles over the long course of their wanderings, but pieces of shell, drops of blood, and membranes from eggs all contain the DNA that records the nesting patterns of generations. Locked in the DNA is a history, revealing that when a female sea turtle matures, she returns to the place of her birth to lay her eggs, to a beach she last saw, touched, and smelled perhaps thirty years earlier. Throughout their long lives, female sea turtles are faithful to the sands of their birth. Green turtles from Ascension Island, from Matapica, Suriname, and from Atol das Rocas, Brazil, all mingle in the rich pastures of sea grass where they feed, but when nesting season begins, they separate, returning to their native beaches, no matter how far away.[5]

On the scale of time's long reach, marking the birth of an ocean or the waxing and waning of the thick ice, sea turtle colonies are short-lived. When ice last covered the northern latitudes and its chill crept into warmer climates, the sands in Florida and Cyprus where green turtles nest today dipped below 77 degrees Fahrenheit (25 degrees Celsius), too cold to incubate eggs. Ascension Island barely rises above the water. A rising sea could drown its beaches, an earthquake ripping through the watery mountains below could destroy the island, and with it the entire turtle colony. Over thousands of years nesting beaches disappear, but new rookeries evolve as occasionally, very occasionally, a turtle strikes out on her own, leaves what is familiar, and builds her nest on a different beach. The green turtle population on Ascension may be descended from a single turtle who somehow found her way there. What brought her? Perhaps her compass failed and she strayed from the path she knew. Perhaps a raging hurricane destroyed her natal beach and forced her to set a course for parts unknown. For us, beached on dry land, her behavior remains a mystery.[6]

Sea turtles move to long, slow rhythms of time, spun out over thousands of years. Over this broad reach, a break in the nesting pattern by a few solitary females enables turtles to adapt to vicissitudes of climate, to rising seas, to cooling currents. In recent years, man has imposed, over these long cycles, his own faster tempo. It may take thousands of years for wind and rain to carve out a new beach, and then thousands more for wave and current to carry away the sand, but man, building on the dunes or redirecting the flow of water with jetties, can drastically interrupt these natural rhythms in only a few years. In this rush, a sea turtle's fidelity to her nesting site could become a liability. In this shorter term, of tens or even hundreds of years, female turtles rarely abandon their natal beaches. They will not swim in from other beaches to recolonize an abandoned or extirpated rookery.[7]

On the eastern side of the Yucatán, in the Mexican state of Quintana Roo, lies one last stretch of undeveloped beach, an important nesting site

for green and loggerhead turtles. A Spanish business conglomerate has recently purchased the beach. Its plan to turn X'cacel into a resort, with restaurants and a large hotel built into the dunes and off-road vehicles marching across the sand, will be incompatible with sea turtle nesting, will jeopardize future generations of turtles. Newly emerged hatchlings turn away from darkened dunes and follow the light of the night sky down to the water. When dunes are brightly lit with artificial light, the hatchlings turn inland, their tracks eventually disappearing in the vegetation. Even if the beach is dark, baby turtles emerging on the sands of a touristed beach are often derailed by the tracks of dune buggies.[8]

Decimation of a population doesn't happen overnight. With a blind and undeviating sense of purpose, female turtles return to their nesting sites year after year, only to have each clutch of hatchlings pulled to its death behind the dunes. After twenty years, when no young turtles have survived to reproduce, the population crashes. Man's impulse to possess the beach is ongoing. Green turtles once thrived in the Mediterranean, but beach development has pruned the population, reduced it to a few hundred turtles nesting on the sands of Cyprus and neighboring Turkey. Each rookery is genetically distinct; extirpating colonies in Cyprus and X'cacel would remove one third of the genetic diversity in the Atlantic green turtle. Narrowing the gene pool would further weaken this remnant population.[9]

Occasionally the fate of an entire species rests on the safety afforded by one single beach. The Kemp's ridley, critically endangered, swims in from all over the Gulf of Mexico, the Caribbean, and the eastern seaboard of the United States to leave her eggs on a hundred-mile-long beach strewn with bits of seashell at Rancho Nuevo, Mexico. Turtle meat, though, is a delicacy, and the eggs are prized as aphrodisiacs. For many years *veladores,* the turtle hunters known as night watchmen, staked out the beach, seizing turtles and plundering nests, thoughtlessly extinguishing the lives of thousands of animals. Turtles were oblivious to the danger. They couldn't know the fate of last season's nesters; they couldn't know they were crawling to their deaths. Year after year they came to Rancho Nuevo, and each year their numbers diminished. In

1985, when the nesting season began, only two hundred turtles arrived on the beach. When the future of the Kemp's ridley was on the line, we withdrew from the onslaught. Now, each April, when the first turtles appear, another kind of watchman patrols. Volunteers, under the aegis of the Mexican and U.S. governments, dig up freshly built nests and move newly laid eggs to protected corrals, safe from all predators. This strenuous effort has begun to pay off. In 1997, the number of nests increased to 2,000, bringing the population up to 2.5 percent of what it once was.[10]

In 1947, approximately 40,000 turtles swarmed ashore to nest at Rancho Nuevo. They came in concert, in an *arribada,* nesting en masse, thousands at a time. When the adult turtles disappeared into the sea, crabs appeared on the sand and coyotes came out of the grasses behind the dunes. They descended on the beach and preyed upon the eggs, but the *arribada* offered herd protection. When few turtles come ashore, many nests are attacked. During an *arribada,* the predators quickly become glutted, leaving a higher percentage of nests undisturbed. Though today's crowd at Rancho Nuevo hardly constitutes an *arribada,* the number of Kemp's ridleys is increasing. We maintain vigilance over the beach, guarding the turtles as the population recovers from our excesses and hoping for a time when they will once again be wild, when thousands of turtles will come ashore to nest undisturbed, when there will no longer be a need to incubate eggs in pens, and when nesters will once again be so many that losses from heavy rains or high tides or the feasting of crabs will make no difference at all.[11]

And yet, even the best of nesting conditions—the safest beach, the most auspicious weather—cannot fully safeguard a population. The neighborhoods of sea turtles, defined by feeding grounds, nesting beaches, and juvenile nurseries, are separated by hundreds, often thousands of miles. The slope of beach, the expanse of sea pasture, the drift of current: these edge turtle neighborhoods. Their community extends beyond our political boundaries. Though we may be divided politically, we fish in the same waters. The survival of sea turtles is threatened, both on the shore where they nest and in the water where they spend most of their lives. Atlantic, vast as it is, is no refuge.

Endangered hawksbills roam the Caribbean, but for them the pellucid waters of this tropical sea are far from safe. As desirable as the ivory tusk of an elephant or the horn of a rhinoceros, the "tortoiseshell" of a hawksbill is a precious commodity. We, who might not identify this small oval-shaped turtle with its curved beak and amber, black-and-brown-streaked shell if we saw one in the flesh, recognize it in eyeglass frames, combs, jewelry, and guitar picks.

Not always was the hawksbill an endangered species. Once hawksbill shells filled Colombian middens, were built into the shields of Yucatec warriors. Receipts from British and French traders record the exchange of large numbers of hawksbills. Our vanity has nearly destroyed this turtle. Fishermen and *veladores* have picked thousands of them off beaches where they nest and coral reefs where they feed, selling them in a lucrative market that reaches around the world. In three generations, the population of hawksbills has declined 80 percent. Although international trade in hawksbills was banned in 1975, Japan continued to import thousands of Caribbean hawksbills each year for another seventeen years. Today, approximately 5,000 hawksbills nest in the Caribbean each year, mostly in Mexico. Other traditional nesting beaches, once crowded with 200 or 300 turtles, are empty.[12]

No single country can claim this turtle. The whole Caribbean basin is their realm. Hawksbills forage on the Doce Leguas Cays, a reef off southern Cuba. Many nest there as well, but others lay their eggs on beaches in the U.S. Virgin Islands, Belize, and Mexico. Turtles foraging on the Puerto Rican reefs of Mona Island build their nests on sands in Puerto Rico, but also in the U.S. Virgin Islands, Cuba, and Mexico. Turtles hatched on the beaches of Mexico swim to the Dominican Republic and the Bahamas to feed. Long-distance to us is local to them; turtle neighborhoods extend from St. Croix and Jamaica to the Miskito Keys in Nicaragua, from the Yucatán to the Dominican Republic. In order to bring this turtle back from the edge of extinction, each beach, each reef, must be a haven.[13]

While the Caribbean is home to the hawksbill, the loggerhead is a more international traveler, nesting in the sands of South Carolina, Georgia, and Florida but spending much of its life across the sea. After the hatchlings scramble into the water and set their course for the Gulf Stream, their fate lies in the hands of distant peoples, distant nations. The turtles drift away in the currents, only to reappear years later on the other side of the ocean, near Madeira and the Azores. There they can be found, basking in the heat of the noonday sun and eating jellyfish. Almost all juvenile loggerheads feeding here were born in western Atlantic waters, in the United States or Mexico. The young turtles follow currents circling Atlantic, drifting from the Azores and Madeira down toward the Canary Islands and then back across the ocean, roaming for eight to twelve years before they return home. The exact routes of their wide-ranging migrations are still unknown, but many detour into the Mediterranean, where as many as half the young loggerheads hail from across the sea.[14]

Appearances deceive. While 90 percent of Atlantic loggerheads nest on U.S. sands, protecting their nesting areas from erosion and encroaching development only stems the decline of endangered populations of turtles. These efforts will not, of themselves, save turtles from extinction. Even if somehow every egg hatches, and every hatchling successfully runs the gauntlet to the sea, an endangered population of turtles will not recover. Even under the best of circumstances, few baby turtles live to reproduce. However, cutting the death rate by 50 percent among older juveniles, among those turtles on the verge of maturity, can bring back a population. Man and turtle cross paths not only late at night, in the darkness of a sandy beach, but out on the water, where as many as 40,000 turtles each year have drowned in the nets of shrimpers. In 1994, after years of debate and dissension, the U.S. government required all shrimpers to install grates in their nets to allow captured turtles to escape. The devices work well enough to give beleaguered turtle populations a chance of recovery. In the South Carolina shrimp fishery, turtle excluder devices (TEDs) reduced turtle strandings by 44 percent. If shrimpers comply with the law, allowing more and more young turtles to reach maturity, the number of turtles may increase substantially at the end of fifty years.[15]

While some shrimpers have accepted the new regulations, others have not. When law-enforcement officers boarded shrimp trawlers in the spring of 1999, they found violations ranging from nets without any turtle excluder devices to nets where the escape openings had been sewn shut. Many turtles continue to wash up on Texas beaches during the breeding and nesting season, but in 1997, when shrimping was banned for eight weeks along the Texas coast, the number of turtle strandings declined 90 percent. In the coastal waters of Rancho Nuevo, Mexico, commercial fishing is banned during the turtle breeding season. For many years, people tried to reestablish a second nesting site, by incubating eggs from Rancho Nuevo and allowing hatchlings to imprint on Padre Island. Now, Kemp's ridleys are beginning to nest in this nascent Texas rookery. Given the number of Kemp's ridleys drowned in fishing lines and shrimp nets or severely injured by boat propellers, a prohibition against shrimping off Padre Island may be necessary to safeguard the breathing room we are finally affording endangered sea turtles.[16]

Further offshore, other threats are pressing in. While shrimpers have been required to modify their gear to save sea turtles, other fishermen have not. Longliners, with miles of line and hundreds of hooks set for swordfish and tuna, snare sea turtles on both sides of Atlantic, in the Caribbean and in the Mediterranean. In the Mediterranean alone, longliners hook 20,000 turtles a year; between 20 and 50 percent die. The U.S. National Marine Fisheries Service, in an effort to reduce the number of turtles killed or injured by fishermen, has temporarily closed portions of the Grand Banks to tuna, swordfish, and shark longliners.

While the behavior of fishermen is clearly linked to the survival of sea turtles, the responsibility for other threats is more diffuse, harder to isolate. Caribbean coral reefs, an essential feeding habitat for hawksbills, are dying, bleached by warming waters and weakened from nearby coastal development. Over the last twenty years, increasing numbers of green turtles have developed multiple tumors around their eyes and mouths. While the tumors are benign, they nonetheless threaten the turtles' well-being, making it difficult for them to see and to eat. The disease occurs worldwide and is now appearing in loggerheads and Kemp's ridleys. In Atlantic, fibropapillomatosis has particularly affected turtles living in the lagoons of the Indian River in eastern Florida. Its cause is unknown. Some scientists suspect algal toxins nourished by nitrogen and phosphorus running off land.[17]

When Columbus arrived in the Caribbean, the sea was awash in turtles, but now only remnants of once robust populations remain, leaving most sea turtles threatened with extinction. Four hundred years ago, the population of Caribbean green turtles may have numbered in the millions, but entire rookeries in Bermuda, the Cayman Islands, and elsewhere throughout the Caribbean have been wiped out, leaving only a shred, perhaps 1 percent, of the population. Where we have identified major threats—shrimping gear and the destruction of nests and nesting habitat—and acted with restraint, some of the tiny turtle populations have begun to stabilize. Numbers of Kemp's ridley nests at Rancho Nuevo, green turtle nests in Tortuguero, Costa Rica, and hawksbill nests in Mexico's Yucatán are finally increasing. We have far to go, but now we can hold a fragile hope.[18]

At the break in the continental shelf, the edges of what once was land plunge into the deep sea, and cool and fertile coastal waters meet the clear blue of open ocean. It is a busy place. Where the *Cramer* passes through, fishing boats gathered at this boundary light the night. They hunt tuna and swordfish feeding in rich waters, fish of great strength, power, and endurance. When the season changes, the huge fish, fattened with food for the coming winter, disappear from America's fishing grounds, embarking on a long and incompletely understood journey through the sea. In the spring, some bluefin resurface to spawn in the warm waters of the Gulf of Mexico. Afterward, they swim with the current through the narrow Straits of Florida, shunting through clear waters off the Bimini coast known as Tuna Alley. Then they race north to feed in the shelf waters off Cape Hatteras, in the undersea canyons off New

York, on the shoals of Georges Bank or the Grand Banks. Others may spawn in distant offshore waters, in the middle of the Sargasso Sea, between Bermuda and the Azores. In Atlantic's eastern waters, bluefin spawn off the coast of Africa and in the Mediterranean, then fan out along the continental coast. Swordfish, too, cover great swaths of sea, roaming the length of Atlantic between Canada and Argentina, between Ireland and South Africa.[19]

Relative newcomers to the marine world, tuna and swordfish have evolved into some of the sea's most highly developed fishes. While the cod, haddock, flounder, and plaice who dwell year-round in the North Sea and the Gulf of Maine are cold-blooded, their body temperatures rising and falling in synchrony with the surrounding water, thus limiting their geographic range, swordfish and tuna, exquisitely adapted to live in the vastness of the sea, are free from the boundaries imposed by temperature. The swordfish who surface at the shelf edge have swum up from the depths, rising hundreds of feet through the water each evening as the sun sets, following their prey of fish and squid. A temperature difference of 34 degrees Fahrenheit (19 degrees Celsius), as great as the swing between winter and summer, night and day, separates cold deep from warm surface. Swordfish exit one realm and swim into the other in under an hour.

Moving between such extremes would stun the nervous system of a cold-blooded fish, but these ocean princes make their own heat, warming themselves in the deep cold. The burner of the swordfish lies behind its eyes, below its brain, a dark mass of tissue surrounded by insulating fat, heavy with blood, and loaded with energy-producing mitochondria. With warm brain and eyes, swordfish can chase their food in waters deep and shallow, near and distant. By night, they feed at the surface, at the edge of the deep water. By day, they move onto shallow banks, like Georges or the Grand Banks, and dive down to feed, slashing through schools of menhaden and mackerel with their long, sharp swords.[20]

Bluefin thrive in waters as cold as 43 degrees Fahrenheit (6 degrees Celsius) and as warm as 75 degrees Fahrenheit (24 degrees Celsius) but unlike swordfish, they do not possess organs whose chief function is to produce heat. Instead they retain the heat they generate swimming. Other bony fish quickly lose their heat to the sea, for their red muscle lies near their skin, close to the cold water. In bluefin, who can weigh as much as 1,000 pounds (450 kilograms), red muscles are housed deep within the body, near the backbone. Warm venous blood flowing away from muscles heats cold blood coming in through the arteries, enabling tuna to retain 98 percent of their body heat, giving them free rein to forage in cold waters and to dip in and out of the Gulf Stream, where sea temperatures plummet as much as 27 degrees Fahrenheit (15 degrees Celsius) across one nautical mile. In cold water, the bluefin, separated from the chill by only a taut skin, maintains an internal temperature of 80 degrees Fahrenheit (27 degrees Celsius).

Coincident with the relocation of its red muscle, tuna developed the unique style of swimming for which they are so aptly named (*Thunnus thynnus,* from the Greek meaning to dart or lunge forward). While the bodies of other fish undulate through the water as they swim, the crescent-shaped tail of the tuna propels its rigid body forward. Retractable fins, small scales, and recessed eyes further enable tuna to thrust quickly through thick and heavy seas, easily overcoming water's drag and resistance. With their warm bodies, rapid metabolism, and sleek design, tuna excel at both short sprints and long-distance travel. They zoom in on prey in short, quick bursts of speed, and they can cruise at two body lengths per second, easily making long-distance endurance swims along an entire ocean basin. Engineers who design underwater robotics dream of replicating the sleek body of this 8-foot (2.4-meter)-long, 700-pound (320-kilogram) fish who rushes without ceasing through the breadth and depth of the sea.[21]

Swordfish and bluefin travel throughout Atlantic with tremendous speed, but from moment to moment, day to day, month to month, their migrations are not well charted. In the winter of 1997, when the warm Gulf Stream edged shoreward toward the coast of Cape Hatteras, pressing against cold water rushing south in the Labrador Current, giant bluefin gathered in the warmth along the boundary. The following year, when the Gulf Stream moved offshore and the chilly Labrador Current filled the waters of coastal Cape Hatteras, bluefin wintered in waters unknown to men. Some bluefin, fattened in American coastal waters during the summer and fall, follow the currents across the sea during the winter. How they navigate, no one really knows. They could be guided by internal compasses of magnetite chips embedded in their skulls, by the warmth, salinity, or motion of the current, by patterns of polarized light received by the pineal window in their heads, or by the scent of prey leaving an oily, odorous slick in the water.[22]

Tagging allows us to observe their transoceanic prowess. Giant bluefin are known to swim between the Bahamas and Norway, between the eastern coast of the United States and the Bay of Biscay. Just how many make the full transit is a subject of controversy. Although fisheries managers have long believed in the separation of eastern and western populations, and in minimal migration (no more than 1 or 2 percent) across Atlantic, genetic analysis and satellite tags may soon show the bluefin to be a true transoceanic swimmer. New and sophisticated tags, beaming the whereabouts of migrating bluefin into the e-mail of scientists at their computers, call into question the notion of separate, reproductively isolated stocks. The new data suggest that the line drawn to separate the populations—a line running through the central Atlantic at 45° longitude, between Bermuda and the Azores—may be arbitrary. Bluefin may not recognize the lines we have drawn. These data, though still preliminary, suggest that mighty bluefin traversing Atlantic heed no political boundaries. Data collected between 1997 and 2000 indicated that each year, between 30 and 70 percent of bluefin leaving their winter feeding grounds off the United States and Canada cross the supposed stock boundary line.

Further studies will reveal the reach of bluefin journeys. The migration routes of bluefin are important to know, for giant bluefin crossing Atlantic to feed or breed are susceptible to intense fishing pressures in the waters off Europe and in the Mediterranean. It may be that fisheries managers on one side of Atlantic who ignore the recommendations of their scientists jeopardize all Atlantic bluefin.[23]

In the twenty short years I have lived by the edge of the sea, the rhythms of life in offshore waters have subsided, the vitality of great fisheries has ebbed away. Swordfishermen who once reeled in 300-pound (140-kilogram) fish just off the coast of Block Island, Rhode Island, now must motor offshore, days away from the sight of land, to catch fish one third the size. Tuna fishing is big business. Today, when a customer is willing to pay $75 for a serving of bluefin in a Japanese sushi restaurant, hundreds of New England fishermen are out on the water hoping to land the $10,000 fish. The price of tuna fluctuates, but even at its lowest, sushi markets pay well. It used to be that Maine canneries purchased tuna for five cents a pound and turned it into cat food.

Only sound management can protect tuna and swordfish, since the hunt itself has become so one-sided. Not that long ago, when great fishermen tested their mettle on swordfish and tuna, they couldn't always find fish, and when they did, they couldn't always hook them. Fishermen, exhausted by the struggle, worn out by the strength and prowess of the prey, often lost to the fish. Today, when the season opens, as many as twenty-five boats will crowd into a single area, led by spotter planes that circle overhead, scanning wide stretches of sea, homing in on schooling bluefin. The gear is solid, the lines strong, and hooked fish are often landed.[24]

The sea empties, for men's sport, and the memory of its fertility disappears. In the past twenty years, fishermen have run down the numbers of breeding tuna, landing fish barely old enough to spawn. Almost two thirds of the bluefin caught in the United States are immature. These fish are taken for pleasure by recreational anglers who year after year exceed their quota. By the time regulators, using an inadequate monitoring system, finally close the fishery each year, fishermen have frequently taken almost twice as many fish as are allowed. Plundering future generations of bluefin has taken its toll. A ripe female tuna produces forty million eggs. During spawning season, dense larvae once clouded the Gulf of Mexico, but over the years the larvae have dwindled and the cloud has thinned. Older generations of fishermen remember the fecundity of the Gulf of Mexico, but many of them have died, and younger scientists tagging bluefin now have trouble identifying the Gulf of Mexico as a significant breeding ground.

Swordfish fare no better than tuna. Longliners set their hooks, and more than half the catch consists of fish too young to spawn. Female swordfish mature at 150 pounds (70 kilograms) and tuna at 235 pounds (110 kilograms), but it is perfectly legal to catch 44-pound (20-kilogram) swordfish and 70-pound (30-kilogram) tuna. Fishermen have taken them in the Gulf of Mexico when they are spawning, and in East Coast nurseries when juveniles gather to feed. Harpooners, who target adult fish swimming on the surface, are more discriminating than anglers, whose gear takes fish of all ages and sizes, but their numbers are small. Unable to compete with fishermen using less selective gear, harpooners have been shut out of the swordfishery. As for their share of the bluefin quota, they are granted a scant 5 percent.[25]

Time winds on, relegating great fish to the past. If the sea were lit and we could see how full it once was and how empty it has become, we might act more quickly, more decisively. Bluefin tuna fishermen, ignoring the abundance of thirty and forty years ago, have set their sights low, lobbying for quotas designed to maintain but not rebuild the current meager population. If tagging studies continue to suggest that bluefin tuna dwell on both sides of Atlantic, quota cuts in the United States will be insufficient unless they are matched in Europe. It may be possible to partition lands and people, but our waters are shared, and we cannot divvy up Atlantic's great long-distance swimmers. It may be that conservation efforts on one side of Atlantic are being undermined by quotas set too high on the other. In Atlantic's eastern waters, regulators, ignoring the recommendations of their science advisers, set eastern Atlantic bluefin tuna quotas at nearly twice what the population can sustain. If large numbers of bluefin cross Atlantic, European fishermen may be taking fish that Americans have left in the sea to spawn.

Europeans have flagrantly violated their all too generous quota, hauling in record-breaking catches. Flouting the regulations, Italy exceeded its 1997 quota by 39 percent, France by 45 percent. In addition, many of the fish are undersized. In 1998, 51 percent of North Atlantic bluefin and 83 percent of Mediterranean bluefin landed in the fishing ports of Spain were below minimum size and weight. Thirty-seven percent of North Atlantic swordfish and 86 percent of Mediterranean swordfish were undersized. Tuna and swordfishing are regulated by ICCAT, the International Commission for the Conservation of Atlantic Tunas. At the 1999 meeting, the U.S. delegation condemned this abuse, and the Europeans walked out.

Not every country belongs to ICCAT, and those who don't, further undermine the quota system by taking what fish they want. Driven by the high price of bluefin in Japan, as many as fifty vessels flying flags of convenience from Panama, Belize, and Honduras come to the Mediterranean each summer to pirate spawning bluefin. They land an additional 30 to 50 percent of the quota.

Commissioners did agree to reduce the swordfish quota, but again not as much as science advisers suggested. To protest the taking of immature swordfish by foreign longline fisheries, the United States has banned the import and sale of swordfish less than 33 pounds (15 kilograms), dressed weight (headed and gutted). The ban helps, but does not prevent the taking of swordfish too young to spawn. At the same time, regulators in the United States, responding to lawsuits, have decided to close swordfish nurseries between Charleston, South Carolina, and the Florida Keys and in the western Gulf of Mexico to long-lining. These closures should reduce by 31 percent the number of young swordfish cast back into the sea. A similar proposal working its way through Congress will, in addition, buy out vessels adversely affected by the closures. At any point, these regulations could become bogged down in political controversy or undermined by lax enforcement, but they are crucial steps, essential to restoring a vanishing fishery.[26]

Though it is in our common interest to replenish these great fish, retrieving them from the recesses of memory, timid regulators, resistant fishermen, and skittish politicians diminish their prospects. Those who govern our fisheries are stuck in a quagmire, caught in quarrels over numbers, arguments over the steepness of the declines, the degrees of decimation, and the apportionment of blame. They are embroiled in controversies over money, over the value of today's paltry catch versus the value of a bountiful future that seems too distant to touch. They are stymied by the political challenge of sharing fish who may take the whole of Atlantic to be their home. Drowning in the pressures of the moment, these stewards of our resources have trouble sustaining the long view. But only in modern man's narrow notion of time is ten years of rebuilding so very long.

We can rebuild swordfish and tuna fisheries to the robust populations of twenty and thirty years ago in less time than it took to destroy them, but because we have tarried so long, the necessary quota cuts are high. In addition, for a rebuilding plan to succeed, we must come to understand the biology of bluefin and recognize that they may be truly cosmopolitan fish. Nonetheless, we are blessed. Large numbers of bluefin born in 1994 survived. Large numbers of swordfish born in 1997 and 1998 survived. These young fish are tomorrow's hope. With adequate protection and high enough quotas on both sides of Atlantic, these fish could bring back what we have destroyed. Over and over again we work the sea until our fisheries collapse, forced to the brink of extinction; yet, in this case, despite our excesses, we have still to tax the ocean beyond its capacity for rejuvenation. Its waters are still forgiving, resilient, and for swordfish and bluefin, there is still time.[27]

Earth has harbored multicellular life for at least 600 million years. Each species of plant and animal appears but once, and when its course is run, it disappears. We, relative newcomers to earth's family, have chosen to impose on these natural rhythms, to determine the tenure of earth's offspring ourselves. We, among the youngest but also among the more aggressive and acquisitive of earth's inhabitants, are choosing when to extinguish the light of other species. We have toyed with the lives of sea turtles, played with the future of bluefin tuna, and all but decided the fate of the North Atlantic right whale.

Millions of years ago, when India and Asia collided, the sea separating them receded. The floor of the old ocean was shoved high into the Himalayas, where it lies today, in northern Pakistan. Layers of silt and mud that once lined the floor of this sea have been squeezed into rock, raised into mountains. Scattered within are imprints of leaves and fossils of seashells once strewn about in the shallow water. Also embedded in the rock are the bones of a 600-pound (270-kilogram) animal, *Ambulocetus natans,* a walking-swimming whale, ancestor of today's cetaceans. *Ambulocetus* walked on land and swam in the sea, flexing its spine to propel its enormous feet through the water. *Ambulocetus* evolved from a sea-dwelling animal, and eventually its descendants returned there, their long tails turning to flukes, their hips and legs shrinking to a few tiny vestigial bones.

Early whales hugged the shore, drinking fresh river water, but as their kidneys adapted to the salt, as their jaws evolved to sense the vibrations of sound, and as their sinuses grew to withstand the pressures of diving, they moved out into distant, deeper seas. Unlike sea turtles, who remained connected to land by coming ashore to nest, whales abandoned terrestrial life completely, continuing to breathe air, but giving birth at sea. We may never know what possessed these animals to cast off from shore. Perhaps the food was more plentiful, the water more sheltering, the predators fewer. Whales have endured in their liquid home for over 50 million years, but now, for some species, such as the North Atlantic right whale, time is running out.[28]

We hardly know this whale. It is rare to see the entire animal all at once unless it has washed up dead on a beach. It rarely lifts its enormous bulk, 50 feet (15 meters) long, 70 tons, above the water. Instead we catch glimpses of parts, the fluke dark against the sky as the whale makes a deep dive, the massive head as it surfaces to feed, or the broad, flat back as it glides away. What spirit, what impulse lies within the roughened skin, the thick blubber? There is beauty in the seasons of a right whale, and rhythm to its days, but our understanding, like our image of the whale itself, is partial.

For years we saw the right whale only with the eye of a predator. Long before seamen settled in Nantucket and New Bedford, Basque whalers killed right whales in the Bay of Biscay and then moved on to more profitable waters, to pristine eastern Canada and to the Cape Farewell hunting grounds in the Irminger Current between Greenland and Iceland. In Canada, they established a large and remote whaling station in Red Bay, Labrador. Red Bay empties into the Strait of Belle Isle, a narrow passage between Newfoundland and Labrador, often shrouded in fog and filled with drifting pack ice. Four hundred years ago the strait teemed with migrating whales, but the Basques slaughtered thousands of them. What remains of the New World's oldest and busiest whaling station is a sailors' cemetery on Saddle Island and several sunken galleons in Red Bay harbor. Similar remnants mark other ghost trails of right whales: an abandoned whaling station in Blandford, Nova Scotia, once launched whalers bound for the hunting grounds of nearby Sable Island. Occasionally the sign is a solitary whale cutting through the Strait of Belle Isle, or a lone pair feeding in the waters of Cape Farewell.[29]

American whalers finished off what the Europeans began, reducing the population of right whales to perhaps as few as fifty. However well-adapted right whales were for a life at sea, they were ill prepared for an encounter with us. For a human hunter, a right whale was the "right" whale to take. The whales swam close to shore, where, buoyed by a generous endowment of blubber, they often appeared at the surface, feeding, courting, resting. Once dead, they floated, and were easily towed ashore. People turned the oil into soap and fuel for lamps, the baleen into umbrellas, corset stays, and whips. They appear to have thrown away the meat. For this, the whale was hunted down to a population with only a few unrelated females.[30]

Today, the North Atlantic right whale, numbering about 350, is the world's most endangered large whale. Its home once reached all along the edges of Atlantic's northern waters, on both sides of the ocean basin. As whalers extended their range, that of the right whale contracted, shrinking to the waters between Florida and Newfoundland. Following the trail laid down in logbooks left from the heyday of the whaling industry, scientists returned to long-abandoned hunting grounds of colonial whalers and found, in coastal waters between Brunswick, Georgia, and St. Augustine, Florida, calving grounds of the North Atlantic right whale.

In January, February, and March, right whales swim into the warm shallow waters off Cumberland and Amelia Islands, where they give birth and nurse their young within twenty-five miles (forty kilometers) of shore. Mothers and babies, occasionally accompanied by schools of bottle-nosed dolphins, are scattered throughout the 1,000 square mile (2,600 square kilometer) calving grounds. The devoted mothers stay close to the calves, lolling in the water, giving piggyback rides, nursing. Though their broad black backs barely break the surface, their distinctive V-shaped spouts rise high in the air. Generations ago, as the whaleboats closed in, right whale mothers refused to abandon their harpooned calves and so succumbed themselves.

As spring approaches, the whales swim north to feed in cooler, nutrient-rich waters, sometimes swimming in close to land. Upper Cape Cod can be a desolate place in winter, when chilly winds sweep over the dunes to meet an icy sea, but for the watchful and patient, there is company. One raw March saw thirty whales in Provincetown Harbor, feeding only seventy-five feet (twenty-three meters) from shore. Some were recent arrivals; others had wintered in the bay. The Great South Channel affords no such easy view. A narrow cut in the seafloor, scoured by an ancient river, it is further offshore, separating the treacherous shallows of Georges Bank from Nantucket Shoals. In the spring, when the sea is rich in food, as many as 180 right whales may gather in the deepest part of the 50-mile (80-kilometer)-wide channel, skimming slowly through the surface water, their mouths open, gorging on the spring bloom of tiny copepods, each whale filtering through its baleen over a ton of plankton per day.

As summer nears, the water warms and the whales move on, many to the fertile waters off Grand Manan Island in the Bay of Fundy, whose extraordinary high tides are said to have begun flowing long ago when a whale slapped its tail hard in the water. Mothers and calves, young juvenile whales, and adults gather here, feeding in nutrient-laden water brought in by surging tides. Adults come together in large groups of thirty or forty to court. They sing, roll in the water, rub their huge bodies against each other, mate. Many right whales, but not all, complete the 1,800-mile (2,900-kilometer) journey from the calving grounds to this bay. We recognize but few neighborhoods frequented by right whales, and their migratory patterns by and large elude us. While some mothers and calves summer in the Bay of Fundy, 25 percent gather elsewhere, in another nursery unknown to us. We have not fully identified right whale courting grounds.

Once adult whales frolicked in the shallows of Browns Bank, off Nova Scotia, but by 1993 they had abandoned the bank to join families feeding in the Bay of Fundy. In 1997, over 160 right whales summered there, but in 1998 those who came left early, only a few weeks into the season, for reasons we have yet to decipher. Calving whales winter off the coasts of Florida and Georgia, but no one knows where the rest of the population spends the year's coldest months. Perhaps it is a quiet place, undisturbed by the comings and goings of men. Or perhaps, like the calving ground that existed for so many years without our knowing, it lies before our unseeing eyes.[31]

First the League of Nations and then the International Whaling Commission banned commercial whalers from slaughtering right whales. Despite more than sixty years of protection, the North Atlantic right whale remains on the brink of extinction, while its southern cousin, dwelling in the coastal waters of Argentina and South Africa, is recovering at a rate of 7.6 percent a year. Scientists wonder whether, in the northern population, enough newborn whales are born and survive each year to sustain the tiny population, to replace those who have died, or whether the right whale has begun a slow descent toward annihilation. For this depleted population, the line between the two is fine. The species is poised between life and death; the premature death of only a few reproducing females could throw the balance. During the 1980s, the population grew at an annual rate of 2.5 percent, but now, at the turn of the century, even that minimal growth rate appears to have ceased. If the death rate is not slowed, the population will become extinct within two hundred years.[32]

Natural losses to a population are inevitable; other losses are not. Although whalers no longer take to the water, we are still killing whales. Our actions, less conscious, less deliberate, are still fatal. As our economy expands, we edge further into the sea, encroaching upon the habitat of whales. We crowd the water, and whales cannot navigate safely. Three congested shipping channels cut across the right whale calving grounds. While mothers tend their young, seemingly oblivious to the traffic around them, car carriers, tankers, and containerships pass through en route to the busy and growing ports of Brunswick and Jacksonville. Volkswagens and Saabs come in; Fords and Saturns are shipped out. Wood, wood pulp, and paper products leave. Chemicals and machinery arrive. Dredging equipment operates twenty-four hours a day, keeping the water deep for freighters. Nuclear submarines, missile carriers, and destroyers from the Kings Bay and Mayport naval bases move stealthily through the water as officers practice tactical maneuvers, test weaponry, and try the mettle of submarines under siege. During the 1997–98 calving season, observers sighted forty-eight whales and over five hundred commercial and military vessels.

Shipping lanes also cut through right whale feeding grounds in the Great South Channel and south of the Bay of Fundy. The Bay of Fundy shipping lanes service more than one thousand ships each year, including tankers weighing as much as 400,000 tons and high-speed passenger ferries that fly through the water at 40 knots, twice the speed of their predecessors. Collisions occur. In 1994, a disoriented young calf, barely a year old, swam up the Delaware River to Philadelphia, where he was hit by a tugboat whose captain couldn't see him in the darkness. When the whale reappeared in the Bay of Fundy three years later, he was healthy, but still scarred by the boat propeller.

This whale suffered only superficial wounds; others are not so fortunate. Hanging in the New Brunswick Museum is a skeleton of a young right whale killed in the Bay of Fundy by a passing ship. In August 1997, another young whale died, its jaw broken from a collision in the bay's shipping lanes. In the spring of 1999, another ship struck another female.

Staccato, as she was called, was found off the coast of Cape Cod. She had five broken vertebrae and a broken jaw. These deaths add up. Between 1970 and 1999, sixteen right whales died from ship strikes. The number, though small, accounts for 35 percent of right whale deaths. When the population is in such dire straits, when it must be rebuilt whale by whale, calf by calf, every premature death jeopardizes the entire population. The three whales struck and killed by ships were females. When they died, the hope for fifteen to twenty-one new calves died with them.[33]

Ships crowd right whale feeding grounds. So does fishing gear. Each year when right whales arrive in northern waters, some are snagged in the thousands of lobster traps and miles of gill nets lining the sea between New England and the Bay of Fundy. In 1997, scientists reported five entanglements. The numbers were the same in 1998 and 1999. Entangled whales thrash around in the line and then swim on, dragging the gear, hundreds of feet of line crisscrossed around their flippers, caught in their mouths, wrapped around their bodies. While some whales shake it off, others can't. Whales entangled in fishing gear are susceptible to serious injury: the line can cause infections, impede swimming, and inhibit feeding.

Between 1970 and 1999, fishing and shipping activities seriously injured fifty-six right whales, about 18 percent of the population. Fishing gear caused at least half the injuries. Gillnetters and lobstermen, depleting their own fisheries, are harming this highly endangered whale as well. In 1999, a nine- or ten-year-old female right whale, just old enough to bear a calf, was killed by a polypropylene gill net. The line, wrapped around her body and flippers, had cut through her blubber and into her muscles. She dragged the net through the Great South Channel, up into the Bay of Fundy, and then back before she finally died off Cape May, New Jersey. More than 60 percent of the right whale population has been caught in fishing gear at least once. To what degree entanglement in a lobster or gill net has prevented a whale from carrying and nurturing her unborn calf, we may never know.[34]

Three million years ago, a change in the climate forced our forebear

Australopithecus down from the trees. Our ancestors hit the ground running, fleeing for their lives from swift predators prowling the savanna. The ones who made it survived not by virtue of speed, which they lacked, but by their wits, making fire and assembling tools to fend off their enemies. We are still running, seemingly impelled by the same urge to grab what we can, even when such impulses are an odd and self-defeating anachronism in these modern times of relative comfort and abundance. We are awake to the threat we pose, but we choose not to back off in any serious, significant way.

Each year a host of observers flies over the calving grounds and the Great South Channel, alerting transiting vessels to the presence of whales. Aerial surveys are not infallible. They cannot take place over rough seas, in fog, or in darkness, and they did not save Staccato. Slowly, ever so slowly, regulators are restricting fishing in areas where right whales gather, but given the numbers of entanglements, these efforts are inadequate. Fisheries officials in charge of implementing the Endangered Species Act have determined that any human "take" of a right whale, accidental or not, further endangers the population.

Year after year we exceed that limit. In 1999, we killed two right whales. The sun had hardly begun to shine on the millennium when, on January 20, another dead right whale, a three-year-old female with fishing gear wrapped around her tail, was sighted off the coast of Rhode Island. In 1980, the life expectancy for a female North Atlantic right whale was forty-eight years. By 1995, it had dropped to sixteen, cutting the number of calves she could produce by two thirds, from nine to three, edging the population toward a size where extinction will be inevitable. Before the die is finally cast, we can ban fishing, shipping, and high-speed ferries where right whales are feeding or transiting. It would not be impossible, or unseemly, for us to slow our pace, to share the ocean with those whose home it was long before we arrived.[35]

Sharing the sea is a partial but relatively straightforward solution to what is becoming a complex problem. Ship strikes and gear entanglements account for some but not all of the difference in growth rates between northern and southern right whales. Inbreeding may be part of the problem. When whalers decimated the right whale in the North Atlantic, they may have left behind only five groups of unrelated females. The shrunken gene pool and resulting inbreeding may produce fewer healthy calves. In the last ten years, only 38 percent of female North Atlantic right whales have reproduced successfully, compared with 54 percent of female southern right whales. Yet this kind of genetic bottleneck has not thwarted the recovery of other endangered marine mammals, such as the elephant seal. Analysis of DNA in several hundred right whale bones found in the ovens and along the wharves of an old Basque whaling station in Red Bay, Labrador, may suggest how much genetic variation this population contained before it was exploited.[36]

In the northern waters of Atlantic, the number of breeding female right whales is increasing while the number of new calves is not. In calving grounds off the coast of Georgia and Florida, researchers sighted only four new calves during the 1999 season. This was the lowest season on record until 2000, when researchers could find only one calf. Right whales in the southern Atlantic give birth to healthy offspring every three to three and a half years, but in the north the time between viable births is inexplicably lengthening. Within the last twelve years, the calving interval has undergone a disturbing trend, increasing from three to five years. As man edges further and further into the ocean, altering her watery habitat, the sea's abundance may be fading.

Our continuous trespass may have tired Atlantic so she can less easily care for her children. Each year, the sea's great fertility washes the water green, grows thick pasture for tiny grazers who fuel the marine food web, feeding all from the tiniest microscopic animal to the largest whale. If the sea becomes less fertile, from natural causes or as a consequence of our own activities, whales may go hungry. Right whales are large animals requiring massive quantities of food. In the clear waters of the calving grounds, food is sparse. Female right whales feed in fertile northern waters, fattening up on plankton, building energy reserves to sustain themselves and their nursing calves during the long winter. Where they feed, the sea is heavy with copepods. The patches are thick, the densest we know, averaging about 100 copepods per cubic foot (3,700 per cubic meter) and growing to levels ten times and even nearly one hundred times higher. For a whale, the effort of swimming through the water and feeding is costly. Swimming with a large open mouth creates drag, requires additional expenditure of energy. When food is scarce, when the density of copepods drops below a certain level, right whales cease to feed; the water is too thin to make it worth their while.[37]

Whales have a keen sense for the sea's riches, which scientists, with all their probes, cannot yet copy. To find the densest plankton, scientists follow whales. What guides these animals, we simply don't know. Perhaps the whales remember, from year to year, generation to generation, those places in the sea where plankton is lush. Perhaps mothers teach the young calves. Mothers and calves roam Atlantic for hundreds of miles, back and forth between the Bay of Fundy and Cape Cod Bay, filtering out swelling populations of copepods in choice feeding grounds. Perhaps they taste the fertile water, turning in toward the Great South Channel when they drink from a plume of fresh water running from New England into the sea. Right whales still grow some hair, a few strands around their mouths, left from their long-ago sojourn onshore. With this hair, richly supplied with nerves, they may sense the thickness of the plankton stew.

The tracks of whales record the invisible lushness of the sea. When the waters empty and food is scarce, whales leave. In March 1997, right whales disappeared from Cape Cod Bay one month early. Thick swarms of copepods had dissipated, crowded out by a bloom of nuisance algae that colored the sea brown. Whether the slimy algae, which has increased in intensity in recent years, signals a decline in the health or fertility of the water, no one can yet say, but in the springs of 1998 and 1999 no mother-calf pairs came to feed in Cape Cod Bay.

Whales abandon the Great South Channel when there is no food for them. Almost every year, in late spring and early summer, right whales gather there to feed, zigzagging through the water, ferreting out dense swarms of copepods swept in by currents. In 1992, spring came and went, and then summer, and the whales did not appear. Neither did the copepod swarms. Pteropods, or sea butterflies, filled the sea instead, in concentrations too low to entice the whales to stay. The water was cold that year, 4 to 11 degrees Fahrenheit (2 to 6 degrees Celsius) colder than usual, the copepods reproduced later, and currents whisked them away before they grew large enough to eat. Perhaps the whales swam farther east, to feed at the edge of Sable Island, where copepods maturing later might by carried by currents. Perhaps they went hungry and, as a result, were not able to breed. Nine right whales gave birth in 1998–99, but none were whales who frequented inshore habitats. Studies are beginning to show an association between the reproductive success of right whales and the richness of their feeding grounds.

As man grabs from the sea and strews his waste there, the waters' riches may dissipate. Where the water warms, plankton decline. Where excess nitrogen pours into the sea, blown east from power plants and car exhaust, leached into rivers from farms, fertilizers, and septic systems, algae thrive, cutting off the oxygen supply for fish and tiny air-breathing copepods. We fish out the cod and haddock and halibut, and their prey thrive. Herring, mackerel, and sand lance compete with whales for the tiny copepod. Natural, cyclical variations in the fertility of the water, compounded by our assault, may make it harder for right whales to find the food they need. Undernourished, they may have difficulty reproducing. Scientists, measuring the thickness and lipid content of the seven inches of blubber insulating the right whale, are hoping to learn whether right whales are sufficiently nourished to reproduce. By comparing southern to northern right whales, they hope to discover whether malnourishment in the North Atlantic right whale might explain the lengthening interval between births.[38]

Other scientists wonder whether sublethal poisoning impedes successful reproduction. Right whales eat low on the food chain, fueling their massive bodies with millions upon millions of tiny copepods. Copepods in two important feeding grounds, Cape Cod Bay and the Bay of Fundy, contain toxics, so along with their food, right whales ingest, metabolize, and excrete dioxins, dibenzofurans, crude oil, and combusted fossil fuels. Unlike some organochlorines, these chemicals do not accumulate in body fat, so their presence cannot be measured directly. As the chemicals circulate and pass through the body, they leave behind a marker, a specific protein, cytochrome P450 1A (CYP1A). The presence of this protein, induced in the skin and membranes by exposure to certain chemicals, correlates with damages to the thyroid, immune, cognitive, and reproductive systems.

Exactly how these toxics might compromise right whale fertility is at the moment unknown. What is known is that CYP1A abounds in the skin of northern right whales when they are feeding, and it occurs in significantly lower amounts in southern right whales, who feed in the presumably cleaner waters off Georgia Island in the southern Atlantic. Man is messy; his debris spreads far and wide. Diesel exhaust and power plant emissions all waft eastward and fall into the sea. CYP1A is a nonspecific marker; it cannot distinguish between truck exhaust and power plant emissions, it cannot separate dioxins from dibenzofurans. Identifying a specific pollutant and a specific source of pollution requires an additional level of sophistication, one that we have yet to develop. Undoubtedly we can do so; the question is when, and whether it will be soon enough. The stakes are high, for everyone. On a molecular and cellular level, we have much in common with the animals with whom we share this earth. If the planet is growing too dirty for whales, it may be for us as well.[39]

To us, the sea is a blank slate, unsigned, unposted, but where we are blind, other animals see, and where we are deaf, other animals hear. We, who are so dependent on vision to make our way, cannot comprehend how whales navigate so easily among dark neighborhoods thousands of miles apart. We guess, but our guesses take us into unfamiliar, uncharted realms, distant from our own ways of knowing. The magnetic pull in the rock of the seafloor varies, depending on its makeup and size; whales, sensing these lines of magnetic variation, may travel through the sea guided by them. Possibly it is the voices of singing whales, echoing off coastal contours or rises and dips of the sea bottom, which illuminate their path. Sound rushes through water at speeds of almost one mile (1,600 meters) per second, but the exact pace varies with the temperature of the water. Perhaps the voices of whales identify oceanic thermal fronts, boundaries between cold and warm water where the sea is rich in food and where the speed of sound varies with changing temperature. Sound waves can travel hundreds, sometimes thousands of miles through the sea before they dissipate. Perhaps whales call to each other, their voices leading the way to distant places.

We dive into the sea and water fills our ears. Hearing nothing, the sea seems silent, but in fact it is a raucous place, filled with the sounds of cracking ice, rumbling underwater avalanches and earthquakes, raging storms, breaking waves, and the voices of the animals themselves; snapping shrimp, barking seals, and singing whales. In recent years, the cacophony has increased. Seismic blasts from oil and gas exploration, oil drilling, the sonic boom of jet aircraft, and the roar of ship engines all add to the din. Science and military activities contribute as well, with noise equally loud. Because sound travels more quickly through warm water, scientists plan to document trends in global warming by measuring changes in the speed of sonar traveling though deep-ocean sound channels. The U.S. Navy detects stealth submarines by sending loud pulses of sound through the water.

What of this roar do whales hear? Anecdotal evidence is piling up, while absolute proof is wanting. Two sperm whales stranded off Great Britain during seismic exploration off the Shetland Islands, and underwater air guns blasted during oil exploration silenced dolphins in the Irish Sea. Across Atlantic, humpback whales feeding in Newfoundland's Trinity Bay became tangled in fishing nets while workers blasted a nearby channel. Two whales, once freed, became trapped again, possibly because their ears were damaged. Detonation of explosives in the same area deafened and killed two whales the following year. In the Mediterranean, twelve Cuvier's beaked whales stranded along the coast of Greece at the same time as NATO was testing a new sonar system. In

March 2000, while the U.S. Navy tested its sonar, fifteen whales stranded on the islands of the Bahamas, their ears hemorrhaging.

The din of men, as loud as the whales themselves, carries hundreds of miles through the sea. At what point does the clamor we have added to the oceanic world—from passing freighters, from oil and gas exploration, from naval operations, from science experiments—begin to distract the whales, block their channels of communication, interfere with their ability to find or call to each other, to sing of fertile feeding grounds, of rendezvous suitable for courtship? We already know that the songs of male humpbacks are lengthened in the presence of sonar, presumably in response to the interference. We may not have the time to wait for further proof, if such a thing even exists. Rather we might ask, as we approach the twilight of the North Atlantic right whale, whether our needs for economic and political preeminence are so very pressing.[40]

On hot summer afternoons, the multilane coastal highway out of Quebec is congested with bumper-to-bumper traffic. The road edging the St. Lawrence narrows as it leaves urban Canada behind, while the river, rushing to meet the sea, widens, pushing the Gaspé Peninsula on the opposite shore further and further away until it disappears from view. At the mouth of the Saguenay River, the coastal road abruptly breaks off. Linked by ferry to the other side, it continues on, passing through the vacation village of Tadoussac and the industrial towns of Baie-Comeau and Sept-Îles. By then the road is quiet and empty, rising high above the Gulf of St. Lawrence, squeezed between steep sea cliffs and dark unbroken forest. Somewhere near Anticosti Island it ends. Peat bogs break up the woods; the estuary opens into the sea. The trees are impenetrable, the bogs impassable. The few who dwell here, in poor fishing villages scattered along the shore, look to the sea.

A visitor, I too look to the sea, hoping to find *Balaenoptera musculus,* the blue whale, the largest animal ever known to inhabit earth. In the northern waters of Atlantic, this whale, larger than the largest dinosaur, once graced the waters of the Hebrides and Faeroes, of eastern

Greenland and Iceland, of the Davis Strait and the Labrador Sea, numbering between 12,000 and 15,000. Whalers killed most of them, emptying seas where they once roamed, shrinking the population to 4 percent of its original size. Today sightings are rare, even for those who spend their days at sea. For the most part, the whales live way offshore, in waters beyond the purview of man. A New England fisherman might come across one or two in a decade. Each summer between 100 and 200 come to feed off coastal Iceland, and over the years, another 300 have been photographed in the Gulf of St. Lawrence and the Labrador Sea.

We know next to nothing about this remnant population, about where the animals give birth, nurse their young, find their food. Sometimes blues come into the Gulf of St. Lawrence in the winter, to feed at the edge of the pack ice. Sometimes they swim up the estuary earlier, in late summer or fall, arriving from parts unknown en route to destinations unknown. Today, a blue whale venturing near the coast is often solitary, traveling alone, rarely staying in one place for more than a few days. When the whale moves on, it covers great distances quickly, three or four hundred miles (five or six hundred kilometers) in two weeks. Occasionally, one or two might linger for a day, perhaps two, in waters around the Mingan Islands, a tiny archipelago off the remote village of Longue-Pointe.[41]

Few trawlers work out of Longue-Pointe, and no sportfishermen or party boats scan the seas and radio in the whereabouts of whales. This sea is not for the taking; air and water conditions frame our days, wind and wave determine our itinerary. Calm, windless weather bodes ill for whale watching, as hot air resting on cold water spawns thick, close fog. Embraced by dense, inscrutable gray, unseeing in air filled with light, we drift in this place with no form, no dimension. Suspended in the silence, we listen for the *pff, pff, pff* of the blue whale's blow, and hear instead the raspy breath of minkes. Late in the afternoons, if the fog lifts, our little rubber boat hurtles over flat, glassy water, skimming over the reflected purple and rose of fading sunlight. Porpoises dive through the mirror sea, and seals laze about, eyeing us from afar, and then, as we approach, slip silently down into the darkness.

When stiff winds drive away the fog, clear the air and whip up waves, we pull out binoculars and watch for the mottled back of the blue whale, its notched fluke, or high slender spout, white against the brown cliffs of Anticosti, gray against the blue sky. We see puffins, comically skittering through the water, stumbling into the air, and razorback gulls, guillemots, cormorants, eiders, and gannets, but no blue whales. Often the waves are high, close to nine feet (three meters), and we can't maneuver safely. Wind and heavy thunderstorms force us ashore on islands where giant limestone sculptures carved by wind and wave line sandy beaches, or where insect-eating plants and leathery, fragrant Labrador tea spring from soggy inland bogs. In the 1500s, the Basques paused in these landscapes now uninhabited by man. On Île Nue, in a meadow above the rocky sea edge, whalers who came from across the water built ovens from sea clay, boiled blubber in copper pots, cooled it in barrels of cold water, and waited for the precious oil to rise. Only the clay ovens and a few Basque roof tiles scattered in the bleak grasses tell of this time gone by.

Basque whalers came to the Gulf of St. Lawrence five hundred years ago. Etched into the rocks of the Mingan Islands are fossils 500 million years old. The small animals once dwelled in a sea whose waters lapped on another shore, a sea that flourished long before glaciers iced over eastern Canada, and long before spewing volcanoes ripped open the Gulf of St. Lawrence. When the continents danced, the waters receded and then returned, recording their history in rock. Tracing the coil of an ancient shell, I feel the spaciousness of time, its distant reaches revealed by the outline of an animal whose years on earth were spent long before man or his forebears were conceived or even imagined.

I like to take comfort in the great arc of time that spans millennia, knowing that when we are gone, the sea will endure and give birth to new life. Spending eight to ten hours a day in a tiny boat on the Gulf of St. Lawrence, in waters empty of other vessels, speeding over great distances that barely amount to a ripple in this broad estuary, I'm tempted to take comfort in the expanse, the openness, the limitlessness of water. I know better. While we have yet to rein in time, we have found the edges of earth. The sea is smaller than we imagine, its fluid waters more connected than we ever knew. We have closed in and shrunk this sea, disturbed the balance of life there in ways we have only begun to understand.

The whale I find during my visit to La Minganie is not the whale I seek. On a day when rough winds and frothy seas have penned us to land, I take a walk through the woods outside the village. Hidden amidst the trees is a small clearing piled with huge and ragged bones. I could sit on thick joints, lie along the length of vertebrae, walk under the curve of ribs. Here I am finally in the company of a blue whale. It is dismembered, rotting. What human act washed this whale ashore? Was it impaled by a boat, choked by fishing nets? Was it weakened by hunger, deafened by blasting? The stench of decay is nauseating.

I leave in the afternoon and stay the night in Sept-Îles at the home of a man who spent his childhood killing whales. He and his father used to shoot belugas, day after day, whale after whale, just for fun. Today, genuinely appalled by his lack of awareness, he remembers that back then whales meant nothing, they were just things to play with. We still toy with the lives of whales, perhaps just as brazenly. An article in a business magazine titled "Don't Tell the Whale Lovers" describes a burgeoning business started by an entrepreneur hoping to tap a vast marine resource. He hopes to make money processing the shells and meat of krill, a tiny sea animal, and selling the protein-packed product to salmon and trout farmers. Blue whales feed almost exclusively on krill, a small shrimplike crustacean inhabiting polar seas, filtering between three and five tons through their baleen each day. Scientists know that the reproductive health of fin whales is related to the abundance of food, to fluctuations in the supply of herring and capelin, yet left unconsidered is the question of whether there is enough krill to go around, whether the blue whale will go hungry as we, having swept the sea of wild fish, further ransack its waters to feed fish we now must farm.[42]

We are like the digital readout on a ship's GPS, focused on the present moment, looking neither forward nor back. We assume an average swordfish weighs 90 pounds (40 kilograms), forgetting the 250-pound (110-

kilogram) fish lining the docks of Tiverton, Rhode Island, not that many years ago. For us, as few as twenty turtles make a crowd on a nesting beach, when thousands once defined an *arribada*. We accept the hawksbill as rare because its numbers were low when we started counting. We rejoice to see a few whales, for we cannot imagine a time when the sea was full of whales. Like the GPS, we have no memory of what has come before, no dream of what lies ahead. If we would place ourselves in a longer continuum, holding on to memory before it fades, before we become any more satisfied with our attenuated lot, we might better appreciate the magnitude of our losses and more easily imagine, and replenish, our future.

For me, the GPS with its flashing points of latitude and longitude illustrates a worldview in which the time is always now, the place always here, and in which we are the sole point of reference. This singularity enables us to open a krill fishery that disregards the hunger of whales, operate a swordfishery that ignores the vulnerability of sea turtles, condone industrial port expansion that ignores the breeding habits of right whales. Isolated in this narrowly circumscribed center, we assume we can light every darkness, know the bounds of every realm, compute every cost and benefit.

A less egocentric view is represented by those sailors who, navigating by starry skies, sense our tiny place in the arc of time, our tiny place in a vast universe of life intricately and intimately joined. It is a view bolstered by the existence of the blue whale, who, large as it is, eludes our understanding, thwarts our efforts to track it, disappears into distant waters on journeys we cannot follow. Its presence is humbling. Navigating with a longer, wider, more encompassing and generous view, we may see that while there is still time, it is possible to keep company with those who share our earthly dwelling and return the sea to its inhabitants.

CHAPTER 7

Wide, Wide Sargasso

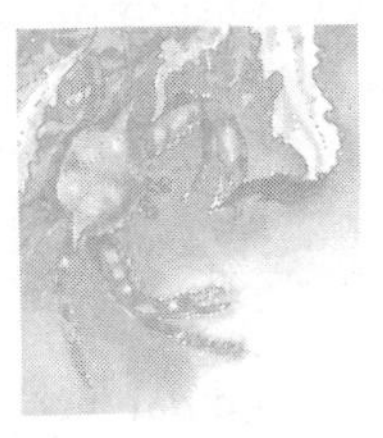

TO the east of the thick green waters of the continental shelf, to the east of the cobalt blue of the Gulf Stream, lies the Sargasso Sea. Bounded on all sides by swiftly moving currents and rushing winds, the Sargasso is the calm, warm interior of Atlantic, 2 million square miles (5 million square kilometers) of intensely blue transparent water touching Bermuda to the west and the Azores to the east. When the *Cramer* passes through, she is fully 1,000 miles (1,600 kilometers) offshore. We are alone out here, where few have reason to be, far beyond the shipping lanes, far beyond the swordfish and tuna boats hugging the edge of the Gulf Stream. No tanker, trawler, or pleasure boat crosses our horizon. The air is still, the sails flap idly, the boat drifts. We are too far east of the Gulf Stream to feel its northward pulse, too far north to feel the lift of the trade winds. Dry air rising off the equator gave off its rain, and now it descends gently back toward earth, becalming us. It is possible to sail out of the Horse Latitudes, carried on the infrequent storm, but if we waited for wind, we might tarry for months.

In the afternoons, under the hot sun, with a lookout posted on the rigging to watch for passing sharks, we dive off the caprail, swim in the warm clear water. Nothing mars the dome of sky, or rim of unending sea reaching around the curve of earth. In turbid coastal waters where sunlight is soon scattered, quickly disappearing beneath the surface, swimmers give themselves to darkness. Here, light falls undiminished through hundreds of feet of pellucid sea, casting long shadows through an immensity whose twilight edge I can only imagine. The unlimited expanse is comforting and intimate. Buoyed by the water, I sense my cares and concerns fade away. Gently rocked by the swell, caressed by the waves, I am instilled with a deep sense of peace, and of possibility. Fully embraced by this water, I easily imagine that man can live in concert with his surroundings. As my bobbing shipmates disappear behind the advancing waves, the sea seems spacious enough to absorb our extravagances. The sparkling water holds promise, but despite its clarity and the breadth of the sky, reality falls short. Even these remote waters are not untouched by the hand of man, so many are the elements joining us to this faraway place.

The Sargasso is a realm unto itself, distant and unlike more familiar coastal seas. Its waters are layered, a lens of warm, depleted sea resting above the cold, rich darkness below. Compared with coastal waters steeped in nutrients pouring in off the land in rivers, rushing in on currents, or welling up from fertile depths, the Sargasso is a desert. Because its layers are separate, unmixed, its waters are meager, supporting few floating plants to color the sea green. Satellites spinning high overhead, tracking the effusion of chlorophyll in the ocean, record that life-giving green fades in the Sargasso, thins to 4 percent of the amount found in coastal waters.

Yet even this desert blooms. The blossoming and withering of floating meadows is sparse, minute, subtle. The cycle of life in earth's barren seas, despite its small scale, produces 30 percent of the sea's carbon. Tiny cells, single chloroplasts visible only through fluorescent microscopes, capture the energy of sunlight to make carbohydrate. Nutrients are few in this unstirred sea, but the cells absorb what little comes their way—bits of iron dust blown in on the wind, and nitrogen spun in from fertile North Atlantic on a Gulf Stream ring, or left in the wake of a passing animal. Scarcity has led to economy; the cells cycle and recycle the same nutrients again and again, losing very little organic matter to the seafloor. There is no waste here. Nitrogen used by one cell is reused by another when the first cell dies. Scientists straining seawater through filters fine enough to trap all but the tiniest, tiniest particles find that in clear impoverished waters like the Sargasso, most of the fertilizer consists of disintegrating fragments of individual cell walls once belonging to earth's tiniest photosynthesizing organisms.

Though these bacteria existed long before we sought to identify the ingredients of seawater, we have been blind to their presence, oblivious to the idea of a realm so tiny, so beyond our ability to see. Few of these microbes can easily be cultured. Only when we have stripped them down to their very essences, extracting their nucleic acids and cloning their RNA, have we come to see the ubiquity and diversity of this hidden world of water.[1]

From so little, from the tiniest of cells making do with the scantiest of resources, is built a web of life, feeding larger animals who thrive in the gentleness of the open ocean, but who could not survive the turbulence, siltiness, and congestion of coastal waters. Small pelagic snails spin gossamer mucus veils six feet (two meters) across, spread wide to catch what little food floats through the clear water. The surfeit and coarseness of food in waters of the continental shelf would choke this fragile animal and many others, such as the shimmering comb jellies pulsating through the water or the ribbony chains of barrel-shaped salps. Watery, transparent, insubstantial, these animals live out their lives in an unbounded sea, never touching a harsh edge. Buoyed by dense seawater, suspended in a three-dimensional world far from the impenetrable bottom or the crash of breaking waves at the surface, they need no skeleton to carry or brace themselves, no protective coloring to block the sun's damaging ultraviolet light. When they brush against the hard foreign surface of a collecting net, they often collapse and fall apart.

Somehow, a salp survives one of our tows. It is a delicate, jewel-like animal, not more than an inch or two long. Despite its unusual appearance, it is a chordate, a distant relative. Bilateral symmetry and the appearance of a notochord or internal skeleton at some time during our lives join us to this seemingly alien animal, give us a common ancestor. Inside the salp's crystal-clear body lies a small crustacean, an amphipod, who has nibbled away its host's innards. In this formless, seamless sea, which offers no shelter, no dwelling place, this animal has found its home in another's body.

Out on the deck, in the air, the salp will soon shrivel. Unperturbed in the blue water, it is a model of elegance and efficiency. A solitary salp replicates itself again and again, forming large circles wheeling through the sea, or long undulating chains linking as many as one hundred animals. Each slender individual continuously filters gallon after gallon of seawater through its barrel body, imbibing at one end and forcing water out through the other, gaining sustenance and achieving locomotion in the same swallow. As the water moves through, the salp strains the sea's scanty bits of bacteria through a sticky mucus net, which it simultaneously consumes and replaces. Hitchhikers move in. Some live at the edge of the salp's throat, catching pieces of food floating through; others lay their eggs in the barrel, leaving their offspring with nest and nourishment. As insubstantial as salps are, they provision a few blue-water animals, contributing to the diets of migrating sea turtles. Jellyfish drifting in this barren sea may not eat for days, but if they encounter a chain of salps, their stomachs swell to consume the colony in its entirety.[2]

The fragile salp, inhabiting the warm Sargasso where life is not abundant, is part of a living, breathing sea, part of an immense exchange of carbon dioxide and oxygen between atmosphere and ocean. Each year, billions of tons of carbon dioxide pass from air to sea and back again, flowing easily between the two realms. Along the coast, where currents mix the water, and nutrients run off the land, the meadows of the sea bloom in profusion, drawing carbon dioxide down from the atmosphere. As plants are grazed by tiny microscopic animals and by copepods, who in turn are consumed by larger and then still larger animals, the carbon passes through the food web. Though much is recycled, some eventually comes to rest on the seafloor, where it is decomposed by bacteria and turned back to carbon dioxide and nutrients to begin the cycle anew.

In the Sargasso, the pulse of life is more muted. Exactly how this sea breathes is less well understood. No one really knows which is greater in the open ocean, production or consumption, the growth rate of minuscule plants or the respiration rate of salps and other animals. No one really knows whether this sea absorbs or releases carbon dioxide. Scientists think that oceanic and terrestrial plants reabsorb approximately half our carbon dioxide emissions. If, as earth warms, fertile coastal waters heat up and become layered like the Sargasso, the sea as a whole will absorb less carbon dioxide, accelerating the rate of global warming.[3]

Salps, plentiful in the warm waters of the Sargasso, were until recently unknown in the cold, fertile waters near my home. The Gulf of Maine, fed by ocean current and estuary, is usually dominated by copepods and other herbivores. Three years ago, as the ocean put forth her last burst of life before the onset of winter, it filled with masses of salps. The local newspaper carried first-person accounts of strange alien invaders even fishermen could not identify. Swarms of salps, thousands of them, clogged bays and tidal creeks, coming in so close it was possible to wade out and catch them in buckets. They looked like clear jelly beans, each with a royal blue dot, the stomach, inside. When the tide ebbed, they stranded and died. No longer cradled by the sea, the water quickly drained from their delicate transparent bodies, leaving only the casings, sticky chains of flattened barrels imprinted on the rocks. They dried in the sun, and washed away with the rain. Why had they come so far afield? Perhaps an eddy of the Gulf Stream, wheeling north with its pinch of the Sargasso, washed the animals inshore. Perhaps coastal water temperatures, abnormally high that autumn, spun a new web of life. When, after a few weeks, water temperatures dropped, the sea cleared and the animals disappeared. The next year they returned, and the year after.

No one can say whether salps washing into New England's waters are transient visitors, their appearance here a passing vagary of nature, or whether, as waters warm, they herald a subtle, and perhaps more enduring, shift in the ecology—a shift that may reverberate through the food web. The scanty meadows of a salpy sea absorb less carbon dioxide, support less life. Dense meadows of floating plants are required to support burgeoning populations of krill, while salps tend to inhabit waters where chlorophyll concentrations are low. In terms of energy provided, watery salp are no substitute for meaty krill. Krill, not salp, feed whales and seabirds: few are the sea's inhabitants who grow and thrive on a meal as insubstantial as that provided by the gossamer salp, whose body is 98.5 percent water and salt. In the Southern Ocean, in seas off the Antarctic Peninsula, penguins and whales thrive on a hearty diet of krill, who in turn subsist on algae growing beneath the sea ice. When the water warms and the ice recedes, rich sea meadows disappear and the numbers of krill decline. In Adélie penguin rookeries on King George Island, the number of surviving fledglings has declined in recent years. Although nearby waters are rich in salps, they lack krill, keystone of the Antarctic food web.[4]

In the Sargasso, though, where resources are scarce, an animal with but a trace of organic matter is not anomaly but norm. Each day, as the light dims and night moves in, salps come to feed in the surface waters, swimming up from the depths as much as half a mile each day. They are among a host of animals rising from the deep water when the nights are black and the moon dark. These creatures come from another world, where day blends in with night, where the pressure of the sea weighs heavy, and where what little food comes their way drifts slowly from the surface. This unfamiliar world of the mid-water, so little studied, so poorly understood, is earth's most common, most widespread habitat. More than three quarters of earth's living space belongs to the deep sea. Many of its dwellers, like those in the lighted water above, are too fragile to wrest from their watery homes. Boxy medusae three feet long by three feet wide (one meter by one meter), hanging motionless, awaiting their prey; delicate siphonophores over one hundred feet long; deep-sea jellyfish, transparent except for stomachs colored red to block out the bioluminescence of the shrimp and fish they consume—all fall apart when jostled into captivity.[5]

Our night tows pick up a few of the hardier animals. Though strange and unfamiliar, these denizens of the mid-water exemplify life in earth's watery realm. Fat deep-water shrimp come up in the nets. Bright red on the deck, they are black in their world, where sunlight's longer red wavelengths are rapidly absorbed. Blue wavelengths penetrate the farthest into the sea. Their red coloring, like that of the jellyfish's digestive system, absorbs the faint blue light permeating the deep sea, blending the animals fully into the darkness. In a world where food is in meager supply, advantage does not necessarily accrue with size. The fish who dwell here are tiny, no longer than an inch or two, but their mouths are disproportionately large, and full of sharp, pointed teeth, designed in a number of bizarre and ingenious adaptations to a life of scarcity. The thin viperfish we catch from the depths has a jutting jaw and wide mouth whose long fangs shut like a trap, leaving prey little chance of escape. Inert in the net, the viperfish displays little of its versatility as hunter and trapper. Food is scant in the mid-water, and meals few and far between, but with its hinged and distensible jaw, the viperfish opens its mouth wide, wider than its own body, to swallow fish larger than itself.

Mid-water fish have overly large eyes, to take in as much of the dim light as possible, but their sea is not always dark. As many as 90 percent of the fish living here emit their own light, from specialized organs located along their bodies. The hatchetfish glistening on the deck has large eyes, a wide gaping mouth, an almost vertical jaw, and many pale photophores along its underside. It lies limp in the air, its life running out, but in the sea it is a skilled stalker. Fish of the mid-water have slow metabolisms. Up near the surface, fish chase their prey, swimming at high speeds, expending tremendous energy, but deep in the darkness, the pace is slower, and the hunters lie still, waiting. The hatchetfish hovers in the water, looking upward, scanning for fish passing overhead, for silhouetted shadows, for an ever so subtle gradation of black in a dark sea barely illuminated by distant light filtering down from the surface. Vulnerable to similar surveillance from below, the hatchetfish adjusts. Emitting a faint light downward from photophores along its belly, adjusting its intensity to match the background, the hatchetfish erases its own shadow.

The mid-waters twinkle with the light of bioluminescent fish. The lights of lanternfish shine and then fade in the darkness, as the tiny fish rise from the depths with the setting sun and then sink back into the deep as day breaks. Their nightly migration once confounded ship pilots, appearing on sonar screens in the shape of a field of cobblestones known as Alexander's Acres, a false seafloor with a disconcertingly inconstant depth. For bioluminescent fish of the mid-water, light is language, the voice of lovers seeking mates, stalkers seeking prey, the hounded seeking refuge. Lanternfish, like most mid-water fish, emit and receive light in the blue part of the spectrum. Their light shines for all to see, while a type of dragonfish attracts mates and hunts with a beacon of red light, its signal invisible to other mid-water species. A deep-sea anglerfish awaits dinner with a lighted lure, a vestigial dorsal fin elongated and converted into a fishing line and bait suspended near, or occasionally in, its mouth. A squid, startled by a potential predator, flashes to distract, throwing off a lighted cloud of ink as it disappears into the darkness.

Bioluminescence evolved independently at least thirty times, as squid, medusae, jellyfish, fish, and many other mid-water animals came to produce light in their darkness. There are few fossils to document the transition, but the glowing light of a deep-sea octopus living in the waters off New England hints at one line of evolution. Photophores extending in a line down each arm of the octopus look very much like suckers, but the muscle has atrophied, lacking the strength to grasp prey. Perhaps, as the animal left well-lighted surface waters for its new home in the darkness, and as it came to live on a diet of tiny organisms trapped in its webbed arms, the suckers, no longer necessary, were used to produce light for attracting prey and communicating in the blackness. Marine animals may once have used luciferins, the chemical compounds required in bioluminescence, to rid their bodies of harmful oxygen derivatives. When they ventured into deeper, darker waters where life slowed and living demanded less energy and less oxygen, they may have put these chemicals to another use, to meet the emerging and more pressing need to produce light. Where luciferins come from and how they are transferred through generations of animals is poorly understood. Particular luciferins are common to so many deep-sea dwellers, scientists suspect they are ingested. And yet, bioluminescent marine animals use more luciferins than their meager mid-water fare provides, suggesting that perhaps at some point the animals learned to synthesize them themselves.

Bioluminescence, a chemical reaction enhanced by a catalyst, is stunning in its efficiency, its economy. A luciferin, reacting with oxygen in the presence of an enzyme, luciferase, creates lively, unstable molecules that, as they come to rest, emit a cold light. We, who both consume massive amounts of energy to light our darkness and produce much wasted heat along the way, cannot duplicate this clean conversion of chemical to radiant energy. Dwellers of the mid-water create cold light in several ways. Some animals come equipped with the critical elements: their photophores contain the necessary luciferin and luciferase. Others purchase them, harboring specialized light-emitting bacteria in their light organs, serving their guests a rich diet of nutrients. The bacteria are generously provided for, supplied with nutrients they would otherwise have to scrounge or manufacture themselves. In exchange for fully assembled meals and room and board, the bacteria multiply and produce light.

Animals of the mid-water have acquired sophisticated means to adjust their light, to screen or shade it when they need to be in darkness. They can focus it on a mate or a meal, filter it to produce a particular color, reflect it. We, bathed in a sunny atmosphere, have difficulty comprehending the myriad ways dwellers in deep-sea darkness harness their light. On land, the most familiar analogue might be fireflies, whose twinkling lights on a warm summer night hint of a world to which we are not privy, a world beyond our ken. How much more distant, more enigmatic are the flashing lights of bioluminescent animals of the mid-water. On dark, cloudy nights when they appear at the surface, it's difficult to read their signals, to know whether the sparkling lights denote a lively conversation we cannot understand, or whether they shine in reaction to the nudge of waves or the prospect of food. These denizens of the sea shine their beacons in the darkness, marking a luminous trail we have only begun to interpret.[6]

Deep as the twilight zone is, there is more below, a layer of blackness thousands of feet thick blanketing some of the deeper parts of Atlantic's basin. We have seen and sampled less than one tenth of 1 percent of the deep sea, this cold and forbidding place where we, whose bodies are filled with air, cannot withstand the weight of water. Water pressure increases with depth, beginning at 14.7 pounds per square inch (1 atmosphere) at the surface and adding the same again every 33 feet (10 meters). When the last rays of light finally disappear, more than half a mile down, the pressure has mounted to an excruciating half a ton per square inch, anathema to those whose bodies contain large pockets of air. Yet the press of the sea can be surprisingly gentle. In the *Cramer*'s science lab we cast collecting bottles into the depths from time to time to measure the water's oxygen content. For one hydrocast, we send Styrofoam cups down as well. I cover mine with drawings, of a whale, of the *Cramer*'s sails and helm, and I cram the remaining space with a list of all of us who serve on B Watch. During the descent, pressure squeezes air from the Styrofoam, shrinking the cup to less than a tenth its former size but leaving its shape fully intact. The touch of the sea reduces my cup without damaging it in any way. It is a perfect miniature, its lettering and illustration completely preserved and clearly legible.

Water, unlike air, does not compress. Though the sea bears down, animals living in the abyss maintain their shape and size because they themselves are mostly water. Few air-breathing animals are physically equipped to visit these crushing depths, to cross between the realms of light and darkness, but the sperm whale does, regularly, descending as deep as a mile, holding its breath for a full hour and a half. Deep-diving whales save energy on the long plunge. As their lungs collapse from the increasing pressure, their bodies lose buoyancy and they sink gently into the depths. Their heart rates slow to conserve energy, and circulation to their kidneys and digestive tract diminishes. At the same time, a diving whale's heart and brain draw on abundant reserves of life-giving oxygen. Whales maintain an extra large volume of blood, densely packed with large numbers of red blood cells containing oxygen-rich hemoglobin. A generous endowment of deep red myoglobin cells in the swimming muscles supplies additional reserves to sustain them during a deep dive.[7]

We briefly glimpse a sperm whale, its rough corrugated skin barely rising above the surface as it swims through the desert of the Sargasso, perhaps passing through en route to breeding grounds in the Azores or the Bahamas. We watch this whale, one of the sea's largest carnivores, disappear into the deeper reaches of the sea, where it goes in search of food. It feeds in these unlighted depths, most likely on another of the sea's largest carnivores, the giant squid. In the sparsely populated world of the deep sea, where life is dominated by the delicate and the diaphanous, it is hard to imagine a sperm whale stalking a fifty-foot (fifteen-meter) giant squid. We know little about the sperm whale, and less about its large, brainy, and elusive prey. To acquaint ourselves with the latter, we must follow the former, for the sperm whale understands this animal far better than we, who know little of its habits or its habitat, its appearance or its appetites, its birth or its life.

Sailors have found dead and dying giant squid floating in the sea off Ireland and the Canary Islands. Occasionally they wash ashore, on beaches in Newfoundland, New England, Scotland. When whaling was legal, we received further intimations of the giant squid's existence, from harpooners hailing from Madeira and the Azores who found them in the stomachs of the sperm whales they speared. From this smattering of dead bodies, we infer both the animal's geographic domain and the size of the population. Sperm whale bellies are full of fresh squid as well as the indigestible beaks left from earlier repasts: nearly one quarter of these chitinous remains belong to giant squid. Despite the seemingly large numbers of giant squid roaming the sea, we have yet to see one alive and healthy.

Giant squid swim at the top of the food web, hunted only by sperm whales. What happens during this encounter we can only surmise. The giant squid, cousin of the sedentary mollusc but having shed its external shell long ago, is agile and quick, more nimble than its predator. Being a water-breather, it needn't surface for air. Its eyes are as large as dinner plates, much larger than those of the sperm whale. Its vision evolved independently, and differently than ours. Squid focus on the distant and the close-at-hand simultaneously, and detect the direction of reflected light waves, allowing them to identify and evade predators more easily.

Ever hoping to observe a healthy giant squid alive in its home, we follow sperm whales, then send video cameras into the depths where we assume the giant must dwell. So far, our efforts have been in vain. Perhaps squid, with their exquisite vision, notice us and turn away. Sperm whales succeed where we have failed, possibly using echolocation instead of vision to plumb the sea. Perhaps, while scanning the water with high-frequency pulses of sound, they find and stun giant squid, slurping them down, tentacles and all, before the shocked prey returns to its senses. This hypothesis, which has yet to be proved, would explain why giant squid regurgitated by dying whales or recovered from their stomachs bear no teeth marks, even though the sperm whale is toothed. If a struggle takes place as the sperm whale ingests its one-ton meal, the only signs are circular sucker scars on the bodies of captured whales.[8]

An extraordinarily large animal dwells outside our purview; our only access is through infrequent dissection of the dead and dying. The sparse remains offer only a glimmer of an understanding of the lives of giant squid. Distant bystanders to their world, we have only the barest fragments to go on. The tantalizing discovery of tiny packets of sperm found embedded in an arm of a female giant squid snagged by a fishing trawler hints at, but does not describe, a life cycle and social habits adapted to the vastness of the open ocean and the infrequent opportunities for mating. What niche giant squid fill in oceanic food webs we can only surmise by opening the stomachs of dead animals. Time and time again, the stomachs of dead giant squid are empty. Once a trawler fishing off the coast of Ireland hauled up a squid with a stomach full of whiting, prawn, and octopus, as well as other species of squid. The meal hinted at the animal's diet, but indicated little about its style of hunting or feeding. The existence of a giant squid, still wild, promises that we haven't yet completely surrounded, closed in on, its—or our—world.

Though we cannot yet follow either giant squid or sperm whales into their deep-water homes, our garbage can, and does. Incompletely incinerated flame retardants, common in children's clothing and in the covers of computers and televisions, make their way into the deep sea, along with PCBs and pesticides. Blown in on the wind or washed down to the sea in rivers, the pollutants attach to particles sinking from the surface, or they slip down the continental slope into waters far away from industrial production and consumption. They are forgotten, but not gone. Insoluble in water, they move up through the food web, accumulating in animals that feed upon them. Sperm whales stranded on the beaches of the Netherlands show traces of toxics in their blubber.

Though we ourselves cannot travel great distances into the deep water, we have laced the food of deep-dwelling animals with our waste. Where does the buildup begin? In tiny photosynthesizing bacteria? In ribbony salps filtering streams of water? In sluggish fish nosing in the bottom mud? The avenues are many. While the *Cramer* drifts through the Sargasso, we use a number of science watches to inspect the stomachs of tiny lantern fish, hoping to discern their feeding preferences. The lantern fish are tiny, the stomachs tinier, and the animals they ingest minute. Only a high-powered microscope brings this world into focus. Magnification unexpectedly reveals that our waste, invisible to the naked eye, permeates this sea. Lantern fish dine on copepods and slivers of plastic.[9]

Harpooned whales may provide tantalizing glimpses into the unknown world of squid, but whales dying at sea in their own time, according to their own rhythms, and sinking to a final resting place on the bottom, bring the water's life-giving fecundity full circle. Whales, living at the top of the food web, spend their lives consuming, taking from the sea's riches, but in dying, they yield what they have received. A dead whale, sinking to the abyss, provides a surfeit of food, an oasis in a desert. The carcass feeds a host of solitary bottom dwellers who never see the light of day, never leave the blackness of the seafloor. These residents of the ocean bottom keep little company; hermaphroditism eases their search for suitable mates. In this place of unending darkness, slow-moving fish rest, propped up on the seafloor by extra-long fins, ready to catch bits of food wafting down from the surface. Weak, toothless gulper eels wait, ready to wrap their large mouths around any prey who unwittingly swim near. Rat-tailed fish lie at an angle, their gently waving tails inclined upward to maintain position, their heads bent down in the mud rooting for food. In the blackness of this abyss, day blends fully into night; the shorter circadian rhythms of lives turning around a twenty-four-hour day disappear.

Yet even this depth is affected by the longer rhythms of the seasons. In spring and summer, when abundant light produces more greenery at the surface, the supply of food drifting deep down into the darkness is measurably greater. Dwellers of the deep sea, far removed from the surface, never seeing the play of light on the waves, nonetheless depend on the bounty of floating plants for sustenance, on the slow but assured descent of bits and pieces of cell walls and other organic debris to furnish their meals. Eking out their existence on the frigid seafloor, they cannot feel the temperature changing at the surface 13,000 feet (4,000 meters) away, cannot feel the warming of waters. Distant perturbations, though, resound at depth. When surface waters warm and settle into layers, nutrients are not replenished. Lush seas thin; there are fewer floating plants, fewer grazers, and fewer leftovers to provide a deep-sea meal. Where the sea has warmed, the amount of food reaching the deep has decreased by 50 percent. If such warming continues, famine may come to the deep sea, to the viperfish, sea cucumbers, and crabs, and to the millions of other species constituting the world of deep water. Persistent food shortages, year after year, are bound to have ramifications. As an already grim struggle for survival becomes even grimmer, and as life's necessities grow ever more scarce, the dwellers of deep-sea communities may not have the energy to grow, to reproduce, or even to sustain themselves.

The deep sea is the end of the line, biologically. Here the last pieces of carbon are decomposed, and the elements synthesized to produce life are returned to the water. The deep sea is also the end of the line geologically, the burial ground for the last bits of dust and sand eroded from continents. Avalanches of debris coursing through underwater canyons closer to shore quickly peter out, leaving only the very fine silt and clay and sand to settle out here on the plains of the abyss. Over time the slow trickle has buried the mountains formed millions of years ago when Atlantic first cracked open, filling their valleys and covering their peaks to make these plains the flattest on earth. When glaciers chilled the planet and soaked up the sea, exposing the broad reach of continents, currents swept away more sediment to feed the abyssal plains, but today the ice is melting and drowned rivers and estuaries trap sand and silt and clay worn off mountains. What reaches the deep sea is Saharan dust blown out to sea in storms. As the wind dies, it settles out from the air and slowly descends to the seafloor as a fine red clay, accumulating at the rate of a tenth of an inch (a few millimeters) every thousand years.[10]

Here, in this remote place, biology and geology are partners. Icy polar currents flowing in over the flat plains carry water saturated with manganese. In a process only dimly understood, the manganese, and other minerals as well, precipitates out around bits of debris lying along the bottom, around shark's teeth or pieces of whale earbone. Shreds of carbon buried in the seafloor eventually turn to oil hundreds of thousands of years later: these remnants of the living become the grit of valuable metals. The manganese nodules grow slowly around the hard pieces of shark and whale, like pearls. The mineral rings grow as much in one million years as the deep plains do in one thousand. The nodules litter the floor of the sea, somehow escaping burial in the rain of dust. At some point, man may find it worth his while to scour the seabed and retrieve these lustrous balls of minerals, many as big as potatoes or grapefruit, but for now the cost of doing so is prohibitive and the mineral pearls remain undisturbed in the deep water.[11]

Some change comes slowly to the deep sea. Sediments and minerals accumulate almost imperceptibly, and animals live unhurried lives, growing little over long periods of time. The dead disintegrate slowly. An oceanographic research submersible accidentally sunk in waters off New England still harbored, when it was retrieved eleven months later, along with all the equipment, someone's barely decomposed lunch. The refrigeration and slow rate of decay left bouillon, apples, and a bologna sandwich with mayonnaise virtually intact. In the landscape of the watery realm, coastal waters appear as busy metropolises, pulsing with life. For the longest time the deep water of the abyssal plains seemed a sluggish, sparsely populated rural province. Baited cameras lowered to the deep floor of the sea attracted crowds of fish and shrimp who gathered for a few hours and then dispersed. Unbaited cameras left on the bottom for days and weeks captured far fewer images of live animals. Many were of sea cucumbers, sucking up mud from the bottom as they fed, slowly plowing a long trail on a seafloor paved with the tracks of burrowing animals who may have last crossed these muds weeks or months or even years ago.

From this we infer barrenness in the deep sea, but perhaps we cannot see in this darkness. If the continental slope, also assumed to be desolate, can yield a trove of life as diverse as a rain forest, there is no telling what profusion of life may inhabit the flat plains of the deep Sargasso. In this barely explored, barely described place, in these waters whose chemistry is poorly understood, in this seabed about which we know so little, some respected American scientists would like to see us bury spent fuel rods from nuclear power plants and radioactive material from decommissioned nuclear weapons. We have little idea whether the canisters will hold up under the pressure and corrosion, how heat emitted from fuel rods will alter muddy sediments or affect water quality. Should these highly radioactive and carcinogenic wastes leak into the seabed, we cannot say how quickly they would migrate up through the layers of clay and work their way into the food web.[12]

Given that we are exhausting our onshore options for safely disposing of radioactive wastes, the distance and darkness of the Sargasso have appeal. Yet as far away as this sea seems, as otherworldly its color of clear, deep indigo blue, and as elusive the life of its mid-waters, the rhythms of the Sargasso are touched by the pace of our life onshore. The connecting threads weave through the food web on a tangled path linking discarded plastics to the diets of sperm whales. They follow the complicated breathing of land and sea, linking terrestrial carbon dioxide emissions to the geographic range of oceanic salp, to the amount of food drifting down to dwellers in the deep abyss. The life journey of one wide-ranging aquatic animal that dwells in both the salty Sargasso and the fresh water onshore further joins these two realms, brings their deep connection into sharp relief.

I live in a neighborhood built on swamp and bedrock. Dark, narrow creeks slip down from hidden woodland ponds, cutting through granite ledge and emptying into soft meadow, spilling into the ebb and flow of tidal rivers fed by the sea. Not far from my house, the thick mud flats of the Little River swell when the tide rises, flooding the marsh, staining freshwater creeks with salt. One small creek rises up into the woods, up a series of mossy granite steps to its source, the Lily Pond. When shadbushes burst into white flowery spring bloom and slick black cormorants come to feed in the marsh, this creek rushes with life. Alewives swim in from the river, returning from the sea to spawn in the freshwater pond. On rainy, moonless nights, when the tide is flowing and the meadow is wet, thousands of tiny glass eels, swept in by the sea, work their way up the river. They run up the creek, swimming up over rocky steps, slipping through wet grass, heading into the pond, where they will settle in nooks and crannies among the rocks on the bottom. Ultimately, they will darken and grow to be three feet (one meter) long, but now they are the small and slender, their bodies thin and transparent, their eyes black. Wading through the waterlogged marsh, standing at the edge of the river, I catch them with my flashlight, darting through the swirling tide. They have traveled 1,000 miles (1,600 kilometers) from the waters of their birth, in the wide expanse of the Sargasso Sea, to reach this little pond.

In my neighborhood, the journey is relatively short: the creek finds its source in a pond not far from the sea. In other places eels swim hundreds, thousands of miles up wide, strong rivers into remote tributaries. They swim through the St. Lawrence estuary and into Lake Ontario. They climb the Mississippi and branch into the Missouri, swimming upriver until, thwarted by dams, they can travel no further. It is an epic journey rivaling that of migrating birds in distance traveled.[13]

Only two species of *Anguilla* dwell in Atlantic. *Anguilla rostrata,* the American eel, lives in ponds and streams draining into the sea all along the continent, from the Gulf of Mexico to the St. Lawrence estuary. Parents of glass eels gathering at the head of the Little River may have lived in icy tidal creeks of Labrador, in marshy bayous of Louisiana. A mother may have made her way up the Mississippi into Minnesota, while the father lived his life in tidal creeks along the Chesapeake. *Anguilla anguilla* inhabits freshwater streams and rivers all over Europe. Only slight differences in DNA and musculature—the European eel has more vertebrae—distinguish *Anguilla anguilla* from its American cousin. After fifteen or twenty years of a reclusive freshwater life, eels on both sides of the sea mature and return to the place of their birth. On rainy autumn nights they reverse the long migration that first brought them inland. Sliding down the creeks and rivers into the tide, they head out into the Sargasso, guided by senses to which we are stunningly oblivious. Whether they steer toward the spawning grounds by the light of the stars, by contours in the seabed shaped by magnetic variations, or by the smell of salt water, we can only surmise.

We cannot even follow eels along their watery lifeline to the Sargasso. Oceanographers from both sides of Atlantic, equipped with sophisticated sonar and submersible cameras, have invested substantial time and resources in the attempt, vainly combing the Sargasso in search of spawning eels. Scientists have netted leptocephali, the larvae, and assumed that the smallest, 0.25 inch (6 millimeters) long, are nearest the spawning area, which they place south of Bermuda. The larvae, shaped like willow leaves and endowed with long, oversized teeth whose purpose is unclear, look so unlike their parents it was thought for many years they were a different animal entirely.

The Sargasso still holds the secrets of eels. European and American eels leave their freshwater homes in the fall and gather in this warm sea, perhaps at the same time, in the same waters, perhaps not. No one has yet found an adult eel anywhere in this sea. We imagine that, exhausted by their long journey, they spawn and die. American larvae drift with the currents; European larvae may even swim. What energy reserves sustain them on their journey we don't yet know.

Nourished by the sea, perhaps absorbing tiny bits of organic material in the water, they grow—American eel larvae to 1.8 inches (45 millimeters), European larvae to 2.4 inches (60 millimeters). Eventually, the two species separate, and when they cross the waters of the continental shelf, they metamorphose into glass eels, their flat, willowy bodies becoming cylindrical, their large, long teeth disappearing, replaced by smaller and finer versions. By the time baby eels have floated out of the Sargasso, dispersed all along the continents, and reached the head of the tide, a whole year has passed for American eels, and two or three for European. Eels inhabit a broad geographic range on each side of the sea, but they seem to form only a few breeding populations. Though separated by great distances, their destinies are linked.

On rainy spring nights in Gloucester, eels swarm in the flooding waters of the Little River and wind their way into freshwater creeks and ponds undisturbed, their passage through the dark woods marked only by the splashing of rain on wet leaves and the spring song of the peepers. In Maine, where catching glass eels is legal, spring migrations can be noisy and well attended. In years when eels were plentiful and prices were high, fishermen crowded the narrow creeks, jockeying for position, clogging the running water with fine-mesh nets set to catch elvers swimming upstream. Altercations broke out, and occasional violence. Fishermen cut each other's nets, resorted to fistfights, and fired guns, all to protect turf.

On the other side of the world, across the Pacific, the health of Japanese banks determines the level of squabbling on Maine creeks and the fate of the American eel. Eel is an expensive delicacy in Asia, where wild populations have long been exhausted. Japanese, Chinese, and Taiwanese fish farmers pay good money to import young glass eels, who still retain their transparency. Farmers raise eels in captivity, fatten them on fish meal pellets, and then sell them for large sums to restaurants. In a strong economy, prices rise. The glass eel fishery, known only to those who take to the woods on moist spring nights, is worth millions of dollars; glass eels are among the world's most valuable marine commodities. When prices are high, a fisherman might net $350 for every pound of eels. One cold night spent tending gear in an icy stream might bring in $1,000, more than enough to cover expenses. In 1997, this lucrative fishery was worth $6.3 million to U.S. fishermen.

When prices rise, men are tempted to poach. A judge recently jailed a dealer for buying $43,000 worth of illegally netted eels. Soaring prices also bring more fishermen onto the creeks. Between 1997 and 1998, the number of Maine elver licenses doubled. When the 1998 season opened, more than two thousand fishermen held licenses to set four thousand nets. Hopes were high, and dark night saw Maine's coastal creeks crammed with nets, but the gold rush ended abruptly when Indonesia's economy crashed and Japan's soon followed. With Japan's banks and securities industries reeling from bad loans, the value of the yen fell and businessmen could no longer afford luxury meals of eel. Demand fell, and the price of eels dropped by two thirds. In 1999, an unexpected surge in the number of Japanese eels depressed prices further, to a low of $10 to $20 per pound. Frustrated American fishermen dumped their catches.

The plummeting Asian economy and the plethora of Japanese eels gave some breathing room to the stressed American fishery. Although tidal inlets teemed with fishermen, the boom was over. Halfway through the 1998 season, the catch was down 75 percent. Creeks that once swarmed with glass eels were virtually empty. Possibly the breeding population is under siege. Fishermen measure their catch by the pound: each pound consists of at least 1,500 glass eels. Each year millions of American eels are caught and sold when they are mere wisps, long before they have attained the size and color and shape of adult eels, long before they are even remotely ready to spawn. For most of its life, *Anguilla* lives hidden in rocky recesses at the bottom of ponds, but on the rare occasion when it comes up into the water, we have intercepted it, threatening its well-being. For many years, the eel fishery was virtually unnoticed and unregulated. Eel fishermen come and go in the darkness of the night, seeking a fish few of us have ever seen or eaten or even imagined others would desire. We have no notion of how rich the population was or might have been; by the time we began counting, the numbers were already declining.[14]

In Canada, in the upper reaches of the St. Lawrence River and Lake Ontario, eeling once was big business. Here older eels are caught and exported to Europe, where they are smoked, jellied, and marinated. Thousands of eels used to swim into the Gulf of St. Lawrence, winding their way through and around the locks and dams of the St. Lawrence Seaway. After living for as long as fourteen years in Lake Ontario, they returned to the sea, coming out through the locks, swimming through the turbines of hydroelectric dams. In the St. Lawrence/Lake Ontario commercial eel fishery, catches declined 50 percent between 1985 and 1997, even though prices had tripled and demand was still high.

An eel ladder installed at the entrance to Lake Ontario in 1974 allowed scientists to observe the decimation of the population. Placed at the Cornwall Dam, it eased the final push upriver. Eels coming into the mouth of the St. Lawrence take three or four years to reach the fish ladder, and when they arrive, they are already juveniles, pigmented yellow from their diets of copepods and plankton. In the decade 1975 to 1985, between 600,000 and 1,300,000 eels ascended the fish ladder each year. In 1986, the number dropped to 200,000, and in 1997 it dropped even further, to 6,700.

Overfishing stresses eel populations in the St. Lawrence. So does pollution. Scientists have yet to tease out how each contributes to declining numbers of eels. *Anguilla*'s life turns in a circle, beginning in the salt water of the Sargasso, when leptocephali are born, and ending there years later, when as adults they return to spawn and die. Record catches in 1975 may have thinned the population of spawners. In addition, those who made it back were marked by their freshwater days, with PCBs from industrial discharges and pesticides from runoff. Mirex contamination temporarily closed the European market, giving eels a respite from fishing pressures but stressing them in other ways. Although government regulation had reduced PCB and pesticide levels in the estuary by 1990, eel catches still plunged. St. Lawrence waters are cleaner, but industrial pollutants are long-lived. They still persist, working their way through the food web into the bodies of eels. In the Netherlands, 2,700 contaminated eels were removed from the Rhine, a river—like others in northern Europe and America—polluted with industrial organochlorines. These pollutants, which accumulate in the bodies of eels, are neither easily nor quickly excreted. When eels were placed in a clean lake, some pollutant levels dropped substantially after four years. Eight years later, others still remained.

Eels pass their contaminants up through the food web. Endangered beluga whales live in the St. Lawrence estuary, at the mouth of the Saguenay River where food is plentiful. These small white whales, easy to glimpse from the river ferry, have had a tough life. They suffer from high cancer rates, low birth rates, opportunistic infections, and endocrine disorders. Their bodies are so riddled with PCBs, pesticides, and mercury that the Canadian government classifies dead whales washing ashore as hazardous waste. For years the waters of the St. Lawrence have been polluted by chemical plants and smelters discharging their wastes into the river. The pollutants are distributed all along the reach of the estuary.

The village of Longue-Pointe, way out toward the sea, is tiny, consisting of a few roads and a few blocks of houses. While I was in Longue-Pointe, I did a lot of walking, particularly when it was too foggy or windy to look for whales. All along the shore, in this remote, isolated place, there were signs prohibiting the taking of shellfish, because they were contaminated. Humans may abstain, but whales cannot; some of the pollutants they consume come from far away. Until 1975, one company manufactured pesticides way upriver, where the water is fresh, near Lake Ontario. Though the company closed in 1975 and the pesticide, Mirex, has been banned, high concentrations still appear in the blubber of beluga whales. Eels swimming up the St. Lawrence accumulate pesticides and PCBs, and belugas gorge on them.

Eels making their way down the St. Lawrence or the Rhine, beginning their descent to the sea, carry the burden of living the greater part of their lives in polluted waters. This burden is evident in the fat stores needed to power their long journey to the Sargasso and to produce their young. As eels begin their final passage to the sea, they undergo one last transformation. Born in the sea as tiny, transparent elvers, they adapt to a youth spent in freshwater. When they are fully grown, they metamorphose one more time, readying to give their lives for the next generation. As they move down creeks and rivers to the salt water, they darken, their yellow skin turning silver and black for camouflage in the ocean's depths, and their eyes doubling in size for better vision in dimly lit water. They have but one purpose. Undistracted by food, they cease to feed, living off stored reserves of body fat. Whatever energy is left goes into egg production. The fat disappears, but the pollutants stay behind, passing into eels' muscles and eggs. The spawning area of *Anguilla* is far away in the distant waters of the Sargasso, where we cannot see what toll pollution takes fifteen years later, twenty years later, on an eel's ability to produce large numbers of hardy offspring.[15]

The waters of the Sargasso may be distant, but the fate of *Anguilla* lies ashore, with the number of fishermen granted licenses, with the cleanliness of fresh water, with the rise and fall of Asian markets. International commerce in eels not only raises the demand for American eels but poses another threat as well. Without human intervention, *Anguilla* would never have contact with Asian eels. Dwelling in two separate oceans, in freshwater streams separated by vast mountain ranges, Asian and American eels have no natural occasion to mingle. International trade has brought them together, in aquaculture tanks and fish markets, with detrimental consequences. Asian eels have long harbored a parasite, *Anguillicola crassus,* a small worm about 2 inches (5 centimeters) long. The worm lives in the eel's swim bladder, where it lays thousands of eggs that are hatched and released into either the eel's digestive tract or the surrounding water. The larvae are tough, clinging tenaciously to the floors of aquaculture ponds and to the sides of tanker trucks. They are highly versatile, living comfortably in salt water and freshwater creeks. They are eaten by a number of animals, by copepods, crayfish, and snails who in turn are eaten by eels.

Sometime in the 1980s, parasites from Asian eels appeared in the Ems River in Germany. No one is quite sure how they arrived. Perhaps a trucker carrying eels imported from Asia hosed out his tanker, releasing worms into a nearby river. Perhaps a merchant carrying Asian eels cleaned his tank, discarding parasite-ridden water in a drain discharging into a river. Or perhaps an Asian eel fell from a crate in an open fish market and sought refuge in a nearby stream. Whatever the initial point of contact, the infestation rapidly spread. Within ten years the parasite had infected between 70 and 90 percent of European eels.

Anguillicola lives uneventfully, harmlessly in Asian eels, but its effects on European eels are not so benign. The walls of an eel's swim bladder, normally thin and transparent, thicken and cloud over, and fill with dilated blood vessels when the parasites feed. Swim bladders give fish buoyancy in the sea, enable them to withstand the pressure of water. A parasite-ridden swim bladder works less efficiently, with the proportion of oxygen decreasing between 35 and 60 percent, depending on the number of parasites. Scientists wonder how adult eels can complete an already taxing spawning migration with malfuctioning swim bladders.

In an era of ever-increasing world trade, where markets are truly global, where great distances are easily traveled, time became the only barrier separating *Anguillicola* from American eels. That buffer has now disappeared. Not many years after the European infestation, an imported eel brought *Anguillicola* across Atlantic. In 1995, infected eels were found in aquaculture ponds in Texas and in a feral eel from a South Carolina Bay. By 1998, infection rates had soared, to 67 percent of eels in North Carolina's Cape Fear River, to 50 percent in some regions of Chesapeake Bay, and to 20 percent in portions of New York's Hudson River.

I have no doubt that *Anguillicola crassus* is well on its way into New England. In my own neighborhood, striper fishermen using live bait often purchase eels or menhaden: at one of the local bait stores, the eels come from Maryland. I went into the store one day to take a look. Standing by the tank watching a mess of eels circling through the water, I wondered how many might carry *Anguillicola crassus.* Gloucester has worked hard to protect eels in her creeks and streams, but the shellfish warden, vigilant as he is, cannot cordon off Gloucester's tidal creeks. At this point, there is little he can do to keep such parasites out. By the time the fishing season ends and the stripers have come and gone, it is quite likely that *Anguillicola crassus* will have moved into the bay. Further spread up the eastern seaboard is inevitable. What kind of obstacle this will present to eels whose migration out to the Sargasso is already compromised, only more time will tell.[16]

No continent edges the Sargasso; it is one of earth's only seas entirely surrounded by water. Many are the ways, though, in which it is touched by land. Migrating eels bring in Asian parasites along with the chemical discards of a consumer society, but eels are by no means the only bearers of onshore tidings. Dusty remnants of eroding mountains, shreds of bone or cells sinking to the bottom: these, too, carry what washes off the land into the deep. Traces of chlordane and dieldrin, insecticides applied on land, lie at the bottom of the Sargasso. Though ribbons of fast-moving currents enclose this sea, giving it a watery edge, the boundary is permeable. The rushing currents separate the Sargasso from the continents but at the same time collect and channel land's castoffs into these clear waters more than 1,000 miles (1,600 kilometers) offshore. Circling currents push the water inward, toward the center of the Sargasso, mounding it into a hill a full three feet higher than the rest of the ocean. What swirls in, stays.

All kinds of flotsam, swept in by the currents, converge in this sea within a sea. Accidental spills and leaks, acts of negligence, or deliberate violations of the law bring oil to the Sargasso, as far away, tankers pump out their bilges, discharge ballast water, and clean out their holds, and cruise ships discharge their oily waste into the water. Oil seeps in and coagulates into soft, sticky clumps of tar. Hatchling loggerhead turtles drift in the Gulf Stream at the edge of the Sargasso, floating amidst clumps of *Sargassum* weed, feeding on shrimp, copepods, jellyfish, and crabs living there. The jaws of almost half these baby turtles are caked with tar. When their stomachs are flushed, it appears that in addition to seafood and tar, 15 percent of the baby turtles have also ingested pieces of plastic or synthetic fiber.

Plastic dumped at sea is virtually indestructible. Noted for its durability, it outlasts our use for it, enduring long after we have thrown it away. On the *Cramer*'s science watch, the tows bring up many plastic fragments, pellets, bits of sheeting, pieces of fishing line. We separate these and place the rest of the catch, mostly gelatinous clumps of scum, in petri dishes. With the aid of a microscope, we identify the first one hundred organisms, the delegation from the minute, hidden world of the surface. In rich coastal waters, this catch is abundant, usually yielding dense concentrations of copepods and amphipods, pteropods and fish eggs, shrimp and snails. In the Sargasso, life is scarce, the catch meager, the counts low. I don't expect to find many animals in this sea, where nutrients are few, but I am surprised to find my petri dish covered with tiny shards of plastic, invisible to the naked eye, hard and sharp to the touch. These shards were once part of larger, recognizable objects, tossed over-

board or washed out to sea long ago. Single pieces disintegrated into many, large broke up into small, but nothing ever really disappeared. Instead their scattered remains were swept into this remote place, where they have entered the marine food web through the diet of tiny lantern fish.

The most visible signs of life in the blue water of the Sargasso are lines of golden brown seaweed floating at the surface, pushed into long rows by the movement of wind and water. *Sargassum* once grew in profusion here. Christopher Columbus noted it in his log, both on his first Atlantic crossing and on his return. He saw rafts of *Sargassum* so widespread they reached the horizon, so thick they actually slowed his ships, so dense it seemed as if the sea had shoaled. When we sail through, the weed is wanting: only a few scattered patches are strewn over the wide expanse of blue. It is possible, and comforting, to believe that Columbus exaggerated, but he was a painstaking, scrupulous sailor, perhaps the finest dead-reckoning navigator the world has known, and his achievement depended on precision.

No one really knows how much gulf-berry or sea holly, as *Sargassum* is also called, floats in the sea, or what contributes to its abundance. Perhaps in five hundred years, currents have changed, dispersing the weed to other places. Some fishermen report that over the last ten years, rows of *Sargassum* forty feet (twelve meters) wide, once common, have disappeared. It is possible that the insatiable needs of our economy have reached even this distant, lonely place, as high-seas fishermen target the weed and haul it from the sea. What sells as an additive to livestock feed is also home to several hundred species of animals, and an important nursery for sea turtles. When *Sargassum* is plucked from the sea, a whole community of life disappears with it.[17]

The weed that gave the Sargasso its name is itself a sojourner, a long-distance wanderer hailing from the shores of the West Indies. Ripped from the rocks during the high winds and heavy seas of hurricanes, it drifted out to the Sargasso, where it made its home floating at the water's surface, buoyed by air bladders the size of berries. As it adapted to life in the open ocean, animals sought shelter in its leaves, and each clump of weed became an island of life unto itself. Tiny bacteria and algae cling to the *Sargassum;* larger barnacles, tube worms, and box-shaped bryozoans make a more permanent and more material commitment, cementing their homes to the fronds. Crabs, shrimp, and snails find anchorage here, as well as small fish camouflaged to blend into the thick foliage. *Sargassum* fish lie hidden here, their fleshy fins the shape and texture of their namesake, their mottled brown color a perfect match.

The floating island is self-sufficient, complete, the individual members of the community finding sustenance and shelter from each other. Flying fish spawn here, and chub and snapper larvae grow in the protection of the leaves. Young jacks come to feed on copepods and larval crabs hidden in the foliage. Hatchling loggerhead, green, Kemp's ridley, and hawksbill turtles all find food and shelter here. *Sargassum* weed produces more than half the greenery in the Sargasso. Wind blows in the necessary iron with dust from the Sahara. Small, buoyant photosynthesizing bacteria supply nitrogen. When they bloom, the sea looks as if it is scattered with sawdust. The world of Sargassum is rich, supporting 145 species of invertebrates and 100 species of fish.

Before we learned that *Sargassum* could propagate in the open ocean, away from its rocky origins, we thought it immortal, believing that it lived for hundreds of years, that the same weed scattered across the blue water today once greeted the ships of Columbus. We now recognize that a tangle of *Sargassum* cannot live forever, cannot support itself indefinitely. The community of the weed, like the wide Sargasso, is not one of infinite possibility, limitless capacity. Like the sea itself, the weed can only hold so much. As it ages, what was buoyant becomes ballast. The air bladders thicken and the leaves, heavy with encrusted organisms, sink into the water. Burdened by a weight it can no longer support, by a community it can no longer sustain, the floating island collapses. A few fast swimmers may find refuge on another weed, but predators snatch most of those who flee. Those who stay behind, who cannot swim, or whose homes are firmly fixed to the weed are doomed as the colony is dragged to its death in the depths. There it disintegrates, making a welcome meal for dwellers of the mid-water.[18]

In the wide Sargasso, this is the natural end of a particular community, but it is not the end of a species. During the course of its life, *Sargassum* provides for the future. As the foliage grows lush, the tips of the leaves break off. They drift away and themselves begin to grow, creating new colonies. In the generous waters of the Sargasso, the cycle repeats itself, again and again. Perhaps these islands of *Sargassum* are microcosms of our own fate. Certainly we as a species have prospered even as individual civilizations, taxing the land beyond what it could bear, have fallen, one after another. Looking at our own heavily freighted society, whose excesses have reached, and possibly burdened, even these distant waters, I wonder if the time will come when the sea can no longer care for us.

Part 3

The Empty Basin

CHAPTER 8

The Moving Earth Beneath the Sea

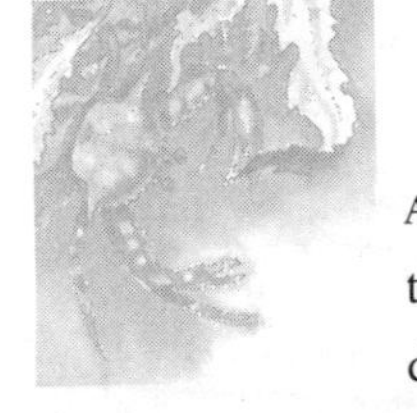

AS the *Cramer* idles through the Sargasso, waiting for the wind to rise, the sea is flat and empty. Nothing demarcates or divides the smooth expanse of water dissolving into the horizon. A thousand miles (1,600 kilometers) away, at the far edge of the Sargasso, the islands of the Azores rise up to meet the sky, but here nothing breaks the endless plane of blue. This vast, unroughened surface, this breadth of uniform sea, deceives. But for a few lonely oceanic islands, the unperturbed surface offers no hint of the grand and sweeping energies hidden below. Though clear, the water conceals potent forces seething in its depths.

Only one thousand miles offshore, the *Cramer* has already sailed through Atlantic's deepest waters. (Only in a trench off Puerto Rico is the sea deeper.) Contrary to what one might guess, Atlantic's deepest waters, like those in other oceans, are along her edges. As we continue east, toward the middle of the sea, the bottom rises. The unmarked plains of the abyss, here flattened by layers of sediment, give way to rising foothills and then to mountains. The first maps of Atlantic seafloor noted, albeit crudely, this rise. Early efforts to plumb Atlantic's depths proved outrageously inaccurate: one naval officer paid out eight miles (thirteen kilometers) of hemp from a drifting ship and concluded the sea had no bottom. Eventually, sailors more or less successfully calculated depth by heaving overboard cannonballs tied to bailing twine. When they hit bottom, the sailors measured and snipped the twine and then moved on, leaving a trail of lead strung out across the seafloor. These crude soundings, forming the basis of the first map of Atlantic's basin, published in 1854, identified a prominent rise halfway between Europe and America.

For many years no one could explain why the basin of Atlantic, unlike a bowl, deepened at its edges and shoaled in its center. People assumed that this "Middle Ground," "Telegraph Plateau," or "Dolphin Rise," as it was variously called, was an ancient and drowned land bridge, or a lost continent, but sailors repairing transatlantic telegraph cable unknowingly produced evidence to prove otherwise. Wrestling with the broken cable, they accidentally twisted off a piece of the "plateau" and dredged up a twenty-one-pound (ten-kilogram) chunk of dense black volcanic rock. It was some of the youngest, freshest rock on earth, and it was torn not from a piece of continent sunk beneath the waves, but from the very foundation of the sea.[1]

Today, highly sophisticated sound waves bring the hazy images of those early soundings into sharp focus, revealing that one of the largest and most salient geographic features on the planet lies on the floor of the ocean. Hidden beneath the waves is an immense submerged mountain range, the backbone of the sea. More extensive, rugged, and imposing than the Andes, Rockies, or Himalayas, it covers almost as much of earth's surface as the dry land of continents. Winding like the seam of a baseball, it circles the planet in a long, sinuous path, running the entire length of Atlantic, slashing the basin neatly in two. Its mountains are stark and black, as black as the sea itself, lit only at their peaks by a thin, patchy covering of white, the skeletal remains of tiny microscopic animals that once lived at the surface. Peaks as high as Mount St. Helens sit in a watery world of blackness, more than a mile below the surface, beyond the reach of light, beyond the sight of sailors.

A great valley, eclipsing any comparable feature on dry land, runs through these mountains. Arizona's Grand Canyon, one of earth's most spectacular places, extends for about 280 miles (450 kilometers). A lesser-known canyon of similar depth but considerably greater length lies hidden in the mountains of the ridge. Although offset in many places by breaks in the mountains, the rift valley, as it is called, extends the length of Atlantic for 11,000 miles (17,700 kilometers). Here in this bleak and forbidding place, where the water is almost freezing, subterranean fires have lifted mounds of fresh lava onto the seafloor. Scientists visiting the rift valley for the first time named the volcanic hills in this otherworldly setting Mount Venus, Mount Mercury, and Mount Pluto, after distant, lifeless planets.[2]

Yet the stillness, silence, and darkness, mistaken for sterility, are the waiting—the quiet that precedes a momentous occasion. What had seemed so foreign is an integral part of earth's very being, for at the ridge our own planet gives birth. The floor of the rift valley is torn; from the gashes has sprung the seafloor underlying all of Atlantic. Here the youngest, newest pieces are made. Earth is still cooling from her tumultuous birth four and a half billion years ago. Heat, leaking from the molten core and from radioactive decay deep inside the planet, rises toward earth's surface, powering the volcanoes that deliver the ridge to the sea.

Early in earth's history, its surface solidified and cracked into approximately twelve giant pieces. Nudged by warm rock flowing below them, the rigid plates drift over the planet's surface, carrying both seafloor and continent. In the hidden recesses of the rift valley, plates drift apart and molten lava ascends to fill the space. What rises, eventually falls. Seafloor created at the ridge drifts with its plate for millions of years, and then, as if riding a conveyor belt, descends back into the depths from which it came. Where plates move apart, continents are torn in two and oceans widen. Where they are squeezed together, towering mountains are raised and entire seas are swallowed. Drifting plates once welded all earth's dry land into one mass, then broke it apart, once gave birth to ancient oceans, then reclaimed their basins and drained the waters away. Today, as India drifts into Asia, the edges of the continents crumple, raising the Himalayas. As the Arabian Peninsula pulls away from Africa, the Red Sea is born. The plates drift, opening one sea, closing another. While the Red Sea grows, the Mediterranean disappears.

Four plates divide Atlantic. The smallest bears the Caribbean Ocean and its islands; the mountainous spine of the Mid-Atlantic Ridge edges the larger three. The American plate sweeps west from the ridge, stretching across the sea, across both continents all the way to the Pacific, except where it bumps up against the Caribbean. There, seafloor fashioned at the ridge and moved across the sea now makes its return descent. Two plates reach east from the ridge, one extending through Europe and northern Asia, the other through Africa. The dividing line slices through the Mediterranean, closing it as Africa moves north to merge with Europe.

These two plates collide and slip east, while bedrock underlying Greenland and the Americas pulls away from them both, sliding slowly, almost imperceptibly toward the setting sun. Each year Atlantic widens by approximately one inch, pushing Europe and North America, Africa and South America farther apart. If Columbus were to make his voyage today, the journey would be longer by the length of a football field. In the last million years, six miles (ten kilometers) have been added to the Atlantic's girth. These tiny motions, accumulated over time, give earth its shape and contour.[3]

In the center of Atlantic, where fresh lava builds new seafloor to fill the gaps between separating plates, the ridge, buoyed and swollen with heat, is lifted into craggy mountains almost two miles high. As young rock is dragged out from the place of its birth, it cools, contracts, and sinks, creating the basin holding Atlantic. The seafloor flattens as the mountains disappear under bits and pieces of plant and animal debris drifting down from the surface. At the edge of the ocean, where rivers wash sand and silt down from the land, the once lofty mountains are now buried under miles of sediment. Earth's heat engine, sculptor of the seafloor, has created a remarkably uniform work; only age and its exposure to the elements distinguish young rocks in the ridge from their 150-million-year-old counterparts submerged off coastal Africa and America.

A drifting plate leaves a marked trail. Throughout history, earth's magnetic field has often reversed direction. Today compass needles point north, while eight hundred thousand years ago they pointed south, and two million years ago, north again. The tracks of the spreading seafloor are etched in rock, in bits of iron that, as the lava cooled, became permanently magnetized in the direction of earth's magnetic field. Like a broken clock, these tiny, stilled compasses measure time passed. Scientists analyzing subtle variations in the alignment of earth's magnetic field have mapped these periodic reversals, which appear as alternating stripes of rock. The stripes, younger in the mountains, grow progressively older the further they lie from the ridge. Mirroring each other on opposite sides of the ocean, they mark the divergent paths taken by drifting seafloor.

The two plates move at different speeds, with more seafloor forming on the plate moving faster. In the center of the ridge, where the rock is youngest, the stripes are broader toward the east, since for the last million years Europe has been drifting at twice the speed of North America. The bands of rock may vary in width, but their symmetry extends all the way to the edge of the continents, where the oldest parts of Atlantic lie. It is a well-signed traverse, where distances are marked and the passage of time is measured, where the history of an ocean basin is charted.[4]

Most of the mountains of the sea lie deep in the water, but occasionally a few peaks reach above the surface. Iceland is born of the sea, anchored to the ridge, pushed out into the air by large plumes of hot rock rising from deep within earth. While other nearby islands, such as the Faeroes and Great Britain, are continental in origin, pieces of granite cut off from the mainland by a rising sea, Iceland is one of the most volcanically active places on the face of earth. In the sixth century, Irish monks fleeing a Europe ransacked by barbarians took to the sea in wicker boats covered with oxskins and discovered Iceland. Seeking a haven of peace and solitude, they instead found a rocky island engulfed in flame, where mountains spewed fiery rock, where burning smoke filled the air, and where the sea boiled.

In accounts of the overseas voyage of St. Brendan, the monks describe thunderous bellowing and hammering emanating from forges on a rocky island where a dark smithy hurled flaming lumps of slag. In all likelihood, they refer to an eruption of Iceland's Hekla, the "hooded one." Often shrouded in cloud and fog, this volcano has exploded violently and repeatedly over the last seven thousand years, frequently covering as much as two thirds of the island with debris. St. Brendan's frightened monks, convinced they had come upon the entrance to Hell, quickly rowed away. Later settlers who built their homes in the valleys around Hekla found that the explosive volcano regularly destroyed their farms and fields, burying pastures and crops under layers of ash, leaving a scorched and desolate wasteland.[5]

St. Brendan understood the power of Iceland's volcanic energies but mistook the source. Iceland may not lead to the dwelling of the damned, but it does expose the undersea world of the ridge, where the seam of earth opens. It is a fiery land, its settlements built and rebuilt on untrustworthy bedrock, on land cut by a broad swath of gaping fissures and smoldering volcanoes. For the last millennium, volcanic eruptions have racked this child of the sea every four or five years, each time adding more rock to the young island. Earth is four and a half billion years old; the scoured rock in nearby Greenland is over three billion years old. Iceland is a late arrival; its oldest rock is a mere sixteen million years old, its youngest has yet to mark its first decade. Iceland sits astride the ridge, where submarine fires are literally tearing the island in two; only flowing lava holds it together. Were the volcanoes to cool and the raging heat to subside, Iceland would split in half, part to drift east with Europe, part to move west with North America, the pieces separating by an average of 0.7 inch (2 centimeters) each year.[6]

In northern Iceland, the roads run through beds of lava to a clear blue lake fringed with lush angelica. Over the past four thousand years, volcanoes have reconstructed this landscape and then repeatedly torn it apart.

Flowing lava dammed a river and then flowed along its riverbed to the sea. The dam created Lake Myvatn, named for the thick clouds of midges swarming over its shores. Bird-watchers from Europe and America gather here to observe the many species of ducks seeking shelter in this important breeding ground sculpted by a volcano. Lava has flowed profusely in the area around Lake Myvatn, filling a dip in the land with a pond of molten rock 65 feet (20 meters) deep. The lava receded and the pond drained, but left behind Dimmuborgir, the "castles," twisted pillars of partially hardened lava rising from burned ground. Three hundred years ago, the land around Myvatn rose, becoming hot and swollen as a restless volcano came to life. Boats moored in the shoaling lake were stranded. Swarms of earthquakes shook the ground, and a line of fissures opened along the shore, flanking Myvatn with a wall of fiery lava.

The land still trembles. Red, orange, yellow, and white minerals streak the ground east of the lake, signaling soft and dangerously unstable terrain. Nearby, a once pleasant bathing pool has heated to a scalding 125 degrees Fahrenheit (52 degrees Celsius). Four miles underground, hot melting rock rises, and as it pours into the chamber of the volcano, the earth swells. Eventually, the roof snaps under the pressure and lava escapes, some exploding in fire on the surface, some coursing underground, stretching the earth, pushing the drifting pieces of Iceland further apart. Earthquakes and rising magma have cracked this ground near Myvatn in over one thousand places. The land moves in fits and starts; one eruption widened the gap by as much as three feet (one meter). During the latest eruptions, between 1975 and 1984, this land widened by twenty-six feet (eight meters).[7]

The ridge cuts through the island, drifting plates wrenching apart the landscape and sometimes poisoning the air. In the south, a fifteen-mile-(twenty-four-kilometer)-long crack known as Laki, or "maw," radiates out from Grimsvotn, a powerful volcano. In 1783, the volcano erupted and a massive lava flow spewed from Laki. Flowing faster than the Rhine, the hot liquid rock swallowed churches and farms, blackened fields and burned crops. A blue haze of poisonous gas settled over Iceland, killing grassland, cattle and sheep. Ten thousand people died of famine. The haze spread across the sea, blanketing Europe and America with fog. When the sulfur sank back to earth, a portion rained into the glaciers of Greenland, where thick ice preserves the record of the outburst.

Old Norse legends, written into Iceland's epic sagas, describe a culture born of ice and fire. According to Iceland's creation myths, life emerged when smoking embers from Muspell, the realm of raging fire, mixed with ice crystals from Niflheim, the land of snow and fog. This ancient myth holds grains of truth which now, hundreds of years later, science has verified and refined. The fusion of fire and ice, which gave birth to the Norse gods and their descendants, continues today, literally, where volcanoes erupt beneath glaciers.

In the fall of 1996, the earth cracked north of Grimsvotn, a volcano hidden under Europe's largest glacier. In a little over one day, hot lava melted ice 1,600 feet (500 meters) thick. For five weeks, rivers of water at 60 degrees Fahrenheit (15 degrees Celsius) ran through the glacier, pouring into Grimsvotn's crater. The volcanic lake filled to overflowing, lifted its heavy, icy cover, releasing a raging flood of ice, rock, and water. The torrent raced to the sea, washing away everything in its path, destroying bridges, telephone lines, and Iceland's main southern road, leaving behind a wasteland of icy rubble and rock. The Norse tales recount that when the gods' long and famous reign ended, earth and sky cracked open and Surt, the fire giant, whose name means "black," appeared, brandishing a flaming sword brighter than the sun itself. With this sword, he vanquished the gods and set earth on fire. From this ruin came a fresh green land, which rose from the sea.[8]

Iceland still rises from the sea, still brings forth new land and life. One November day in 1963, fishermen working off Iceland's southern coast noticed a strange sulfur smell. Within minutes their boat began to lurch, the sea warmed and turned brown, smoke and ash shot from the water. Out of the hissing steam and thunderous explosions a black volcano emerged from the sea, streaming molten lava from its vents, hurling clumps of exploding magma into the air, raining down ash upon the terrified crew. Volcanic convulsions racked the new island for a year and a half, prohibiting visitors from landing on the scorching shore. Migrating birds paused on the hot rock but moved on. Seeds of flowering plants, blown in on the wind, carried in by waves, sprouted and burned. Icelanders named the island Surtsey in honor of the fire giant. When the seas finally cooled, the green earth foretold by the myths appeared. Seaweeds grew along the shore, mosses along the rim of the crater. After seven years, the first birds nested and raised their young. The community is still growing. Today, six species of birds nest on Surtsey, and twenty-five species of plants have colonized the island.

If restless Iceland has anything to say to us, perhaps it is that we are merely passengers riding the drifting continents while other authorities steer. Man has, however, attempted to redirect the flow of earth's energies. Seven thousand years ago, the volcano Helgafell, "holy mountain," created the island of Heimaey, home to Iceland's premier fishing port. After years of quiescence, Surt drew his flaming sword and, in 1973, drew a wall of fiery lava from a flat plain on the outskirts of town. The fires coalesced into Eldfell, "fire mountain," and a tongue of lava bore down on the town. Most of the residents were evacuated. Those who remained made a heroic effort to halt the advancing flow, bulldozing earth dams in front of the liquid rock, then pumping in seawater to chill, harden, and slow it down. The molten rock was doused with over 200 million cubic feet (6 million cubic meters) of seawater, cooling almost the equivalent amount of lava, but Eldfell unleashed forty times more. While seawater thickened the lava into a rigid barrier, the volcano continued to spew forth liquid rock. It piled up behind the barriers and then breached them, continuing on through the town to the sea, burying or burning two hundred homes, the hospital, a fish-processing plant, and a hotel. It stopped just short of destroying the deep-water harbor.

Whether the massive effort actually saved the harbor or whether it merely slowed the flow and postponed the inevitable, scientists can't really say. When the paroxysms ceased, Heimaey had grown by 20 percent, but one third of the town was buried in ash and debris. The eruption, which lasted five months, was small compared with others in Iceland. The villagers who returned repaved their roads with the volcanic rock and piped water through the hot lava to heat their homes, but the thick seawall spawned by the "fire mountain" still stands, a reminder that all this land is leased, impermanent, its inhabitants but tenants on a terrain they ultimately cannot possess.[9]

A plume of melting rock, as hot as 2,500 degrees Fahrenheit (1,400 degrees Celsius), rising from deep within the earth, fuels Iceland's fires. The plume carries its own unique signature, expressed in bits of volcanic glass containing traces of rare elements and unusual concentrations of primordial helium, remnants of primitive earth trapped deep in the mantle, here carried toward light in ascending rock. Earthquake waves map the plume, locating its roots down near earth's core, some 1,800 miles (2,900 kilometers) below the bottom of the sea. Iceland's "hot spot," along with another plume of molten rock rising beneath the islands of the Azores, stoke nearly all the volcanic fires along the northern branch of the Mid-Atlantic Ridge. As volatile as Iceland's fiery lands may be, between 90 and 95 percent of earth's volcanic activity takes place on the ridge under the sea.[10]

Hot liquid rock builds new seafloor, widening the ocean basin much as it does Iceland. Molten lava collects in chambers underneath the rift valley, hardening into heavy coarse-grained rock or breaking through the roof, rising up toward the seafloor in long, thin sheets. Each sheet, or dike, about three feet (one meter) wide and six tenths of a mile (one kilometer) long, adds another plank to the seafloor. Most of the new rock hardens below the bottom of the sea, but it can rise up into the freezing water, through cracks in the rift valley like those near Iceland's, Lake Myvatn, Grimsvotn, or Helgafell. Knolls or hummocks, fashioned by lava, coalesce into low ridges inching along the rift valley floor like caterpillars. Early in an eruption, lava seeps out along the length of a crack, but as it progresses and its plumbing is established, the flow is restricted to one larger hill rimmed by a crater, just as it has done in Iceland.

Whether melting rock rising from the heart of the planet reaches the seafloor in icy waters off Iceland or farther south, near the Azores at the edge of the Sargasso, its face is the same. No violent explosions, no showy pyrotechnics mark these deep-water eruptions, for a heavy sea suppresses a volcano's gas and steam. Instead, lava effuses quietly, like toothpaste squeezed from a giant tube. Draped over the volcanic hills strung out along the middle of the rift valley are rounded pillows of glassy rock, the product of hot lava doused by a chilling sea. Where molten rock emerges into the rift valley and then recedes back below the seafloor, it leaves a hardened line of lava to mark its presence, like a ring on a draining bathtub. Near cracks in the middle of the valley floor sit the youngest volcanoes, pristine and undisturbed. As they begin their journey across the rift valley, an entire hill may go in one direction, or, if it lies across a fissure, the hill may split, part to ride east toward Europe, part to move west with the Americas. The restless plates shift, breaking up the volcanic hills, lifting them up into the walls of the rift valley and out onto the ridge. The symmetry startles. Matching halves of pillow lavas melded into the rising terraces face each other across a valley several miles wide.[11]

From a distance, the mountains of the sea seem to wind their way through the ocean continuously, without interruption, but actually the chain is broken into pieces, each offset by a jagged tear. Each segment, furnished with its own supply of molten rock, works independently, at its own pace, in its own time. Atlantic widens at the ridge, but not all at once. One valley may lie dormant while flowing lava seeps into another. Along one segment, where melting rock has only recently filled fissures in the seafloor, fresh, glassy pillows, uncovered by any sediment drifting down in the water, dress a volcanic ridge. Subterranean fires have burned here within the last few hundred years. They may burn again tomorrow. Nearby, a thin veneer of sediment coats the craters. Here the glassy rock has weathered, lost its shine. Earthquakes have reduced once distinctive lava formations to massive piles of rubble strewn across the valley floor. Thousands of years may have elapsed since the last eruption.

Rock resists a clean tear. Trapped in the undersea mountains, where the seafloor is young and fresh, are pieces of limestone, formed 140 million years ago, when Atlantic was a narrow, shallow waterway. Trapped also are slivers of continental rock, even older, shaved off the land as Atlantic was born. In the fractured seafloor separating the ridge segments lie chunks of earth's mantle, normally hidden below the seafloor, here heaved by earthquakes out into the pile of debris. As moving pieces of seafloor scrape past each other, this ancient rock is stuck, snagged between them. While volcanic eruptions on Atlantic's seafloor may be episodic, delicately tuned seismographs eavesdropping on a tiny segment of ridge register hundreds of tiny earthquakes in only a few hours. Rumbling like distant thunder in a summer storm, they beat the pulse of the moving earth.

Often we can't hear this pulse. Along the rift valley, halfway between Miami and Western Sahara, lie massive heaps of rock, the remains of an enormous avalanche, larger than any ever witnessed or recorded on land. Triggered perhaps by an earthquake, the equivalent of two hundred thousand truckloads of rock hurtled down one mountain, then slid up and over another before coming to rest. The cascading rock jostled the sea, creating a long fast-moving wave traveling through the water at several hundred miles per hour, yet in an ocean so deep, such a large motion would be imperceptible on the surface. What would be catastrophic on land is muted and silenced and rendered invisible in the sea. The age of this avalanche is also blurred, for time as we measure it loses meaning here, eclipsed by the long reach of geology. The rockslide took place within the last half million years, but the time has not been pinpointed, not within a thousand years, not within many thousands of years.

What is of consequence in our own lives is marked in minutes, hours, days, years, decades. The passage of ten or twenty or thirty years is critical in describing our own existence. A span of one hundred years encompasses the lives of several generations, but even this reach is insignificant, irrelevant out on the ridge, where time's brush is so much broader, where geologic timekeepers don't measure increments as small as a human lifetime.[12]

And yet, these smaller measures of time are marked here, evident in biology's faster pace. Dense communities of life, scattered along the ridge, thrive in the depth and darkness. Food in the deep sea is scarce, and concentrations of animals easily seen with the naked eye are rare; such an abundance came as a surprise. The paroxysms of a moving earth spawn these lush communities, whose membership is mostly uncharted and whose rhythms of existence are still by and large unknown. Fueled by the planet's inner heat, they bask in its warmth and wither away when the heat subsides. Nor far from the bleak and barren spot where scientists first visited the ridge lie other valleys, valleys teeming with a life so profuse that the distant and desolate names first given the mountains of the ridge, names such as Mount Venus and Mount Pluto, have come to seem misleading.

Near that first landing site is a segment of ridge with a large volcano prominently rising from the rift valley. Crowning its summit is a lake of young lava, ringed by a field of hot springs. Amidst the fresh, glassy volcanic pillows lie chunks of rock laced with streaks of brown, green, and yellow minerals, an oasis of color in the darkness. Towering spires of these minerals rise as much as 65 feet (20 meters) from the seafloor, spewing scalding water from twisted chimneylike jets. The water shimmers with heat, reaching a searing 620 degrees Fahrenheit (325 degrees Celsius) in the mouth of the chimneys.

Life as we have understood it cannot survive here on Lucky Strike, a vent field located by accident as scientists hauled in a routine dredge. The water is scalding, hot enough to poach any animal venturing too close. It is also toxic, laced with deadly hydrogen sulfide. What little light shines here is dim, too dim to fuel the level of photosynthesis necessary to sustain life. And yet, on the jagged rocky towers—scientists have named them Sintra, the Eiffel Tower, the Statue of Liberty—life abounds. Thick colonies of mussels cling to almost every crack, crevice, or platform bathed in warm water. In the absence of mussels, dense swarms of shrimp hug the vents. Sea urchins, limpets, and crabs all dwell at the edge of the mussel beds. In this rich community, each kind of animal represents a species unknown in shallow coastal waters.[13]

Heat, the force that builds and extends the seafloor, gives birth to deep-ocean hot springs. In the ridge's central valley, the seafloor is porous, cracked by cooling lava and the pull of drifting plates. Seawater percolates down through the cracks, passing through veins in the rock, circulating in the warmth. Along the way, it gives up its oxygen and rusts the lava. Leaving behind whatever magnesium it carries, it turns acidic. What began as seawater turns to another liquid in the baking rock, transformed almost beyond recognition by its journey into the depths. Continuing its descent, it falls deeper beneath the seafloor, becoming hotter and hotter. At temperatures of 660 or 750 degrees Fahrenheit (350 or 400 degrees Celsius), the caustic fluid leaches out metals embedded in the surrounding rock—sulfur, copper, iron, zinc, silver, gold. Gold, one of earth's rarest minerals, was forged at the planet's birth, when the bright, hot stars of supernovae exploded. A heavy element, it sank into earth's inner recesses, where it rested, still and stable, until superheated seawater loosened the inert metal from the rock holding it captive. Here, on the bottom of the sea, is the alchemist's dream. For years medieval chemists sought in vain to turn earth's elements into gold, unaware that the bottom of the sea held both crucible and formula.

The process requires intense heat. Surrounded by hot, perhaps molten rock, the fluid continues to heat until it becomes buoyant. At that point, it reverses direction and rises back through the seafloor. Where the rock is permeable, cold oxygenated seawater seeps through pores and interstices in the seabed to mix with the mineral-laden fluid. The metals precipitate out, and clear, warm water flows back into the sea, creating a "white smoker." Where the rock is less permeable, jets of scalding fluid shoot out from cracks in the seafloor. As they hit the chilly, oxygen-rich sea, the metals rain out in a billow of black "smoke, " the tiny grains building towering chimneys around the vents. Millions of years from now, when the dance of a restless earth has cooled the hot springs, carried the chimneys across the sea, and squeezed them up into the continents, these veins of precious metal may surface. Some already have. The rich mines of Cyprus, Oman, and Newfoundland, formed long ago in seafloor hot springs, were thrust up on land when ancient oceans closed.

Hot springs at 26° north have already deposited 1.2 million tons of metal sulfides beneath the seafloor and another 2.7 million tons above, comparable to the massive ore deposits of Cyprus. This hydrothermal vent field, known as the Trans-Atlantic Geotraverse (TAG), is the size of the Houston Astrodome. It is located on the eastern wall of the ridge's rift valley. Perhaps a young volcano two miles (three kilometers) away, or a lens of molten rock lying somewhere below the surface, supplies the heat. Fueled by the as yet unidentified furnace, high-temperature black smokers spew out riches of copper and iron, rapidly building tall chimneys of metal sulfides. Hydrothermal-vent fluids are so rich that chimneys can grow as much as twelve inches (thirty centimeters) a day. Not far from TAG's billowing chimneys, lower-temperature white smokers, rich in zinc, have built the Kremlin, a mound of chimneys capped with onion-shaped domes. The vents vary in shape and size, reflecting the temperature and mineral content of the fluids. Hot springs on the seafloor have built tall spires and round domes, shelves that hold the warm water, and structures resembling beehives. Whatever the design, the seething heat that creates it eventually cools. When the heat subsides or the plumbing clogs, edifices collapse and break. Warm water wending its way beneath the rubble precipitates additional sulfides, cementing the pile of broken rock. Over time, the metals accumulate. In the TAG field, hot springs have built metal deposits on and off for over twenty thousand years.[14]

To whom do these rich deposits belong? The Dutchman Hugo Grotius, defending his countrymen's right to sail trading ships through waters claimed by other nations, argued against private ownership of the ocean in a 1609 publication, *Mare Libernam,* "Free Seas." The idea that earth's oceans should be accessible to all, free of the claims of individual nations (except for a small strip along the shore), was modified in the twentieth century by a series of unilateral decisions, led by the United States. In 1945, when President Truman extended U.S. jurisdiction out to the edge of the continental shelf, other nations followed. In 1976, when the United States claimed exclusive jurisdiction over all fishing activity within 200 miles (320 kilometers) of its shore, other nations did the same.

Between 1973 and 1982, the United Nations developed a lofty and ambitious treaty to establish the deep sea as the "common heritage of mankind," to be managed for the world as a whole and conserved for future generations. The UN Convention on the Law of the Sea gave coastal nations exclusive rights to the resources in the sea and seabed for two hundred miles beyond their shores, but it also required them to guarantee freedom of navigation on the high seas and through narrow straits, to take responsibility for reducing and preventing marine pollution, and to manage fisheries sustainably. Finally, the convention required industrialized nations interested in mining deep-sea minerals to share their technology and their claims with the world's developing nations via an international seabed authority.

The U.S. government, displeased by the idea of sharing, refused to sign the treaty. At the same time, it seized the exclusive economic zone and encouraged industry to mine the deep sea according to U.S. law. Even though the treaty has since been amended to eliminate the mandatory transfer of technology, and even though enough nations have ratified the treaty for it to go into effect, the United States has yet to sign. As of this writing, the treaty languishes in the Senate, where lawmakers have allowed corporate interests to supersede the public good.

The far-reaching vision of the Convention on the Law of the Sea offers a powerful opportunity for the nations of the world to cooperatively manage and conserve the ocean's resources, and the United States' refusal to participate is an embarrassment. For the moment, depressed metal prices and the high cost of extracting ore from the seabed have taken away the incentive to mine the ridge's hot springs. Investors are turning their attention closer to shore, to rich lodes of gold and silver, copper and zinc, deposited near hydrothermal vents off the coast of Papua New Guinea, where Pacific seafloor is beginning its descent back into the depths. Papua New Guinea has already issued exploration leases for these deposits, which are well within its exclusive economic zone. With seabed mining, we stand on the brink of another frontier, open to the possibility of acquiring even more wealth in metals. This time, though, the mining will take place out in the water, in the dark, where we will not see the effects. The United Nations Convention on the Law of the Sea offers one means to protect and regulate mining in international waters, but only if all industrial nations join the world community and participate.[15]

The long delay in the exploitation of the deep-sea bed has given scientists time to understand more fully the significance of hot springs, in cooling earth, in balancing the salts of the sea, in nourishing the planet's early life and sustaining it in the absence of sunlight. Plumes of hot water, altered by their descent into the rock beneath the ridge and marked on their return, leave a distinct trail of manganese and methane, in concentrations up to one million times greater than those in the surrounding sea. The telltales signal the presence of Atlantic's hot springs, revealing that these oases of the deep, once considered oddities, are a common feature of Atlantic's underwater mountains. Some fields, like TAG, are built along the walls of the ridge's rift valley. Others, like Lucky Strike, lie in the center of the valley, resting on high volcanoes. Still others flourish where scientists least expected to find them, on mantle rocks strewn in the offsets connecting pieces of ridge. The Rainbow vent field sits in such an offset at the Azores Triple Junction, where Atlantic's three large plates meet. The hot springs are nine to twelve miles (fifteen to twenty kilometers) away from the nearest volcano. Wherever seawater can sink through the bottom of the ocean, and capture the heat lying below, vent fields can form, helping to cool earth's feverish core. We once believed that these hot springs of the deep were anomalies; now we have located them along the length of Atlantic—north of Iceland, south of Argentina, and clustered in numerous places in between. Where we cannot see them, it may be because we do not know fully how to look.

Over the course of ten million years, an amount of water equivalent to all that is contained in earth's oceans is routed through the hot springs of the deep sea. It is enough to affect the chemical makeup of the ocean profoundly. Most ingredients of sea salt are worn off the land and washed in by rivers, but deep-sea hot springs bestow their own ingredients as well. All over earth, rivers wind their way into the sea, washing in four billion tons of salt each year, and deep in the darkness of the ridge, hot springs release mineral-laden water to the sea. Water has flowed into Atlantic for millions and millions of years, and yet the sea is not growing appreciably saltier. The sea does more than merely receive salt; it is a simmering stew of chemical reactions, recycling the salts, removing them as they are added. Phosphorus nourishes the plants and animals who dwell in the sea; potassium binds with other minerals on the seafloor and turns to clay; magnesium is carried into the deep rock by hydrothermal fluids. Some minerals move through quickly; others have long-term residencies. Iron cycles through the sea relatively quickly, in about 150 years, while calcium stays for millions of years.

The salts held by the sea are different in proportion from those it receives. Rivers are rich in calcium and carbon; the sea is rich in sodium and chloride. While river-born calcium enters the sea in abundance, it is used to build the shells and skeletons of animals. The path of minerals entering the sea is far from clear, the journey still poorly understood. Man has not yet identified all the sea's sources and sinks for salt; nor have we balanced the sea's complex chemical equations. Yet in what is perhaps a great miracle of marine chemistry, the living waters of the sea are balanced, each salty element remaining in constant proportion to the others, unchanging over millions of years and across thousands of miles.[16]

One of the most startling and far-reaching discoveries at the hydrothermal vents overturned our assumptions about where life can flourish. For many years we had assumed that without light, there would be no life, yet life abounds here, in an abundance no scientist ever imagined. Millions of shrimp cluster around the black smokers. Mussels, thick in shallow estuaries nourished by the tide, dwell in comparable densities here, showered with an abundance of food manufactured far from the light of the sun. They cover every surface bathed by warm water. Hydrogen sulfide dissolved in the vent water enables this community to flourish. In this dark place, far away from the light of the sun, the foul-smelling gas is the bread of life, its presence awakening masses of dormant bacteria to a frenzy of production.

Here tiny flakes of bacteria soak up hydrogen sulfide as surface organisms soak up sunlight, multiplying and filling the envelope of warm water surrounding the chimneys. Their contribution to the carbon breadbasket of the sea pales compared with that of drifting, photosynthetic plants, a mere .03 percent, but concentrated at the hot springs, it supports vibrant communities of life. Like floating plants at the sea surface, blooming as the spring sun rises in the sky and winds stir nutrients from the depths, sulfur-loving bacteria blossom in

hydrogen-sulfide-laden hot springs, creating a blizzard of white in the black water, a feast for shrimp and mussels living at the vents.

Like their sunlit counterparts, deep-sea mussels filter water, but they receive substantial nourishment from within as well. Vent bacteria live not only in the water, but also in the gills of mussels. It is a satisfying partnership. Mussels draw in seawater through their gills, supplying hydrogen sulfide and oxygen to the bacteria. Bacteria use hydrogen sulfide to manufacture a generous carbon repast for mussels. At Lucky Strike, where some chimneys spew water steeped in methane, the mussels have adapted. They host two kinds of bacteria; one produces carbon from methane, the other from hydrogen sulfide.

For most animals, the deep-sea hot springs, laden with hydrogen sulfide and heavy metals, would be one of earth's most toxic environments. Animals that ingest too much hydrogen sulfide die for lack of oxygen. The gas prevents life-giving oxygen from binding to hemoglobin, which carries oxygen throughout the body. It poisons the enzymes necessary for cellular respiration, and yet dense populations of animals flourish in these waters, taking in the poisonous gas along with life-giving oxygen. Vent mussels survive the usually lethal exposures, detoxifying hydrogen sulfide by converting it into a form less harmful and finding the means to store high concentrations of insoluble metals in their gills and digestive systems.

In this hostile environment, levels of dissolved metals are one thousand times greater than in other bottom waters, yet the mussels at deep-sea vents have a tenacious grip on life. Exposure to heavy metals, to radon, and to high levels of arsenic and mercury damages DNA. Vent mussels can repair damage to their DNA, but the repairs are far more effective when the mussels are young. The rapid growth rate and early maturation of vent mussels, compared with those of their shallower-living relatives, allow them to reproduce before damage from the toxic substances becomes irreparable. Growing quickly, reproducing early, they have endured for millions of years, finding sustenance in an inhospitable place.[17]

Mussels reveal formerly inconceivable strategies for survival; shrimp remind us of the breadth of our ignorance. At a number of Atlantic's vent fields, where the particle load in the hydrothermal fluid may be too high for filter-feeding mussels, shrimp dominate the fauna. At TAG, Logatchev at 14°45' north, and the Snake Pit at 23°22' north (named for the long, skinny eel-like fish congregating in the warm water), shrimp swarm all over the chimneys, crowding in to feed. They scrape sulfide off the walls with their feet and scoop it into their mouths, packing their guts with bits of bacteria-filled rock. Living so close to the vents, they are drenched in filaments of bacteria; their legs, shells, gills, mouths, even their antennae are covered.

While vent mussels are sedentary, their nourishment home-delivered, shrimp seek out their sustenance, navigating a narrow and perilous plane of existence where to stray ever so slightly, ever so briefly, is to perish in the scalding vent water. Perhaps their antennae, sensitive to sulfide concentrations in the water, lead shrimp on this treacherous path to their food. Seeming to lack the eyes and eyestalks of its shallow-water cousins, one vent shrimp was named *Rimicaris exoculata,* "shrimp of the rift without eyes." Upon closer examination, the eye was there, a large light-absorbing patch on the back of the shrimp. These shrimp eyes have no lenses; they can't form images. Instead their photoreceptor cells have grown large for maximum sensitivity to the dimmest of light.

Scientists returned to the vent fields, seeking light in what they perceived as darkness. Cameras set with long exposures captured glowing chimneys, their intense heat radiating a dim light. Other faint light, also beyond our range of vision, illuminates this darkness, perhaps emanating from gas bubbles imploding in the vent water, or from minerals in the fluid crystallizing and then breaking apart. Scientists can only speculate about what this light might mean to a vent animal. Perhaps it is a beacon shining the way to food and nourishment, or a warning of the proximity of scalding water. Exploring this unfamiliar terrain, carrying assumptions from our own narrow field of vision, can have unhappy results. Vent shrimp have evolved to live in the faintest of light, and we, unaware that black vents shine, that eyeless shrimp see, have blinded them. Caught in the glare of our research vessels, in the blaze we need to light our darkness, their pink eyespots turn cloudy white and they lose their vision permanently. We have learned much from the sea's hot springs. By working to have a number of these areas designated as true marine sanctuaries, it might be possible to preserve both what we have been privileged to see and what we have yet to recognize.[18]

Each visit to the hot springs demonstrates in an unexpected way the tenacity of the sea's living organisms. Life on the ridge is precarious. Lush communities can suddenly disappear as the seafloor shifts and stirs. Upheavals in the mountains or surges of molten lava can incinerate mussel beds, fry shrimp. Earthquakes rerouting seafloor plumbing snuff out entire communities. When the hot water is shut off, animals unable to leave for warmer places succumb. Extinct, crumbling chimneys and broken mussel shells at the Lucky Strike vent field speak of such disruptions in a time gone by. Yet despite the transient nature of individual vent fields, the life they generate endures. For adults dwelling at the hot springs, whose lives are circumscribed by the reach of warm water, it is their wide-ranging, roaming progeny who ensure continuation of the species. Somehow, mussel larvae make the arduous journey between the vents. Perhaps they drift in the currents or ride eddies of plume water. They carry in their young bodies the sulfur- and methane-loving bacteria who will later provide them with nourishment. These bacteria, it is thought, may descend through generations of mussels, passing from the adults down into the eggs and larvae, never entering the sea. How bacteria and larvae survive until they reach the rich water of the hot springs, no one is quite sure. Perhaps they subsist on generous energy reserves inherited at birth, or perhaps their metabolism shuts down during the trip, conserving energy.

Whatever the strategy, young Atlantic hot-spring mussels travel far and wide, finding shelter in scattered oases. A single species inhabits all the vent fields of the Azores Triple Junction, and another lives in the more southern fields of TAG, Snake Pit, and Logatchev. DNA traces a relationship between these species: mussels living at Broken Spur, a field south of the Azores Triple Junction, contain genes from each. As the mussels spread out along the ridge, isolation or subtle changes in habitat split them into separate species. The long-distance travel seems surprisingly easy in these forbidding waters, in these undersea valleys bounded by steep cliffs and separated by wide gashes in the seafloor, and mussel larvae are not the only ones to make the trip. Along with sulfur-loving bacteria, a blood-red worm lives inside the mantle of many Atlantic vent mussels, feeding on the waste of its host. As the mussels spread out along the ridge, so did the worm. The mussel evolved into two species, and the worm, still a single species, lives equally comfortably in both, its range encompassing all the vent fields where mussels dwell.[19]

Like mussel larvae, new generations of vent shrimp travel quite far along the ridge, their choices many when fields inhabited by their parents sputter out. The life at Broken Spur, 29°10' north, one of Atlantic's least-populated vent fields, may be ebbing. Few swarms of shrimp hug the smokers, and mussel beds are thin and scattered. Months go by, animal populations stagnate, scavengers move in. At one chimney the warm water has receded and most of the mussels have died. Other chimneys, no longer active, lie in crumpled pieces on the seafloor, reducing a once robust field to a few thin and isolated smokers. Broken Spur lies in a deep valley, with walls six tenths of a mile (one kilometer) high and sills blocking each end. The active smokers—the Spire, Saracen's Head, Wasp's Nest—and the broad, flat ledges of Bogdanov all spew hot water into the sea, in plumes whose height never exceeds the valley walls.

Despite these barriers, shrimp larvae from Broken Spur somehow disperse, rising above the sill, drifting or swimming into the next piece of ridge and into the offset beyond. The juveniles spread out, often 60 miles (100 kilometers) from the nearest vent. One drifted 600 miles (1,000 kilometers) away, to waters south off Madeira. Upon reaching adulthood, these shrimp end their wanderings. Sensing hydrogen sulfide in the vent water, they follow the trail to a black smoker where they will settle and live out the rest of their lives. If the hot springs of Broken Spur should cease to flow and the life-giving bacteria return to dormancy, the shrimp would carry on, finding warmth and food and shelter in other fields.

As our understanding deepens, the hot springs that at first glance seemed so separate, so removed from the rest of Atlantic, prove to be neither. Details unveil the linkages. The everyday lives of young shrimp, living and growing far away from the vents for months, perhaps years, reveal that the community of life in the sea's deep-water hot springs is neither isolated nor independent from the rest of the ocean. Juvenile shrimp grow six to twelve times as large as the larvae, attaining these sizes without consuming any of the plentiful bacteria living at the hot springs. Instead, they store up food reserves of deep orange and yellow fats—all, it turns out, supplied by a diet of tiny plants and animals once living at the surface in the light of the sun. Their food descending from sunlit waters, vent shrimp are not quite a community unto themselves, fully separate from the rest of the sea. Their dependence on light-driven energies of the surface is just another example of the unexpected but essential connections among the sea's widespread inhabitants.

Vent shrimp live on during the famine precipitated by the loss of their homes because their offspring are free to roam, their energy needs provided for by the constant rain of organic debris from the surface. The ebb and flow of life at the hot springs, once thought to have a rhythm all its own, responds to the ebb and flow of life at the surface. The spawning of adult shrimp is episodic, timed perhaps to the seasonal blossoming of floating plants, when the remains drifting down from the surface will be more plentiful, and food for their young will be more abundant. Two seemingly separate realms—the lush marine food web woven in the waves by energy from the sun, and the exotic life of hot springs, fed by the synthesis of hydrogen sulfide in the darkness—are linked by animals whose very existence depends on their ability to partake of both.

Shrimp travel between these two life-giving worlds, but other bonds connect the luxuriant hot springs of the ridge to the grassy meadows at the sea surface. Vent fluids alone cannot water the oases hidden in the sea's inner valleys. Saturated with hydrogen sulfide, hot springs generate grist for massive carbon production, but some bacteria require oxygen to manufacture this bounty, and animals who consume it require oxygen to breathe. Hydrothermal fluid loses oxygen to the seabed long before it resurfaces in the chimneys. The larger gill chambers of vent shrimp and the bright red hemoglobin packed into the mussel-dwelling worms are adaptations designed to compensate for the lack of abundant oxygen. Whatever oxygen there is comes not from the vent fluid, but from surrounding seawater, and it arrives circuitously.

Oxygen, produced as floating plants turn light into energy, stays at the sunny waters of the surface, riding the currents to chilly latitudes. There the cold and salty water, still aerated, plunges to the depths to begin a long, slow journey along the seafloor, crossing flat plains, dipping into basins, cutting through gaps in the ridge. Hundreds of years later, it passes through hot springs, mixing with the chimney water. Oxygen-laden water from the surface breathes life into the vent animals of the hot springs. In this fluid world of the sea, where time and distance are no barrier, coastal meadows of floating plants and cold polar water nurture mussels and shrimp growing amidst the newest, youngest pieces of earth.

Seemingly disparate worlds—one sustained by sunlight, the other by the planet's inner heat—are inextricably joined by the sea's indivisible waters. Should catastrophe occur in the waves, deep springs are not immune. Ancient hot springs have left fossils, their remnants cast into the uplands of continents. Some descendants of vent animals whose histories are recorded in this rock still thrive in hot springs fed by contemporary oceans. Other fossils are of animals long disappeared, their lives snuffed out millions of years ago in a mass extinction whose reach extended onto the deep and distant ridge.[20]

We barely discern the rudiments of existence on this plane where our basic assumptions are so often overturned, but each stumble through unfamiliar terrain broadens our vision, expands our understanding of how and where earth supports and nourishes life. The sulfur-loving bacteria that sustain life at the hot springs live in mixed waters, in hydrothermal fluid chilled by oxygenated seawater. Hidden in the recesses of chimney walls reside other vent dwellers, microbes that prefer their vent fluid less dilute. These heat-loving organisms dance where others venture only at their peril, thriving in scalding fluid whose intense heat kills most life. At temperatures where others languish, these microbes awaken, making their homes in water as hot as 230 degrees Fahrenheit (110 degrees Celsius), their presence yet another illustration of earth's manifold capacities to support life.

Given such names as *Pyrolobus fumarii, Pyrococcus furiosus,* the "smoking" and "flaming" fireball, these organisms have an affinity for hellishly hot water. Their worth may be as great as that of the gold and silver ore deposited by the vent fluid. Science and industry have spent millions of dollars identifying, extracting, and cloning the enzymes that enable vent microbes to flourish in extreme heat. Until these microbes were discovered, the technique for making multiple copies of a piece of DNA was time-consuming and cumbersome; enzymes propelling the reaction couldn't withstand the heat and so needed periodic replacement. *Pyrococcus furiosus* loves heat. It thrives in waters that vaporize at the lower pressures of the surface, so its enzyme, marketed as Deep Vent, is well suited for high-temperature duplication of DNA. Deep Vent may not be a household word, but its uses are familiar, aiding criminal prosecutors in identifying the DNA fingerprints of possible suspects and medical researchers in finding the genetic components of disease.[21]

For biologists, the existence of heat-loving microbes holds the hope of life in environments as seemingly hostile as ocean-ridge hot springs. Sulfur-loving bacteria, providing vent dwellers with an abundance of carbon, require oxygen delivered from the surface. The world of the heat-loving organism is more local, more self-contained. Everything this microbe needs to satisfy its lust for life may be found inside the black smoker. Ensconced in the walls of chimneys, living without light, without oxygen, it makes energy from ingredients supplied solely by the vent fluids. Its anaerobic metabolism, synthesizing methane from carbon dioxide and hydrogen, hints at the possibility of extensive microbial life beneath the seafloor, housed in dark, water-filled pores of volcanic rock. Surges of lava, not radiation from a sun high in a distant sky, determine the tempo of this life.

How extensive this deep biosphere is and how independent its existence still remain to be seen. Scientists have not yet retrieved a piece of it. Microbes may throng a seabed heated by melting rock rising from deep within earth's mantle, but their presence in hot springs does not prove a seabed origin. They may have arrived from within the seafloor, dislodged by volcanic eruptions, catapulted up to the surface, and washed out in the vent water. Or they may lead a more sedentary existence, lying inert in the chimney walls until the rush of warm water stirs them to life.

These organisms are too few to sustain the mussels and shrimp crowding the hot springs, but their very presence, even in low numbers, has redirected our search for life on neighboring planets. If tiny microbes can persist below the seabed or in the black smokers, they may dwell beneath the frozen seas of Jupiter's moon Europa. These frozen seas are not a solid block. In places they appear to have cracked and pulled apart, then been mended by slush rising from the depths. Perhaps the strain of Jupiter's undulating gravitational pull has generated heat and warmed the bottom of these seas, creating a broth that, though screened from the sun, may harbor life.[22]

The tiny heat-loving organisms of deep-sea hydrothermal vents, so small that millions fit into a few drops of water, have uprooted the tree of life, called into question our ideas about the orderly and hierarchical nature of evolution at the dawn of life. From earth's ancestral seas have sprung the great domains of life, Eukarya, those organisms whose cells contain a nucleus, from which plants and animals descend, and Prokarya, those organisms whose cells are without a nucleus. Bacteria such as *Synechococcus,* one of the smallest and most multitudinous organisms in the sea, are prokaryotes. Initial DNA sequencing indicated that perhaps the prokaryotes should be split, leaving one domain, Bacteria, and creating a third, Archaea. Initially it seemed that many heat-loving microbes of the sea's hot springs belonged to this new group. It consisted of single-celled organisms whose metabolism resembles that of Bacteria, but whose methods of replicating DNA are more closely related to those of Eukarya. *Methanococcus jannaschii,* dwelling in the sunless, airless swelter of hot-spring chimneys, is such a microbe. It was one of the first to have its genes sequenced, and the map of its DNA seemingly confirmed the existence of life's third domain: 44 percent of its genes resembled those of bacteria and eukaryotes; the rest were unknown.

We seek order in evolution, in the clean categories of species and genera, phylum and family, in the neat divisions that distinguish us from other creatures, that set our place on the family tree. Early orderings of creation, described in the Old Testament, evolved into more detailed lineages conceived by Darwin. Retracing our steps through these lineages, further and further back in time to where the separate branches of evolution converge, we hoped to find our common ancestor, the organism giving rise to us all. With the sequencing of the *Methanococcus jannaschii* genome, we thought we were approaching these misty beginnings. Eukarya evolved from Archaea, or so it seemed, sometime after Bacteria and Archaea diverged. Lodged in the tree's lowest branches were the Archaea, the ancient ones, the closest relatives of our common ancestor. The heat-loving microbes, living in waters so inimical to life, may have been among the first to arrive in earth's early seas. However, the sequencing of genomes of other organisms, including another heat-loving bacterium residing in volcanoes off the coast of Italy, toppled this carefully constructed tree with its orderly branchings, leaving behind a thicket in which it is impossible to tease out who evolved from whom, who stood at the first division in the tree of life.

DNA sequencing, laying bare a microbe's genes, blurred the lines of descent. For dwellers in earth's hottest waters, it is difficult to say at this point who is older, who is parent and who is child, who is sister and who is cousin. Two organisms may share genes but not lineage. In any one organism, two sets of genes may trace two separate ancestries, placing that particular organism in more than one species, more than one phylum, more than one domain. If strands of RNA define an organism as belonging to Bacteria, genes regulating its metabolism or cell division may place it equally comfortably in Eukarya or Archaea. Like the mythological Greek chimera, with serpent's tail, goat's body, and lion's head, earth's ancestral organisms defy easy classification, perhaps because they once lived communally, sharing everything, including their genes.

Gene swapping, rampant in earth's early seas, continues today. In the last thousand, perhaps even hundred years, *Pyrococcus furiosus,* an Archaean microbe living in scalding waters off Volcano Island in Italy, has lifted a sequence of sugar-transporting genes from a microbe of another genus living nearby in slightly cooler water. In metabolic matters of sugar transport, the flaming fireball more closely resembles its neighbor than members of its own species that live farther away. Such affinities suggest that the tree of life, with its orderly hierarchy of limbs and branches, twigs and leaves, may no longer be an apt metaphor to describe early evolution. Further research may suggest another, but for now we are left with a close and tangled thicket, its branches tightly interwoven here, loosely spread there, its organization varying with the perspective of the observer.[23]

The presence of heat-loving organisms in the deep sea has both uprooted our understanding of evolution and planted other ideas, particularly concerning the origin of life on earth. It may be that oceanic hot springs, new and unfamiliar to us, spawned early life on the planet. Early in its existence, asteroids frequently bombarded young earth, crushing and burning its newly forming skin, vaporizing its emerging seas. Vent fields deep in the water that cooled the planet's fever offered refuge from the cataclysm. Perhaps life originated at the sea's surface and survived the onslaught of asteroids by descending to the depths and waiting there until the upheaval subsided. Or perhaps the first stirrings of life took place in the quieter water below. Throughout our own lives and the life of our culture, we embrace light, equating it with life and warmth, shunning darkness and its associations with coldness and death. And yet, the pitch-black depths, not the sunny surface, may have been the womb that nourished earth's first living, breathing organism.

Vent fluids, coursing through hot, melting rock in the seabed, provided the critical ingredients, the gases rich in hydrogen and carbon, the metals iron and nickel sulfide. Molecules of hydrogen sulfide, of carbon dioxide and carbon monoxide, may have flocked to the metallic ions. In the infernal heat of the smokers and the immense pressure of the deep sea, they may have jostled against each other, breaking apart and coming together, then breaking apart and coming together again. On the flat surfaces of the nickel and iron sulfides, they may have reassembled themselves in new combinations, sometimes as molecules of ammonia, sometimes as molecules of acetic and pyruvic acid, the linchpins of metabolism. If earth's first intimations of life appeared in the dark world of oceanic hot springs, they may have emerged not as cell walls, separating the living from their surroundings, and not as complex bits of DNA or RNA, replicating themselves, but as metabolism, as maintenance. Their first purpose was basic—to keep themselves going. Spurred on by the pull of metallic ions, the new molecules may have combined with each other and grown more complex. Pyruvic acid may have joined with ammonia to form amino acid, and amino acids, the building blocks of life, may have linked together in long, folded chains of peptides, ultimately evolving into proteins.[24]

Restless undersea mountains built earth's foundation, created the basins that hold her oceans, and filled them with the rich broth that gave birth to life on this planet. At oases hidden deep in valleys stretched and torn by spreading seafloor, the sea nurtures organisms that were present at the dawn of life, organisms for which toxic waters are not death but elixir. They came into being when the planet was young, and as long as heat and water and volcanic rock pass through the seafloor, they will, in all likelihood, persist. They arrived almost four billion years ago—long, long before the appearance of multicellular animals, long before the rudiments of the creature that would become human appeared. When our short tenure comes to a close, they will probably continue on. Today Atlantic's ridge houses the hot springs that shelter these ancient organisms, but the vents themselves, the moving seafloor, and even the basin of the ocean itself are recent additions to earth's seascape. From the deck of the *Cramer,* bobbing about in the windless Sargasso, there is no hint of what came before.

CHAPTER 9

Atlantic's Primordial Ancestors

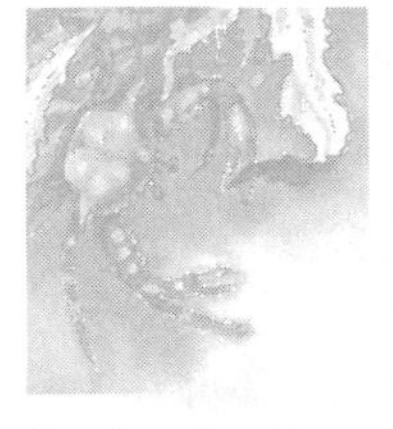

THE island of Newfoundland is cold and remote, and often shrouded in mist. Dark forests of spruce and fir and thick bogs of peat cover the interior; rocky headlands edge the coast. Nearly one thousand years ago, Norsemen, navigating without compasses, steering by the sun and stars and the distant mountains lit by the Arctic mirage, sailed beyond Greenland to reach North America. Impressed with the abundance of natural resources, the forests of timber and streams teeming with salmon, they established a settlement at the gateway of this new world, on the northern tip of Newfoundland, at L'Anse aux Meadows. From their camp at the entrance to the Strait of Belle Isle, they sailed farther down the Gulf of St. Lawrence, gathering wild grapes and butternuts from the banks of New Brunswick rivers. They returned to Newfoundland and wintered there, cutting timber and repairing their ships. After a number of years—no one knows how many—they abandoned the settlement. Over time, the buildings fell apart and lush meadow covered them over, reducing the former colony to a gentle rise in the grass.

Nine hundred years later, archaeologists uncovered the site and found the ruins of eight Norse buildings, including dwellings and workshops belonging to shipwrights and carpenters, and a smithy where bog iron was melted down and forged into nails and rivets. Buried in the peat bog beneath the buildings were old boat nails and trimmings from planked lumber. Inside the dwellings archaeologists unearthed a bronze cloak pin, a stone oil lamp, and a fragment of a bone knitting needle. These few artifacts give insight into how the Norse lived and sketch the livelihood of a people long gone, whose material possessions have by and large disintegrated. This thin story, pieced together from cloak pins and rivets, has led to a revised account of American history, identifying the Norse as the first Europeans to settle on American shores. This story, far from complete, far from definitive, is not the first version of European exploration in America; nor, in all likelihood, will it be the last. Another mound in another meadow on another shore may add another layer, another dimension to a history obscured by the passage of time.[1]

The ruins at L'Anse aux Meadows speak of the comings and goings of men, of the brief sojourn of a group of people in a land ripe with promise, but theirs is not the only story held here. The bedrock beneath the peat records another history, of more ancient arrivals and departures. Underneath the Norse excavation is the seabed of an old ocean, an ocean whose waters caressed earth's first animals and nurtured an unparalleled radiation of life. In its waters, single living cells evolved into complex life-forms, bacteria evolved into plants and animals, and tiny, ambiguous intimations of life grew into large, distinct organisms visible to the unaided eye. In the waters of this ancient sea, populations of animals rose to unprecedented heights and succumbed to extinction, but like the Norse settlers, they left few traces of their passage. The sea itself is gone, its life-giving waters drained away, most of its basin recycled into the depths of earth's mantle. Those fragments thrust ashore to become the bedrock of continents lie scattered, in the remote sea cliffs of Newfoundland, in the Lake District of England, in the foothills of the Andes. In the rock of dry land is inscribed the history of this sea, Atlantic's predecessor, whose great basin filled and emptied long before the first piece of earth cracked to bring in the waters that now lap Newfoundland's shores.

On the scale of time marked by the march of men, Atlantic is ageless. Her sandy edges may shift, but the basin seems fixed, the slow creep of seafloor away from the craggy peaks of the ridge barely perceptible. Years go by, then hundreds of years, then thousands, then a hundred thousand, before older rocks in the ridge are shunted aside, before mountains swollen with heat cool and gently subside, losing their height and grandeur. As seafloor drifts farther away from its birthplace, the continuous rain of debris from the surface softens its hard edges, smooths its contours. Hundreds of thousands of years pass, then hundreds of thousands more, then millions. In this enormity of time, entire mountains are buried by motes of dust and by the minute remains of tiny organisms. When the mountains finally reach the edge of continents, where they are further blanketed by silt and mud washed in on rivers, they are two hundred million years old.

Atlantic is one of earth's oldest oceans, but in the deep time that marks the drifting of continents, the elevation of mountains, and the erosion of high peaks to bare bedrock, two hundred million years old is young—too young, in fact, for this sea to have its history written into the continents. The lands edging Atlantic, whose rivers turn rocks and boulders into grains of sand, are millions, billions of years older. Sculpted long before the first volcanoes split open Atlantic, they will persist long after the sea is gone. This bedrock, the heart of continents, is born of the sea, and in it is recorded the passage of Atlantic's forebears.

Glaciers in harsh Arctic Svalbard have scoured valleys in rock over 800 million years old, rock that once rested beneath salty waves. Vermont's gentle, rounded Green Mountains and New York's lake-filled Adirondacks, which once towered as high as the Himalayas, were fashioned one billion years ago from ancient seafloor. Sheep graze on the desolate, rolling moors of Scotland's Outer Hebrides—old, old mountains thrust up from a departing ocean, then worn down to their roots after two billion years in the wind and rain. Outside Godthaab, the capital of

Greenland, lie ancient exposures of rock formed when earth was in its infancy. Baked and twisted, raised and eroded as continents have come together and fallen apart, they too are of the sea. Continents are a gift from the sea, and these lands have all been bequeathed by Atlantic's ancestors, a long line of oceans extending back to the early days of the planet.

In the beginning there was no ocean. It is the stars moving through the heavens that shine on this dim and hazy past, outlining the moments of earth's birth. Untold numbers of galaxies spin through the sky, blown outward on an intense pulse of heat emitted fifteen billion years ago when the universe blasted into existence. That warmth still guides the nebulae toward an edge we cannot see, countless light-years away. In spiraling galaxies, stars are born, exploding in fiery furnaces where hydrogen fuses into helium, lighting the darkness for millions of years. When combustion runs its course, the star fades. More rarely, it collapses and blasts apart, a supernova dying in a final burst of brilliant light. In 1054 a giant star exploded, eclipsing all but the sun and moon in brightness, broadcasting the sky with carbon and iron, oxygen and silicon. It left an enormous glowing cloud of gas and dust, the Crab Nebula, shining in its wake.

Four and a half billion years ago, such a cloud gathered at the distant, undistinguished edge of a distant, undistinguished galaxy. Seeded by supernovae with all the elements essential to life, it contracted into a flat spinning disk. Pulling in on itself, it collapsed into a burning star, becoming the nuclear reactor that would light and heat our planets. Our sun is in the prime of its life, but the sky is laced with the possibility of other heavenly lamps yet to shine. A new star spawned by a spinning cloud in the constellation Taurus carries that potential. After our sun formed, gas and dust continued circling the blaze, swirling at the outer edge of the disk to form Saturn and Jupiter. On the inner edge, burning dust began to cool, to solidify into grains of mineral and metal. They collided, and stuck, and pulled on each other, coagulating into rocky clumps that would become Mercury and Earth, Venus and Mars. The loose fragments that remained would live on at the outer reaches of the solar system as asteroids, meteors, and comets.[2]

The further back one reaches into time, the more difficult it becomes to trace Atlantic's lineage, but odds are it was a violent beginning. For five hundred million years, earth grew as rocky meteors and chunks of planetary debris slammed into it and congealed. From one such bombardment came the moon. A giant conglomeration of rock and gas, as large as Mars, crashed into earth. Some rubble spun off and coalesced to become earth's satellite, but the rest remained, sinking into earth's first ocean, a molten sea born in the heat of collision. This sea of boiling lava may have been as deep as 430 miles (700 kilometers). Of it, there is no record. Atlantic's earliest ancestors were too hot to harden, too liquid to fashion the rocky evidence of existence.

Heat engulfed the planet from within and without. The constant barrage of meteors melted the surface, and as the planet contracted and pulled together and radioactive elements in its depths began to decay, the interior generated its own heat, carried to the surface in melting rock. Seafloor forming today at Atlantic's ridge cools in the chilling water, leaving solid rock to signal its arrival, but earth's early seas were molten and left no trace. The moon is our window on this distant time. When it cooled, three billion years ago, volcanic activity ceased, preserving in stone the quality of earth's early days. Lunar maria—the Sea of Rains, the Sea of Tranquillity, the Sea of Serenity—are at rest today, but they were born in upheaval, in a flood of lava that poured from early volcanoes into craters dug by blazing meteors and then, when the heat dissipated, hardened into repose.[3]

Cold stilled the young moon while heat rising within earth kept the planet in motion, making an ocean of water possible. Some lava hardened into rock, and water streamed onto its surface, delivered by steaming volcanoes and icy comets. It vaporized immediately. More water rushed in, and that too evaporated. When, after several hundred million years, earth's fever abated, it rained and rained, and watery seas filled the meteor craters. Earth built an atmosphere, its gravity holding fast the gas and water vapor carried in on comets and spewed out by volcanoes. Whatever this atmosphere was—carbon dioxide, methane, and ammonia, or some other combination of gases—it held in heat, warming the planet as the young sun, still faint, could not. This early atmosphere also protected the planet from the fusillade of incoming meteors, burning many as they approached. Comets circling in from the distant reaches of the solar system, meteorites plunging to their deaths in the lava, and hot water spewing from the new seafloor all may have introduced compounds of hydrogen and carbon and caches of amino acids to the young planet. Whatever their origin, earth's seas now held the seeds of life.[4]

The very first moment of transubstantiation, when disparate elements that are the dust of stars came together and lit the spark of life, is far in the distant past. Even if the moment of that first breath spanned millions of years, we can pinpoint neither the time nor the place, for the record, the first line in evolution's biography, is missing, melted away in a surge of rushing lava, cauterized by a flaming meteorite. Perhaps the first living cell was assembled in the inferno of the sea's hydrothermal vents, where hydrogen and carbon compounds joined with nickel and iron ions to initiate metabolism. Or perhaps the union took place in a quiet lagoon where molecules sloshing gently in an oceanic broth were catalyzed into life by a bolt of lightning. Or perhaps a greasy hydrocarbon, launched to earth by a meteorite, slipped into a sea where, nudged by a wave, it circled back on itself to make a membrane enclosing a drop of nutrient-rich water. Each story is possible. The chronicle of this genesis is incomplete, with many missing pieces and only the tiniest fragments with which to imagine the whole. These fragments, though, suggest that life was conceived in an oceanic womb, the enduring legacy of Atlantic's oldest ancestors.

With oceans came continents. Earth's very first lands were probably small islands, like Iceland, built of dense, heavy seafloor, lifted above the water by the heat of a fiery planet. When two pieces of seafloor, steered from their volcanic birthplaces by heat rising from the planet's interior, jostled each other, one sank back into the depths. As it descended, part melted and then resurfaced, taking on a different character, a different consistency. Erupting in violent, explosive volcanoes, it hardened this time, not as the heavy, dense foundation of the sea, but as the stuff of continents, light, buoyant granite riding above water. When continents collided, they crumpled up into mountains, mountains that would later wash into the sea. Along the banks of the Acasta River, in Canada's Northwest Territories, lie scraps of earth's oldest continents. Baked in the pressure cooker of moving mountains, heavily weathered after years of exposure to wind and rain, they retain only a touch of their original identity. Tiny crystals, formed in the rock as it cooled, are four billion years old.

Seafloor born in the heat of earth's interior drifts along the planet's surface, returns to the depths, and then rises again, emerging as continents. The recycling of stone, begun over three billion years ago, perhaps four, continues today, still powered by earth's internal heat. Drifting continents and shifting seafloor may be a phenomenon unique to our planet. The contours of Venus appear to be fashioned only of basalt, the dense, heavy foundation of the sea. While unevenness in the skin of Mars—thick in the south, thin in the northern lowlands—indicates that in the planet's youth, heat may have caused land to drift, today that motion has ceased. Compared with our moving earth, the surfaces of our neighbors are still, and lifeless.

The recycling of stone on an earth continuously re-forming itself may be what sustains the life first cradled in our planet's primordial seas. Carbon, foundation of metabolism, links living being to inanimate rock. It constitutes the starch of a cell, the backbone of limestone. Cycling between the living and the dead, it builds the strength of trees, gives substance to a piece of coal. Along with other elements, it joins the breathing and the still in a continuous cycle, one giving substance, identity, to the other. Cell and sand grain, mitochondria and mineral, are different renderings of the same elements. What dies in the sea gives weight to stone, which, as the seafloor drifts, is pushed up into the mountains of continents. The minerals of mountains, released by wind and rain, give life to those who dwell in the sea.[5]

Nourished by this fellowship, cells manufactured carbohydrate, and life proliferated in earth's early seas. Living, breathing organisms turned out billions of copies of themselves each day, becoming as numerous as stars wheeling through the galaxies. Born in the morning, they died in the afternoon, and came to rest in the mud at the bottom of the sea. Embedded in the mountains of western Greenland is a piece of an ancient sea that spawned this early life. The rock is 3.8 billion years old, preserved when the planet's atmosphere thickened and slowed the barrage of meteorites. Some is lava, the volcanic floor of the sea, and some is granite, formed as the plunging basin melted and resurfaced. Some is mud, compressed and baked on the bottom of the ocean and then heaved up onto shore as shale and slate. Scientists believe that living organisms, dwelling at the sea surface, then dying and sinking beneath the waves, were buried in this mud. Here, at the edge of a remote glacier, on a continent forged as one of Atlantic's ancestors closed, lie the first stone tablets recording the evolution of life. Undoubtedly there are other pieces of primordial seafloor to uncover and decode; we don't know where on the planet they lie.

The Greenland rock holds no fossils, no hardened body parts, only the traces of carbon that signal existence. How these cells lived and breathed, we can only surmise. They may have lived on their own, producing their food with the help of hydrogen sulfide and sunlight, or they may have lived more simply, consuming what the water provided. Their carbon residue, entombed in ancient Greenland rock, suggests that if earthly life originated in the relative calm of deep-sea hot springs, it spread to cooler surface waters almost as soon as the meteorite onslaught subsided. Perhaps water gushing from a hydrothermal plume wafted the cells to the surface, where they retrofitted the tools of deep-sea survival for life in the bright sun. Heat-shock proteins had protected vent bacteria from flickering temperatures at the hot springs, and light-absorbing molecules alerted them to the dangers of scalding water. These, scientists suspect, may have been reapplied to the business of photosynthesis. Who exactly these early surface dwellers were, we may never know. We are blessed to have even these barely recognizable vestiges.[6]

As for earth's early deep-sea dwellers, scientists have found their traces in remnants of 3.2-billion-year-old hot springs thrust up into the rocks of western Australia. Laced in the rock are thin, threadlike filaments of bacteria that once lived in the scalding waters of ancient hydrothermal vents. Far from the light of the sun, protected from the bombardment of meteorites, they thrived in the rich chemical stew of the vent water. In the vent water were many metals: it is possible that we still bear traces of our early ancestry in the molybdenum, copper, zinc, and nickel found in some of our cells.

The give-and-take between land and sea, one creating and nurturing the other, was felt even in this distant time. Moving seafloor shaped the continents, and growing continents relieved the sea of half its salt. Brine accumulated in pools of deep groundwater and in surface basins, leaving behind thick layers of salt as the water evaporated. Neither ocean nor atmosphere held oxygen: cells turning sunlight into energy synthesized dissolved iron and hydrogen sulfide instead. Their breathing precipitated iron from the sea. It fell to the bottom and accumulated layer upon rusty layer, settling into what would become, millions of years later when seafloor was hurled ashore, the world's most economically valuable iron deposits. Layered in the volcanic rocks of western Greenland are thick bands of iron, bred in the sea, then given to the land when ancient pieces of seafloor collided.[7]

With each collision the primal continents grew, as pieces of ocean floor were stitched to their borders. Ur, Arctica, Atlantica, Baltica they are called. They drifted together and were torn apart, only to re-form in other configurations and split again along other lines. Today they are broken into fragments, parceled out among other lands: Ur scattered in the bedrock of Australia, India, and Antarctica; Arctica divided between North America and Greenland; and Atlantica given to West Africa and South America. Baltica rests at the core of northern Europe. The seas that hugged these shores were lively, burgeoning with newly evolving organisms, and each time the drifting lands fused—Labrador with western Greenland, or Svalbard with Norway—stone from the sea recording this riot of life was thrown ashore.

Each time the continents broke apart, new seas opened, each generation with its own character, its own story, its own key chapter in the history of evolution. Earth's earliest seas lacked oxygen, but as life proliferated, that would change. For earth's first living cells, oxygen was anathema. They proliferated without it and perished in its presence. They lived on hydrogen sulfide, plentiful at the deep-sea vents but in short supply everywhere else. In this time of scarcity, the resourceful flourished. A bacterium evolved to wrest hydrogen from a water molecule, to build a sheath to protect its thin skin from damaging ultraviolet light, and to repair its DNA when the sheath failed. Using the abundant supplies of hydrogen, carbon dioxide, and sunlight in the waters of the surface, it multiplied and turned the ocean green, giving off considerable waste in the process. Polluting atmosphere and ocean, it poisoned its competitors, whose numbers diminished as they slipped away into the uncontaminated refugia that remained.

Breath by breath, bacteria imbued the planet with oxygen. Fatal contaminant for some, it became the life force for others, who came to dominate the sea. Pieces of seafloor 3.4 billion years old may contain fossils of these bacteria who transformed earth's atmosphere, but the fossils are tiny and insubstantial, easily confused with ripples etched by a passing wave. About 2.7 billion years ago, in the shale of a younger sea, they left more easily decipherable clues to their identity: molecular markers and lipid fingerprints by which they could be recognized.

Surrounded by abundance, bacteria multiplied and became dominant. Initially those pumping oxygen into air and sea lived as solitary individuals, their fossils consisting of thin threads and strings of spheres laced in the floor of shallow waters. They soon established large colonies, earth's first housing cooperatives, rocky structures with niches to accommodate the diversity of their kingdom. Bacteria that loved sunlight and breathed out oxygen lived at the top, their slender threads woven together in a thin sticky mat. In a layer just below them, green and purple bacteria made carbon with whatever light shone through the mat. They abhorred oxygen, but other bacteria removed it. Below them, still others consumed the remains of the dead. The mat trapped tiny bits of sand carried in by gentle currents. Mud and dust washed in with the rain, building layers of life and grit. The mat thickened, and the whole colony moved up toward the light, building additional layers as abandoned quarters were cemented into rock, into sculpted mounds and domes lining the seafloor. Lying in the harsh Arctic winds of Svalbard are the remains of these domes, three billion years old. Gifts from a departing sea, they hold the memory of balmier days, of quiet waters warmed by equatorial skies and steaming volcanoes where tiny bacteria changed the face of earth.[8]

Stromatolites, as these early bacterial communes are called, determined the atmosphere and dominated the seas, building huge limestone reefs that ran the length of oceans. The remnants of a sea that opened and closed two generations before Atlantic contain fossils of these colonies, once so widespread. I have seen them hidden in the weeds in the woods along the back roads of Saratoga Springs, New York, unsung memorials to a time that has passed. I like to touch these rocks. Stepping off the quiet road, I push the weeds aside and run my hands over the stone epitaph of an organism that literally gave us the breath of life.

Ultimately, stromatolites could not survive in the world they created. Only a few of the once pervasive communities live on today, in the quiet backwaters of Shark Bay in western Australia, and in swift-moving currents off the Bahamas. There, hostile living conditions—brine in one, fast water and steep channels in the other—shelter them from predatory animals. Stromatolites gave animals the air to breathe, and animals in turn grazed them down to nothing. Life endures, its perpetuation made possible by individual species whose days are numbered. These are the impersonal exigencies of evolution. Stromatolites declined, but some bacteria, tough and resistant, glided away from the homes they had built, abandoning hard edifice for liquid sea, trading a permanent address for a wandering life in open water. Unencumbered, they spread throughout the marine realm, into icy polar seas, nutrient-rich coastal waters, salty embayments, and deserts of the open ocean. Today they build no material monument to their productivity. Though invisible, they are the most widespread and abundant organism in the sea.

In the waters of Atlantic's forebears, bacteria gave birth to the first nucleated cell. Single cells of bacteria not only lived communally, sharing the structure of a stromatolite, sharing food and water, but also shared themselves, exchanging their genes, the essence of their identity. They began as individual cells with individual functions but soon engulfed each other, and dwelled within each other, and reproduced together. From these collaborations came new forms of life.

Scientists speculate that in the darkness of deep-sea hot springs, a spirochete—a darting, corkscrew-shaped bacterium—invaded one of its heat- and sulfur-loving neighbors. In the skirmish that ensued, neither was vanquished and neither triumphed, and neither deserted the battlefield. They stayed on, dwelling together, combining motion and metabolism to make earth's first nucleated cell. As other bacteria filled the water with dangerously high levels of oxygen, these cells took in another tenant, an oxygen-loving bacterium whose role became that of mitochondrion, energy producer. Mitochondria floating in the cytoplasm of nucleated cells bear an uncanny resemblance to purple bacteria, their free-swimming brethren. Some of these cells attempted to digest photosynthesizing bacteria and failed. The bacteria stayed in their new home and adapted, becoming chloroplasts, creating the design that would ultimately yield the first green plant. Many scientists believe that each cell distinguished with a membrane-bound nucleus came into being not from the transmission of genetic information down through the generations, but from independent bacteria sharing their resources and living communally in the sea.

The early days of earth are distant and shadowy, and our exploration of this dimness tentative. Just as no one knows precisely when life arose on earth, no one can yet say, even within a million or a hundred million or several hundreds of millions of years, when a cell with a nucleus first appeared. Each time we find a fossil of what we think is the oldest cell, we find another, even older. To date, the oldest evidence comes from molecular biomarkers found in the shales of an ancient ocean; 2.7 billion years old, they belong to nucleated cells. This latest finding pushes back the date of their emergence by at least half a billion years, a poignant reminder of how faint are the outlines of deep time, how dim our understanding of life's beginnings. Our best guess had been off by a length of time that eclipses the tenure of man on this planet, the tenure of animals, the tenure of seas.[9]

The ancient paths of earth's wandering continents are as faint as the traces of life proliferating in the planet's early seas, but as time passed, moving nearer and nearer to the present, the tracks become fresher, easier to read. Approximately one billion years ago, drifting continents fused to raise one of the planet's longest mountain ranges. As high as the Andes or the Himalaya, it sutured earth's lands into one great continent. Here and there, pieces of bedrock expose the ancient seam. It reached through Sweden and Norway, the Scottish Highlands, the edges of Greenland and Labrador. Its remnants appear in the Adirondacks of New York State, the Blue Ridge Mountains, the hills of Georgia. The range ends on an icy beach on the Weddell Sea in Antarctica, and in the bedrock of Peru and Bolivia. Rugged as they were, these mountains were not to last. Today they are but shadows of their former glory, the stone that built them gone to raise new heights on other shores. Wind and rain tore jagged peaks into boulders, dissolved boulders into sand. Running water wore away the towering peaks of Greenland and carried away the debris into neighboring Scotland, where eventually it was pushed up into the stark sandstone hills rising from Loch Torridon, young mountains made from old.

As heat continued to rise from the planet's interior, the huge continent began to fall apart. Earth's skin stretched, and thinned, and cracked as hot liquid rock poured through the rents. It hardened into heavy basalt, the fresh floor of an incoming sea. This sea, unlike the ones that came before, has more clearly defined edges, bounds we can begin to draw. Hardened lava flows, some as wide as 90 feet (30 meters), in Skinner Cove on the west coast of Newfoundland, in Labrador, and in Norway—

all mark the birthplace of the new ocean, Atlantic's predecessor. It would be called Iapetus, after the Greek Titan who fathered Atlas. In Virginia's Blue Ridge Mountains, the old granite rocks of Shenandoah National Park bear stretch marks from this sea's birth. To ride along Skyline Drive or hike the nearby trails is to retrace the birth of Atlantic's parent, for here the ancient granite is torn apart by columns of darker rock, the volcanic tubes that carried hot lava up from the mantle. As molten rock met chilling sea, it hardened into pillows. Shenandoah's pillow lavas, hundreds of millions of years old, are identical to the fresher pillows, hundreds of miles away, filling the dark, watery valleys on Atlantic's ridge to make new seafloor.[10]

Upheaval marked this birth. As the land was ripped apart, the cracks widened and filled with debris washing off eroding mountains. They grew longer and broader, some joining to form jagged rents across a continent, others stopping short, leaving the tear unfinished. The Ottawa River in southern Quebec flows in such a crack, an arm of an incipient sea that failed to materialize, a road not taken. Other tears opening the way for the young sea are still restless. Motions that parted mountains five hundred million years ago to let in salt water still reverberate in our time, as tremors continue to shake the Appalachians. In the winter of 1811, the town of New Madrid, Missouri, was leveled by an earthquake so strong that the Mississippi flowed upstream and church bells rang in Charleston, South Carolina. Scientists do not know when the land will stir again, when the ancient pull of continents will once more tug our present. Land quivers along the cuts that first opened Iapetus, a softer version of energies that splintered an ancient chain of mountains and scattered their fragments across the globe.

As pieces of North America and Europe drifted apart, Iapetus emerged between them, a shallow sea filled with light and nutrients, a sea that harbored an explosion of life. The graveyard on the ocean bottom recorded its arrival, and when continents reunited and the basin closed, it yielded the record to land. Arctic Svalbard is poor in modern amenities but wealthy in stone diaries. Layers of rock 22,000 feet (7,000 meters) thick document both the presence of bacteria and stromatolites and the proliferation of larger and more complex single-celled organisms—organisms whose descendants, coming into their own in another time, another sea, would eventually anchor lush marine food webs.

The layers of rock show single cells growing more elaborate. They grew larger to make room for more DNA in their nucleii, and to house equipment they needed, chloroplasts and mitochondria, to produce energy. As they grew heavier, they compensated for the added weight, pushing against the water with delicate spines, keeping afloat with new body shapes, stars or vases or polygons blistered with lighter liquids, designed for buoyancy. They grew to accommodate the lean seasons in the sea, resting in thick-walled cysts when food was scarce. This ability to hunker down, to wait out times of deprivation, of impoverished waters, would carry these single cells of life through wave after wave of mass extinction.

In southwest China, near the gorges of the Yangtze River, lie deposits of phosphate, washed into a shallow sea after a glacier had come and departed. Death came quickly to those who dwelled here. They were buried by the rain of phosphate and fossilized immediately, preserved in stunning and exquisite detail. Entombed in the sediments are earth's oldest marine sponges, with needles still attached to their cells, and embryos of animals, some still undivided, others split into two, four, and eight cells. These animals lived and died in seas distant from Iapetus, but the legacy of Iapetus is far from fully revealed. As scientists continue to examine the remains of this ancient sea, they may find more windows into the distant past, bringing a hazy and obscure narrative into sharper focus.[11]

The life that blossomed in young Iapetus lurched forward in fits and starts, the opportunities and successes of some organisms made possible by the departure of others. Oxygen winnowed out some; ice vanquished others. Drifting continents opening Iapetus created more coastal waters, more habitat for earth's burgeoning population of organisms. Photosynthesizing bacteria abounded, soaking up carbon dioxide, blowing oxygen into the air. Ancient mountains eroded into young seas, and their sand and silt buried bacteria before they decomposed, before solid carbohydrate reverted to gaseous carbon dioxide. Carbon that would have been breathed into the sky instead drifted to the bottom of the sea. The sun was still faint, and earth could not sustain its benign climate without a blanket of carbon dioxide.

Breath by breath, grain by grain, oxygen-giving bacteria and sand washing into the sea brought on the most severe ice age earth has ever known. As more and more bacteria were buried in the sea, carbon dioxide levels dropped in the atmosphere, and the planet plunged into a deep chill. Almost all earth may have frozen into a thick, icy snowball. Glacial debris from low latitudes is eerie evidence that not even the Tropics were spared. Virginia rested near the equator at this time, but the bedrock of the Blue Ridge Mountains contains a jumble of stone, boulders and cobbles and silt, dumped by a retreating glacier at the edge of a melting sea. The icy breath of glaciers deposited the same rock, of the same age, in the coastal waters of Svalbard and Namibia.

For ten million years earth was nothing but rock and ice, except near the equator, at the edge of the glaciers, where a narrow band of open sea still flowed. In this blue-water refuge, and in volcanic hot springs on the seafloor, the hardy survived. In these hostile conditions, niches filled with a profusion of life emptied. When spewing volcanoes building Iapetus returned carbon dioxide to the air and warmed the planet, some of the surviving animals prospered and left signs of their passage on the seafloor. Resting above glacial stone are layers of limestone, forged from their remains.[12]

As Iapetus grew and the ice thawed, earth's first large multicellular organisms came into being, beginning to fill the open space left by those who perished in the ice. Hard rock has preserved their soft bodies, in the Ediacara Hills of Australia, the sea cliffs of Newfoundland and Ireland, the savanna of Namibia. Their arrival marked the advent of a time when life could be writ large, a time when earth could nurture and sustain organisms of increasing weight and dimension. Their existence is indisputable: these organisms, some as large as three feet (one meter) across, cannot be confused with ripple marks in the sand or minerals precipitated from the sea. The Ediacara, as they are called, represent an important moment in the evolution of life on earth. They stand at the threshold of a new age, and yet, despite their large size, scientists cannot agree on what they are, not even whether they are ancestral plant or animal.

The fossils of these mysterious organisms are beautifully preserved, even though their bodies were soft, easily ravaged by decay or squashed by shifting sands. The Ediacara preserved in the cliffs of southeastern Newfoundland lived in still, deep water, in sands that bore no wave or ripple marks to indicate the slightest disturbance. The sea, lacking oxygen, lacked bacteria to decompose the dead. Their bodies remained intact as lingering currents dropped fine-grained mud and sand to gently bury the delicate organisms. Occasionally a thin layer of ash would settle nearby, blown in from volcanoes building the young sea.

The Ediacara lived in deep water and shallow. Some floated; others clung to the bottom. Their fossils look like feathery ferns, odd coins, pouches. They grew in diameter but stayed thin, no more than one quarter inch (six millimeters) from top to bottom. They lacked the guts and gills of animals and had no lungs or circulatory systems to pump air and energy through their bodies. Instead they soaked up what they could from the surrounding sea. Living by diffusion, they needed as much surface area as possible to absorb the sea's life-giving light and nutrients. Internal walls arrayed in symmetrical patterns propped up their weak bodies: in appearance they are more like a quilted air mattress than any living animal.

Iapetus has yielded a record, but we have yet to decipher fully its meaning. The Ediacara were large, and they left many well-preserved fossils, yet their identity eludes us. The Ediacara in Newfoundland are lifted into remote sea cliffs scoured by wind and rain and often hidden in fog. Human inhabitants abandoned this terrain years ago. The fossils are embedded in a desolate headland jutting out into the sea, named Mistaken Point for the many sailors who, incorrectly believing they had reached the turn for safe harbor at St. John's, met their deaths on the rocky cliffs. Mistaken also may be our perceptions of where we find ourselves in the distant reaches of evolution. Scientists subscribe to many conflicting interpretations of these organisms given by the sea, but no one is quite sure where fact slips into fiction, where the hard lines of reality shade into the softer, hazier realm of imagination. Perhaps one day the discovery of additional fossils will reveal an undisputed link between the Ediacara and living animals. Or perhaps, as some scientists claim, the Ediacara will always stand alone, with no descendants, no connection to the living world, a brave start down a path that ended.

Pioneers in the realm of substance, Ediacarans carpeted the seafloor for over 100 million years. For earth's first large multicellular oganisms, as for the stromatolites, size was both blessing and curse. The Ediacarans' long-lived experiment with amplitude ended when their peaceful, sedentary lifestyle proved no defense against the predatory stalkers who soon made their way across the seafloor. From the time salt water first poured into young Iapetus until the sea closed and the basin disappeared, many populations of marine organisms would blossom and perish, and the rise and fall of the Ediacarans were but one of many arrivals and departures. Layer upon layer of rock fashioned by this sea speak of enduring life in transient forms, of many rising to prominence and few prevailing.[13]

Animals emerged all throughout Iapetus's coastal waters. At first the newcomers left but a trace, a tiny track in the mud, a burrow dug in the sand. The rocks of Scotland's isle of Islay record one of the first appearances of an animal with a digestive tract. Six hundred million years ago, it roamed the seafloor, leaving a trail of its waste. The animal disintegrated when it died, but its excrement was written into the seafloor, turned into fossil as soft bottom compacted into rock. Preserved throughout the sea cliffs of Fortune Head in southeastern Newfoundland are complex tracks left by a worm as it fed and burrowed in the seafloor. The closing sea yielded this signature, this tiny fragment alerting us to a world undergoing rapid change.

No longer did all of earth's inhabitants float passively in the water or rest peacefully on the seafloor. The new generation was on the move, wriggling its way along the bottom, a new generation of homesteaders. It may have been the splitting of a continent, the opening of a new sea, that made the water so inviting. Volcanoes wrenching the land apart laced the atmosphere with carbon dioxide. The gas held in the planet's heat and warmed the young sea. Rain came more often to a warmer earth and eroded the mountains, freeing mineral from rock to build the skeletons of animals.

The animals, if they were animals, absorbed the minerals, turning calcium into hard shell, into body parts preserved in the seafloor. It happened little by little. Initially the newcomers left behind a motley collection of shelly fragments. The scales, spines, and tubes hinted little about the dwellers within. Like the Ediacara, many left no descendants, but during their time they thrived. One, shaped like a heavy insulated goblet, built extensive limestone reefs in shallow coastal waters at the edge of Iapetus, creating quiet lagoons where life could blossom, sheltered from open-ocean waves.

In these peaceful waters began an intense biological arms race. It escalated quickly as animals began to prey upon each other, each round of thrust and parry becoming increasingly sophisticated. Teeth figured prominently in these engagements. When Iapetus was young, the shores of Edinburgh, Scotland, were a shallow lagoon in a warm sea. As mud on the bottom turned to shale, it preserved an eel-like animal no bigger than a tiny twig. The conodont, as it was called, had a tail and a mouthful of cone-shaped incisors. While its forebears filtered out food suspended in the water, the conodont hunted, grasping, cutting, and grinding prey with teeth it would leave behind as fossils. Bearing a stiff notochord along the length of its body, the conodont was one of earth's early vertebrates: a predatory life was encoded in the genes from the very beginning.

The conodont was a snippet of an animal. *Anomalocaris,* the "odd shrimp," was another matter. Almost two feet (sixty centimeters) long, it swam, or perhaps flapped, through the water with primitive fins, grasping prey with two curved limbs and stuffing it into a large circular mouth ringed with thirty-two teeth. Few escaped the attack of this animal, who apparently was not intimidated by the hard exoskeletons that defended others. Fossils of those once beseiged show the war wounds—gouges, large bites taken out of their shells. *Anomalocaris,* along with other animals from earth's early seas, stumped paleontologists for many years. At first the mouth shaped like a pineapple ring seemed to belong to an ancient jellyfish, and the curved appendages used to attack looked like some kind of tail.

Another animal from this time was thought so strange it was named *Hallucigenia.* As it turned out, *Hallucigenia* was not a figment of an imagination run wild, but merely a mistake in assemblage by the scientists who first pieced it together. When the odd protuberances extending from its back were reinterpreted as legs, *Hallucigenia* reentered the realm of the possible. The first of these animals were found in the Burgess Shale, a piece of ancient seafloor upended in western Canada. As more fossils were uncovered, some from pieces of Iapetus plastered onto Greenland, the construction of the animals became clearer. What was considered bizarre became, if not familiar, at least comprehensible. Now it seems that *Hallucigenia, Anomalocaris,* and *Opabinia,* an animal described as a swimming vacuum cleaner, probably belong to Arthropoda, the phylum including crustaceans and insects. As the case of *Hallucigenia* illustrates, the lens with which we look into the past is distorted and dirty, blocking a clear view. Privy only to tatters of earth's history, we can never be completely certain when we have pieced together the whole fabric.

The period when *Anomalocaris* and the others evolved is called the Cambrian explosion, the Big Bang of biological evolution. In retrospect, the word "explosion" may prove inaccurate, the concept an artifact of research. Perhaps the line marking the advent of these animals seems sharp because Cambrian animals left teeth and scales and shells to record their existence, while their soft-body versions succumbed to decay and destruction by waves and currents, leaving little of their history inscribed. Or perhaps we haven't quite figured out where to look. The fossil record reveals the past, but it represents only a small fraction of what once lived. Most of the life that blossoms on earth disintegrates at death, its component parts recycled to nurture the lives of others. Only a tiny fraction of the dead, 0.1 percent, receives burial in the sand and mud that will become rock. And when the rock is heated and twisted and raised into mountains that are then washed out to sea, many of those fossils not altered or crushed beyond recognition are worn away.

The Cambrian explosion may or may not have been an explosion compared with the foggy history that preceded it, but compared with what followed, it certainly was. In twenty million years, an instant of geologic time, almost all of the animal phyla we recognize today emerged. It was a time of seemingly unbounded opportunity; the waters of Iapetus were abundantly fruitful. Glaciers had scoured the ecological playing field clean, and the sea level was rising over continental shelves, opening vast realms to newcomers. Warm rains in a benign climate rearranged hard rock, washing mountains into the sea, and hot springs on a growing seafloor released dissolved minerals into the water, both saturating the sea with life-giving nutrients. Life was young and full of possibility. Genetic sequences were not firmly established, body plans were still on the drawing board, the stages of development were not firmly set. It was a time of experimentation not to be repeated. In the 500 million years that have followed, few new phyla have appeared, a negligible number compared with those that arose in the bustling times of the Cambrian. As we have become, at least for now, more set in our ways, our boundaries more rigid, the pace of evolutionary creativity has slowed, its range narrowed.[14]

During Iapetus's existence, marine animals came and went. Earth's first reef-builders had broken the surge of the open sea, creating calm waters, the cradles of early animal life. They had their heyday, and disappeared. These organisms, known as archaeocyanthans, belonged to their own phylum and had their own unique body plan, which hasn't been seen since. In the quiet lagoons protected by archaeocyanthan reefs, a new hard-shelled animal came into being. When it died, it left behind not bits and pieces of uncertain affinity and identity, but a whole shell.

The animal molted as it grew, leaving many shells to be buried on the seafloor. From them we infer a scavenger creeping through the mud. It had eyes and a mouth with a jaw, and jointed legs for walking and eventually swimming. Some trilobite species were small, no larger than a thumbnail; others were larger than lobsters. They were dominant, successful, ubiquitous. Thousands of species populated earth's seas. They endured for approximately 300 million years, but by the time Iapetus closed, they too had died out.

As the glory of the trilobites faded, others rose. Graptolites emerged, living out their lives afloat in the waters of Iapetus, feeding on plankton. Remnants of these tiny, lacy organisms with serrated edges appear on a piece of shale as faint pencil markings. Widespread, distinctive, they thrived for 100 million years. Then they were gone.

Not all the animals born in Iapetus disappeared, however. Brachiopods, more familiarly known as lampshells, lived on rocky tidal ledges and soft, muddy oozes in deep water. They spent their infancy as free swimmers but then established permanent residency on rocks, reefs, or bits of gravel and shell on the bottom. They stayed there for the rest of their lives, shifting occasionally to avoid burial. Inside their two shells, which, unlike those of clams or mussels, were of uneven size, were a tiny heart and a gut for breathing and feeding. The metabolism of brachiopods was slow; they could survive for up to a year or two without a meal. Their hardiness, their adaptability to diverse habitat, and their unpalatable taste, both to fish and to man, has helped secure their survival. If the fossil record is a measure of endurance, brachiopods rank as earth's longest-abiding large animal. They appeared over 500 million years ago and, after weathering wave after wave of mass extinction, still hold a place in the modern ocean, in the waters off Antarctica, where they dwell in large numbers.

Mostly, though, the record of animal life held by Iapetus is one of impermanence and unpredictability. Populations of animals waxed and waned, coming into great prominence, then dwindling into obscurity and, for most, extinction. Archaeocyanthan and trilobite, graptolite and brachiopod—each had its moment of greatness, and each dominated earth's seas for millions of years. Each declined. Today we can barely pronounce the unfamiliar names, barely imagine seas brimming with animals whose glory is now but a mark on a rock left by a long-gone sea. Who is to say who will live and who will die, who is to be culled out and who will proliferate? The fossil record lists the heirs but does not explain why or how they were chosen.

The flow of water, the building and disintegration of rock, the diffusion of gas into and out of the atmosphere, the living and dying of plants and animals—the call of one elicits a response from the others in an elegant interchange among geology, biology, chemistry, and physics. Together the ensemble creates a living planet. As earth evolves, its inhabitants grow more or less suited to changing conditions. The fossil record suggests that no matter how dominant, how pervasive, how successful a particular organism, its days, in most cases, are numbered. It does not predict who, in earth's next change of circumstances, will be best prepared.[15]

The animals who lived in Iapetus, confined to a particular time and a particular place, make it possible to reconstruct the continents surrounding this sea, even though they now grace other shores. When Iapetus was at the peak of its growth, the line of volcanoes building the seafloor was 6,200 miles (10,000 kilometers) long, almost the length of Atlantic today, and its waters were at least 2,800 miles (4,500 kilometers) wide. Many trilobites couldn't swim, and those who did couldn't venture beyond the shallows of the continents. Their parochialism gave each coast its own community of species, distinct from the others. *Olenellus* lived along the shore of North America, then called Laurentia, a large continent facing the equator. Though *Olenellus* was confined to this coast, its fossils are now scattered, revealing that North America is not what it once was, that when drifting continents come together and then split, they do not always break along the same lines. The fossils of *Olenellus* reveal that the Highlands of Scotland once rested in North America.

Terra firma shifts. With the opening of each new ocean, the dance of continents changes. A large reef that once edged Laurentia, once gave shelter to *Olenellus,* was pushed up into the Appalachians when Iapetus closed. A large chunk is missing. In the hole the Mississippi River empties into the Gulf of Mexico. Scientists had assumed that Africa always faced North America across the sea, but the lost reef and its fossils lie elsewhere, in the foothills of the Andes, in western Argentina. When seas have drained away, the ancient motions of continents are difficult to plot. Perhaps eastern North America and western South America were once joined, or perhaps the two continents stood on opposite shores. Somehow a piece of one broke off to become part of the other. Whatever the mechanism, a block of the North American continental shelf, a piece of limestone 500 miles (800 kilometers) on each side, surfaced in the Precordillera, the steep cliffs at the edge of the Andes. Embedded in the limestone are fossils of trilobites and brachiopods born and buried in an ancient lagoon of the now distant shores of eastern North America. *Olenellus,* a homebody while alive, became a great traveler after it died, taking to the hills of Scotland and the desert cliffs of Argentina, riding the backs of drifting continents.[16]

Fossils reconstruct other lands drifting through Iapetus. Shells of brachiopods and fragments of trilobites identify Baltica, a continent whose lands extended from what is now Norway and Sweden across to the Ural Mountains, south to the Caspian and Black Seas, and then back up north through the Holy Cross Mountains of Poland through Denmark. In the time of Iapetus, Baltica was a southern continent, drifting well below the equator, its travels clocked by the fossils of ancient animals. While its trilobite and brachiopod fossils were distinct, Baltica drifted alone, far from other lands. When Laurentian fossils appeared in the mix, the two continents had drifted within 620 miles (1,000 kilometers) of each other, close enough for floating brachiopod and trilobite larvae to mix. South of Baltica, on the southern boundary of Iapetus at what is today the edge of Antarctic sea ice, drifted Gondwana, an enormous landmass consisting of South America and Africa, Antarctica and Australia, Florida, and pieces of India and the Middle East. For over 200 million years Iapetus separated these lands, but the sea was not to last and the continents would once again converge. Its waters began to close when a piece of land, Avalon, broke off Gondwana, somewhere near Florida and Brazil.[17]

Both land and seascape were transformed as Iapetus closed. Plants slid out of the water for the first time, greening the continents. Lofty mountains rising from Laurentia eroded away, hard granite worn into tiny grains of pure sand. The seafloor off Laurentia buckled, creating a deep trench. The shells of dead animals had built an immense bank of carbonate in the warm sunlit waters of Iapetus, stretching from Newfoundland to Alabama. Earthquakes racked the carbonate bank, and huge pieces broke off. It foundered in the deepening water and grew no more. The seafloor, pressed by approaching continents, began its descent, returning to its origins deep in earth's mantle. Some rock was scraped off and plastered against the side of the trench; some melted and returned to the surface, erupting into an arc of volcanic islands several thousand miles long. Erosion from the rising islands settled into the deep water as mud, and then shale, sprinkled with dustings of ash blown in from volcanoes. Well-oxygenated water yielded red, green, and purple shales; stagnant water produced black. As the seafloor continued its descent, the stack of sediments in the trench rose higher and higher until it appeared above the surface of the sea.

New England's landscape records the closure of this sea. As Iapetus narrowed and the arc of islands approached Canada and New England, the stack of sediment was shoved ashore, over the remains of the carbonate bank, where it became a towering mountain range, the Taconics. Today the Taconics have worn away, but their remnants curve through the gently rolling hills east of New York's Hudson River, up through western Vermont and Canada along the shoulder of the Gaspé Peninsula, and over to the western edge of Newfoundland. Pieces of the carbonate bank have been stitched into the landscape as well, in the quarries of upstate New York and nearby battlefields of the American Revolution. French

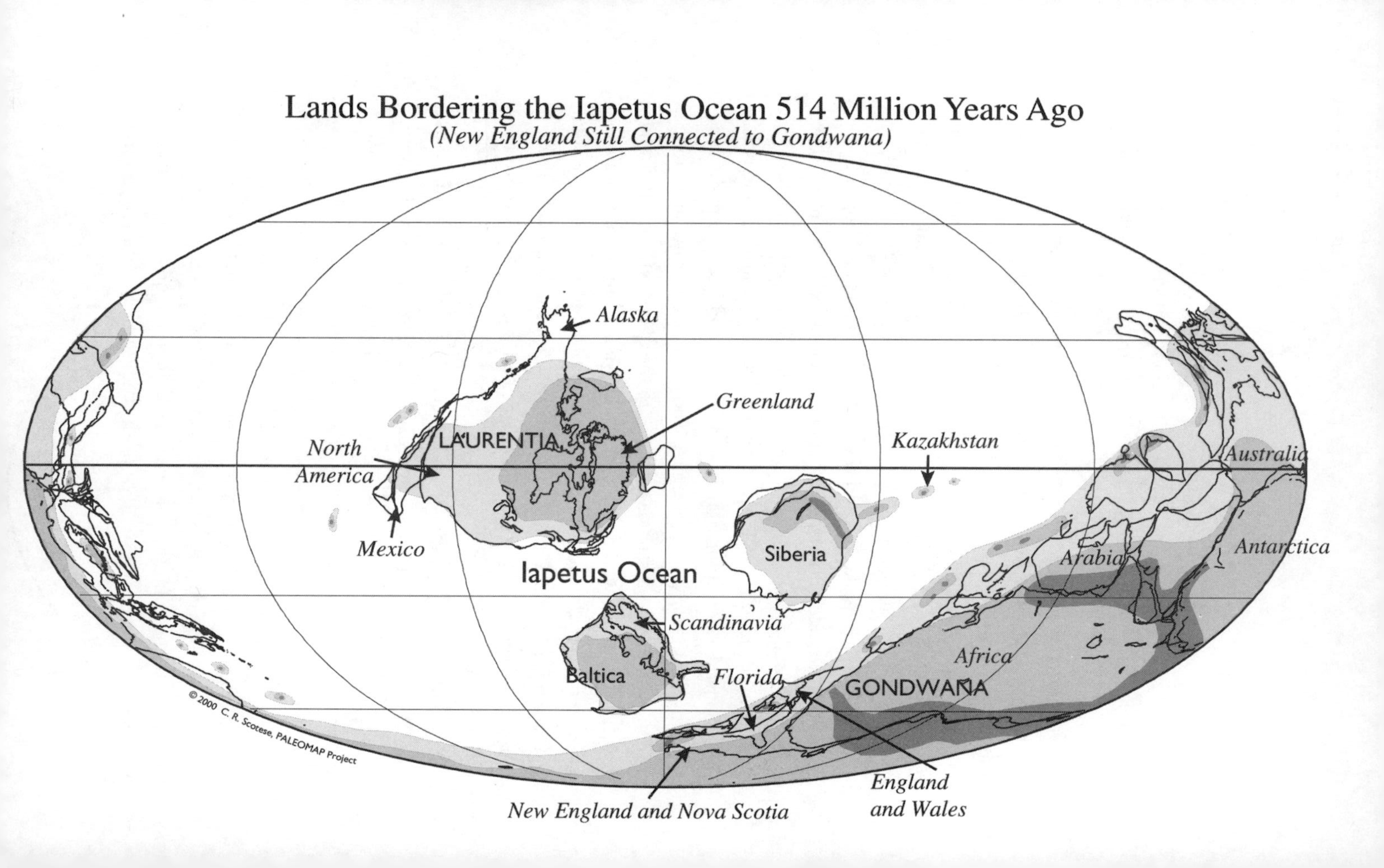
Lands Bordering the Iapetus Ocean 514 Million Years Ago
(New England Still Connected to Gondwana)
Alaska
Greenland
LAURENTIA
North
America
Mexico
Kazakhstan
Australia
Siberia
Antarctica
Arabia
Iapetus Ocean
Scandinavia
Africa
Baltica
Florida
GONDWANA
England
and Wales
New England and Nova Scotia
© 2000 C. R. Scotese, PALEOMAP Project

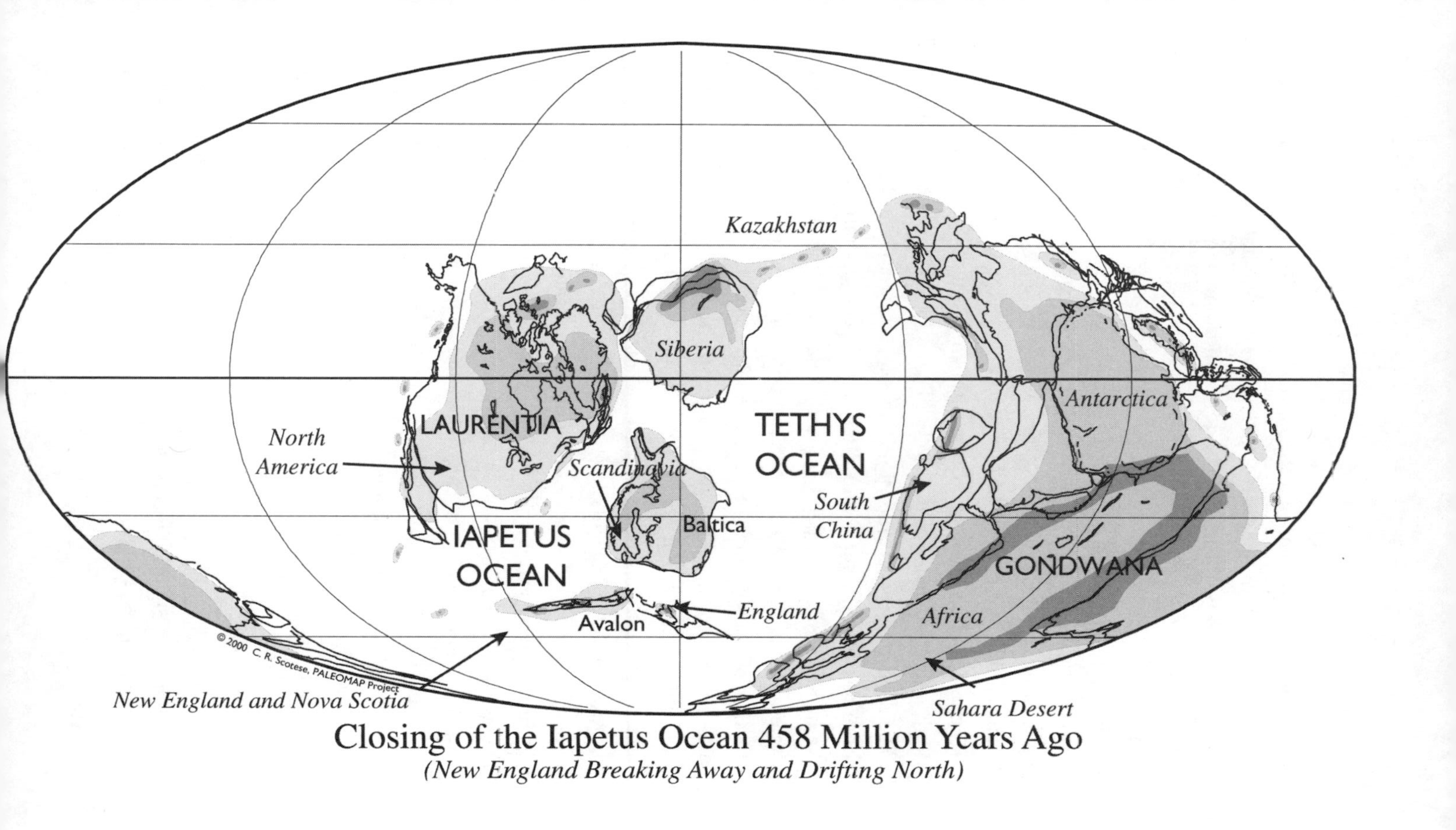

Closing of the Iapetus Ocean 458 Million Years Ago
(New England Breaking Away and Drifting North)

and British forts overlooking the narrows of Lake Champlain, between New York and Vermont, were built on the continental shelf of Iapetus. Amidst the ruins of the old parade grounds and barracks lie pieces of limestone, built in quiet shallow lagoons when the sea was young. A thin layer of shale, sign of deepening waters and approaching continents, caps some of the limestone.

The transition from sea to land was far from gentle. Squeezed by the advancing arc of islands, limestone was baked into marble, shale into slate. A ribbon of pink and white marble, metamorphosed remnants of Iapetus's shallow seafloor, winds through Vermont—ancient seafloor built into the facades of downtown Danby, Middlebury, and Rutland. The deeper deposits, veins of red, purple, green, and black slates, have been quarried and tiled onto the roofs of sleepy country towns on the Vermont–New York border. Even the lava lining the seafloor, beneath the limestone and shale, was baked. As it returned to the depths, slices were scraped off, heated and compressed into veins of asbestos and talc, and heaved up onto land. They were strung out along the length of Vermont and into Canada. For many years a piece of Iapetus, surfacing in Quebec's Thetford mines, supplied Canada with asbestos.[18]

Much of New England's bedrock is of the sea; the rest has emigrated from other shores. The arc of islands docked against Laurentia, leaving its hilly remnants scattered along the western edges of Maine and New Hampshire, the interiors of Massachusetts and Connecticut. Pieces of the arc follow the eastern side of the Connecticut River and continue through Canada into the Gaspé Peninsula, New Brunswick, and Newfoundland. The arc nudged Scotland, still on the edge of Laurentia. It pushed up the Highlands, spitting some sideways, creating a scar running between Inverness and Fort William through Loch Ness—the Great Glen Fault. The crash attached the hills of England's Lake District onto Scotland.

Once the island arc was welded onto New England, continents from the opposite shores of Iapetus drifted in. Baltica hit first, fusing Scandinavia, Svalbard, and Greenland, raising the Caledonide

Mountains, lifting the craggy peaks and steep ridges of Glencoe, Scotland. Avalon landed hard against New England, raising more mountains, Maine's Mount Katahdin, the first place the sun rises in the United States, and New Hampshire's Mount Washington and Mount Monadnock. Monadnock, "the mountain that stands alone," is the hard weather-resistant core of once towering peaks whose softer edges have eroded away.

Avalon collided with Laurentia and split apart, but fossils from its unique community of animals can be used to reassemble the scattered fragments of this enigmatic land torn from Gondwana. Inscribed in the sea cliffs outside Newport, Rhode Island, the abandoned quarries of Braintree, Massachusetts, and the hills outside Harlech, Wales, are fossils of the trilobite *Paradoxides,* named "unbelievable" because it was thought impossible that such a poor swimmer could traverse such great distances. Today Avalon's indigenous fauna identifies a lost continent that once drifted across an ancient sea, a land where England's Shropshire Hills adjoined Maine's Mount Desert Island, and southern Ireland was neighbor to Nova Scotia's Cape Breton.[19]

The waters of Iapetus have drained away, but the history of that sea is indelibly written into the landscape of continents. Nowhere is the impermanence of life and land inscribed more beautifully than on Newfoundland, an island deeply touched by the passage of this sea. Its mountains and sea cliffs record the splitting of a continent as the sea was born, the riot of life its waters once nurtured, and the arrival of distant lands as the water drained away. The Long Mountains, rising from the coast, are the spine of western Newfoundland. Before the days of satellite navigation, Gros Morne, one of the taller hills, guided sailors through the Gulf of St. Lawrence, but more often than not these hills dominate without being seen, from a throne hidden in cloud and fog.

Rounded down, weathered, the hills are the remnants of towering peaks raised over one billion years ago when earth's continents fused. When Iapetus opened, they marked the landward edge of Laurentia, the

North American continent. Long, deep ponds, full of landlocked salmon and trout, cut into the mountains, record Iapetus's birth. The steep granite cliffs rising from Gros Morne National Park's Western Brook Pond are cut with wide stripes of dark volcanic rock, veins of lava that cracked mountains and parted the land for waters from a new sea. Softer than the resistant granite, they have crumbled away, leaving precipitous stairways for tumbling waterfalls and migrating caribou traveling through balsam and fir forest down to the pond.

Newfoundland is a land built and rebuilt as ancient seas opened and closed. Gros Morne, the second-highest mountain in Newfoundland, is crowned with rock formed in the deep water of Iapetus. Sand, eroded down from the mountains, washed into the sea, where, layer upon layer, it accumulated. Under pressure, it turned to stone, and when the sea closed, it was squeezed up to cap the mountain. Gros Morne is a child of the sea, raised over a billion years ago when one ancient sea closed, and refurbished millions of years later as another departed. As Iapetus closed, pieces of the thick limestone reef lining its shores buckled and fell into the deeper water, settling amidst the mud and shale. The approaching land pushed these pieces of the continental shelf up onto the shore and over what was left of the mountains. Limestones and shales carrying Iapetus's history were upended and pushed miles inland, coming to rest on Newfoundland's west coast, virtually intact.

It is possible to hike along the shore, passing through dreary seasonal fishing villages, by the homes of fishermen who can barely wrest a living from the depleted sea, by racks of tiny cod drying in the wind and sun, out to a windswept point containing this ancient seafloor. Iapetus was a wellspring, its life-giving waters home to the very first animals emerging on the planet. The ocean itself has come and gone, once flowing in to fill a small basin, growing and spreading into a vast sea, then ebbing away as continents coalesced. Gone are the waters, and gone are the many organisms that flourished at the dawn of animal life. This stone holds their history. To walk along these cliffs is to retrace the evolution of animals nourished by Iapetus.

The fossils present the long view, a perspective spanning millions and millions of years. In layer after layer, the history is the same: abundance, profusion, and, ultimately, extinction. Even the history itself is crumbling away. Out on these lonely points, surrounded by relics from a dim, distant past, there is only the sound of wind and rain, and rock tumbling down the cliffs, shattering at the bottom. In the rubble are pieces of soft rock containing fossils of graptolites. These animals were raised by the sea; their bodies built up the seafloor and then fashioned the shore of western Newfoundland. Now, worn down by wind and rain, the rock awaits its return to the water, where its minerals will dissolve to nurture another generation of animals.

Iapetus was fed by molten rock rising from deep within earth's mantle. When the sea closed, a large slice of this deep rock was written into the landscape of western Newfoundland, in the Tablelands of Gros Morne National Park. Nowhere else is there such a vast exposure of the rock that built this sea. It is possible to walk for days across the weathered orange rock. Surrounding it are balsam and fir forests and stretches of alpine tundra, but the mantle rock itself is bare, supporting little life. An occasional fox or caribou may pass through en route to more plentiful feeding grounds, but little here beckons them to stay. The soil is depleted; only the hardiest of mosses will grow. Many of the plants are carnivorous, obtaining sustenance from insects instead of earth. The rocks of the Tablelands are rich in iron, magnesium, nickel, heavy elements that sank into earth's mantle when the planet was born. Eerily, here they are, up in the light of day, on a piece of deep earth raised from the darkness of the planet's interior, miles below the sea floor.[20]

The Tablelands made their ascent en masse, virtually unscathed and unaltered, providing a rare view of earth's interior. Elsewhere, along the old edge of Laurentia, up through central Vermont and on into Quebec and Gaspé, tiny pieces of mantle rock appear here and there in road cuts and quarries. These have been squeezed and heated by approaching continents and transformed into soapstone. Man has mined this rock, turning slivers of Iapetus into sinks, stoves, pancake griddles, and foot warmers. The string of soapstone exposures ends in Newfoundland, at the end of the Baie-Verte Peninsula, east of the Tablelands, at a large quarry in the village of Fleur-de-Lys. Two thousand years ago, long before the Norse ever set foot in Newfoundland, Arctic hunter-gatherers worked this quarry, using stone tools to scrape and chisel the softer soapstone. For one thousand years, Dorset peoples living here depended on the sea, hunting seal from the water and carving soapstone cooking vessels, bowls, and lamps from rock that was once the floor of Iapetus.

The Dorset were not the only ancient peoples to settle in Newfoundland. Recent excavation at Fleur-de-Lys has unearthed a soapstone plummet, used to sink fishing nets. It was carved by Arctic hunters who migrated across the strait from Labrador into Newfoundland four thousand years ago, providing for themselves with tools fashioned from an ancient seafloor. East of the Baie-Verte Peninsula, seafloor gives way to the arc of islands that formed as the sea closed. The landscape of central Newfoundland is built from this arc of islands sutured to the mainland as the sea drained away. Here, in the remote islands of Notre Dame Bay, once distant lands lie side by side. Here, the remains of trilobites living on opposite shores of a vast tropical sea now rest together under the shadow of drifting icebergs, distant realms collapsed into a single neighborhood with the closing of an ancient sea.[21]

For the animals who once dwelled in the waters of Iapetus, the sea was immense, its expanse unbridgeable. Today the entire reach of Iapetus is squeezed onto Newfoundland: the edge of Laurentia and its continental shelf run to the north and west, flanked by remnants of seafloor and island arc running through the island's center. Wedged up against them is shelf from the eastern edge of Iapetus and the ancient continent of Avalon. Newfoundland's Avalon was once as veiled as the mythical island behind the mists where the dying King Arthur was spirited away, for it too is of another realm, hailing from the far shore of an ancient sea, carried to Newfoundland by the restless motions of earth.

It is a land layered with history. Archaeologists combing through the correspondence of early colonists relocated the former village of Avalon, settled in 1621 by Englishmen hoping to take advantage of the rich cod fishery nearby. At the site, they unearthed a house and a well, fortifications, and a cobble platform for drying fish. The artifacts the settlers left behind included the crucibles of an alchemist, silver pins and thimbles, pieces of velvet. Digging further down to where the shoreline stood five hundred years ago, archaeologists found the vestiges of an even earlier culture, burnt bones and arrowpoints belonging to the Beothuks, descendants of the Arctic maritime hunters who were Newfoundland's earliest human inhabitants.

Layered below the grass of coastal meadows, below the roads of modern villages, is the history of a people who arrived thousands of years ago. Not a great deal is known about them. They hunted salmon and capelin, seal and walrus, caribou and beaver. They ate the meat, used the skins for clothing and shelter, the bones, antlers, and ivory for tools and weapons. Eventually they disappeared, giving way to new arrivals, with each new wave of immigration ultimately replacing the existing culture in a high-speed version of the rise and fall of animals whose lives are recorded in the nearby sea cliffs.

The passage of peoples, the rise and fall of animals, the opening and closing of seas—these stories share a common rhythm. Earth's primordial seas witnessed the birth and blossoming of multitudes of marine plants and animals, and for many, their demise. Ancient seas filled and drained away, giving girth to the continents, surrendering their record of evolution. It is a record of profusion and decimation, of abundance and scarcity. Hiking along the coast of Newfoundland, walking amidst layers of deep time, along the floor of ancient Iapetus, is a constant reminder that the days of our species are numbered, our time measured. Yet, though our own tenure is evanescent, life endures. The motions of a moving earth, which raise mountains and wash them into the sea, which open oceans and crush continents, which recycle minerals between rock and animal, land and sea, replenish and sustain our living planet, bestowing life on earth. Animals flourished in Iapetus and then faded away. When that primordial sea closed, the continents fused into one great landmass, Pangaea, "all earth." Though its ground was seemingly firm, even this land would split and give birth to a new ocean, and to new life.[22]

CHAPTER 10

The Birth of Atlantic

MILLIONS of years passed after the closing of Iapetus. In that time, earth's continents continued drifting toward each other as lands torn asunder were reassembled and other ancient seas drained away. Africa and South America joined and merged with eastern North America. Morocco became neighbor to Newfoundland; Brazil moved next door to Namibia. The distant and disparate fused. The massive continent of Pangea ran the length of the planet, curving from one pole to the other, embracing a great bay, Tethys. A few continental scraps, pieces of China and southeast Asia, drifted in the bay, named for the Greek Titan who nurtured the sea and who wedded Oceanus.

Pangea, great as it was, would not endure; eventually plumes of heat rising from deep within earth's interior would break it apart, making way for Atlantic. Yet it is worth pausing to consider the time of Pangea, when glaciers melted and the atmosphere enveloped the planet in a warm greenhouse, when reptiles evolved to roam the continent's arid plains, and when seed-bearing plants and trees came to fill dense forests. At this time earth also experienced a great dying: most species dwelling here became extinct, and the course of evolution was reset. Pangea's story, written into layers of sediment laid down beneath its waters, is a story of the devastation and unpredictability of extinction, of the susceptibility of united lands and seas. It is a confusing tale, riddled with mystery, controversy, and speculation, a reminder of how complex and unknown are the workings of earth.

Young Pangea was striking, in height and in girth. As ocean basins disappeared, pieces of seafloor heaved ashore added amplitude to the land. Long chains of towering mountains seaming the island continent added texture. The Urals joined Russia to Europe. A long arc of islands raised rugged mountains along the continent's southern edge. The Appalachians, oriented along the equator, cut through the core. High and steep, they caught moisture from the wind, drenching a tropical rain forest, feeding great rivers flowing to the sea. Spiders and dragonflies lived in the forests, freshwater sharks in the rivers. Some of those rivers are no more. A trail of rock crystals embedded in the dry sands of Nevada marks the ancient course of one river. That river, once as wide as the Mississippi, thundered from mountainous headwaters across land that today is as dry as dust, where streams are ephemeral and water more precious than gold. Pangea may have contained the substance of continents to come, but its singularity dictated a different climate and topography.

Away from the rain forest, in the long shadow cast by the mountains, stretched vast deserts filled with fields of dunes. From time to time, the sea washed in over the desert, spreading across its low plain. More than once it spilled in and receded. It inundated broad lowlands at the base of the Ural Mountains, rinsing land between the Barents and Caspian Seas in a wash of salt. It poured in across Brazil and into Gabon. It flooded western Europe and the American Southwest, its waters rolling inland for hundreds of miles. A few small, salty lagoons, scattered at the edge of continents, are today's equivalent. Water flowed over the plains of Pangea and then evaporated, leaving its salt in thick layers of sodium chloride, gypsum, and anhydrite. The layers thickened, compacting and transforming the masses of reef-dwelling animals buried there into reservoirs of petroleum and natural gas.

Pangea's seas were lively; the warm waters were filled with a panoply of unfamiliar animals who have long since disappeared: coiled ammonites, lampshells, and tiny bryozoans growing in large mossy colonies of hundreds, sometimes thousands of members. Floating foraminifera, single-celled organisms invisible to the naked eye in today's seas, grew to colossal size, almost an inch (two centimeters) across, in the time of Pangea. An abundance of lampshells, in all sizes and shapes, created vast banks of limestone. Meadows of sea lilies carpeted the shallow floor. Sedentary animals easily mistaken for flowers, sea lilies live attached to the bottom, filtering their sustenance with feathery arms waving from the ends of thin stalks. They still inhabit dark waters of the deep ocean, but their numbers have been winnowed away. Those who seek to know the dwellers of Pangea's seas can see the best representations not in the living waters of Atlantic, but up on dry land. Embedded in the Guadalupe and Glass Mountains rising from the desert of Texas are limestone reefs whose fossils preserve a measure of this distant time, a vision of a realm that no longer exists.[1]

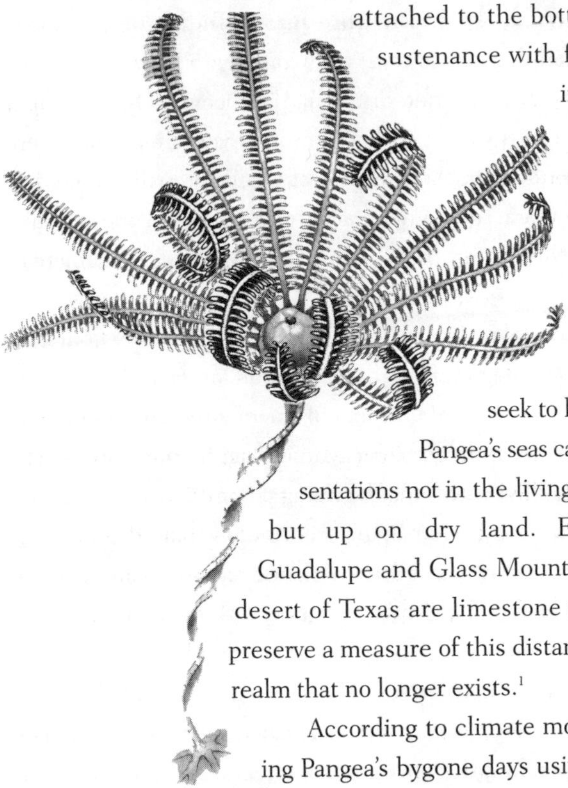

According to climate modelers reconstructing Pangea's bygone days using high-speed computers, interior Pangea should have been a harsh land. The continent was bulky, too bulky in many places to feel the ameliorating effect of the ocean. Land cannot absorb heat as well as the sea, whose liquid reservoir takes the icy edge off winter and breaks summer's swelter. Warm water seeping in over the Poles could soften the seasons along the coast, but in the continent's parched, central core, untouched by the moderating sea, seasonal temperatures should theoretically swing from one extreme to another. However, botanical evidence suggests a more benign environment. Thick, luxuriant forests grew in the distant polar reaches of Pangea, in what are now the icy mountains of Antarctica. Their remains are buried in glaciers where no vascular plant could possibly live. The trees grew on what were once the floodplains of rivers, the edges of lakes. They left behind a dense grove of saplings, and almost one hundred standing trunks from a mature forest. Today the ground is frozen, but back then deciduous trees grew rapidly in the twenty-four-hour sun of the Antarctic summer, dropping their leaves when the light waned. The fallen leaves piled up on the moist ground and later were pressed into seams of coal. In the darkness of winter, growth ceased, but only a few fossilized tree rings show signs of frost. So far, no one can fully describe the flow of wind and water that brought such a gentle climate to such a high latitude.[2]

Evolution may have taken a slower and somewhat narrower course in Pangea, where those who dwelled in contiguous lands and seas were more homogeneous than the inhabitants of fragmented continents and oceans. An ocean is a formidable barrier. Broad, deep water thwarts the migration and mixing of populations, fostering diversity. Compared with the earlier time of Iapetus, when many continents floated separately across earth's seas, or the later time when Atlantic had split Pangea apart, the number of ecological niches in the Pangean world was relatively low. Lands separated by deep water today, Africa and South America, India and Antarctica, were once contiguous. The easy passage created competition among species. In the early years of Pangea, spore-bearing horsetails and mosses dominated the northern continents, and seed ferns the forests of the south.

Over time, seed overwhelmed spore. Horsetail ferns and mosses are fastidious: each stage in their two-generation-long reproductive cycle fails in the absence of water. Delicate spores grow into gamete-producing plants only in moist soil, and gametes wed to form the first cell of the next generation only on wet plants. For them, a dry climate is a challenge. Seeds—more robust, better protected from heat and cold, better able to withstand long dry spells—prevailed, and conifers came to dominate northern forests. Today, seed-bearing trees continue to characterize our forests, dwarfing mosses and ferns, former giants now returned to their original niche, low in the moist ground.

Just as the existence of a single landmass did not encourage the wide proliferation of plant species, neither were earth's animals as diverse as they would become when Pangea broke apart. As Pangea formed, fish swimming in its seas came ashore. Their fins grew into limbs, fleshy and strong. As amphibians made longer and longer forays onto land, their gills disappeared, replaced by lungs. Amphibians maintained their dual residency, returning to water to lay their eggs, until it became possible to replicate and maintain the liquid womb of a swamp within a hard shell. The egg was its own incubator, filled with nourishment, receptive to oxygen but impervious to water, the amniotic fluid safely sealed within. Protected and complete, the parent could leave the egg, without ill effect, in the dry sand.

Pangea's inhabitants, no longer tethered to standing water to provide for future generations, were free to roam. And roam they did. *Mesosaurus,* a small, slender reptile with needle-like teeth, fished along riverbanks in South America and Africa. In the waning days of Pangea, *Lystrosaurus,* a squat, lumbering, mammal-like reptile with jaws to crush and chew plants, ate its way through the continent's lush vegetation, leaving fossils in Europe and North America, as well as in Africa, South America, India, and Antarctica.

Reptiles rose to prominence in this great continent, but even in their heyday they were never as diverse as the mammals who followed. In 200 million years, reptiles gave rise to twenty orders, while in only 65 million years, thirty orders of mammals came into being. Drifting continents may

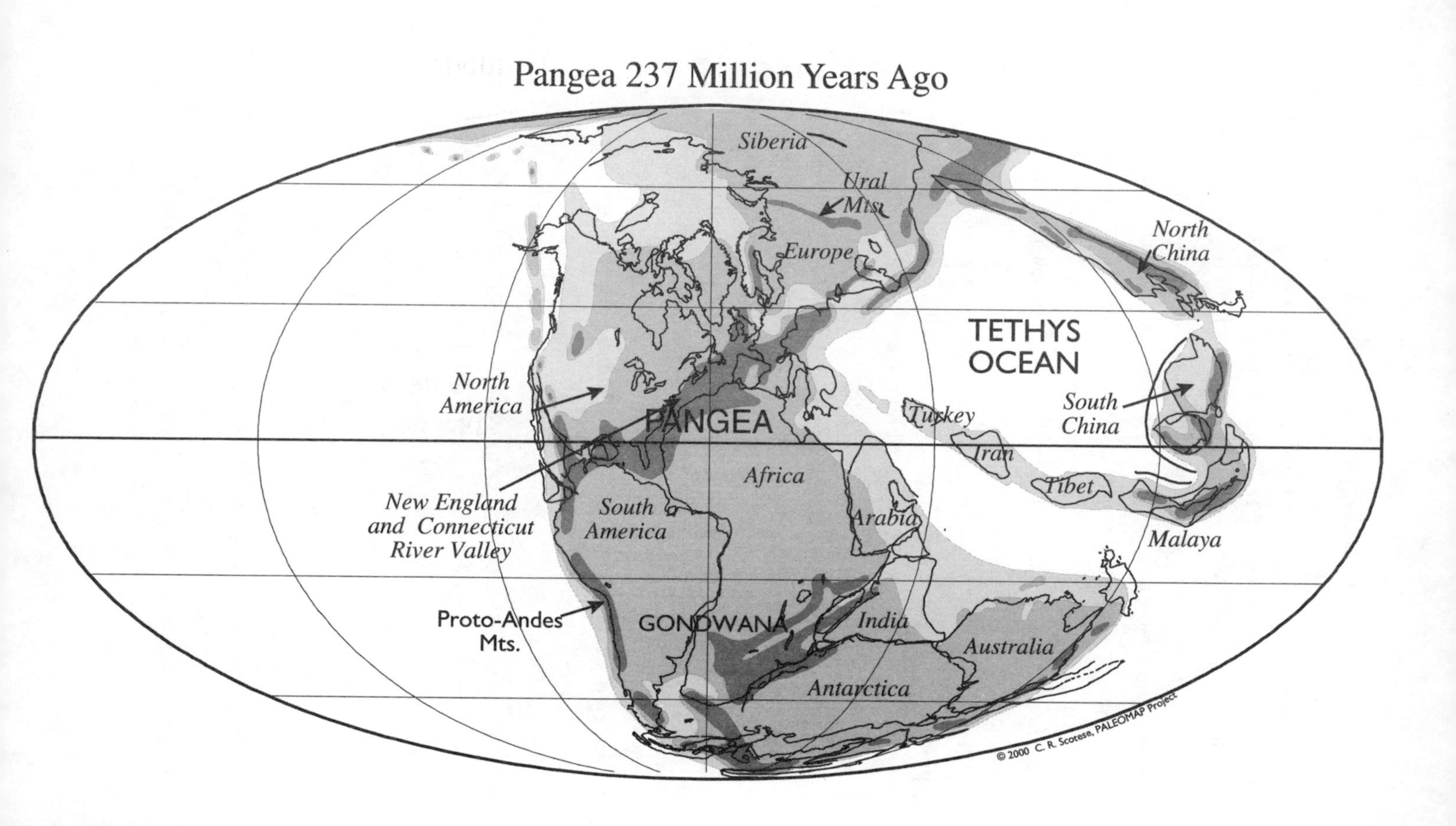
Pangea 237 Million Years Ago
Siberia
Ural Mts.
Europe
North China
TETHYS OCEAN
North America
PANGEA
Turkey
South China
Iran
Africa
Tibet
New England and Connecticut River Valley
South America
Arabia
Malaya
Proto-Andes Mts.
GONDWANA
India
Australia
Antarctica
© 2000 C. R. Scotese, PALEOMAP Project

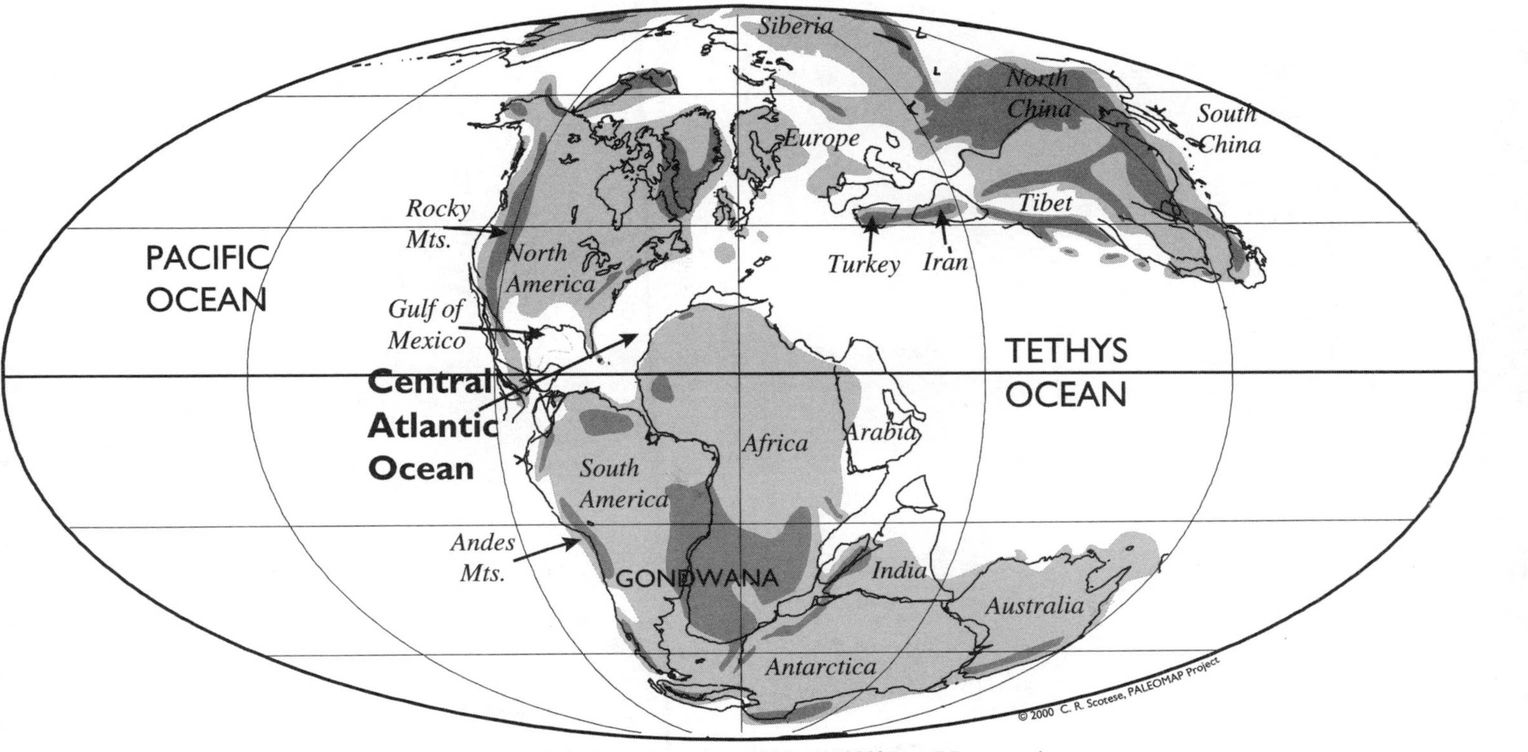

Opening of Atlantic 152 Million Years Ago

help account for the difference. Crocodiles, turtles, and lizards evolved while Pangea was whole. They originated in one place and then made their way throughout the continent. As mammals evolved, Pangea ruptured, creating multiple habitats isolated by impenetrable barriers of water. Today, four orders of aardvark, antbear, and anteater feast on the termites of South America, Africa, Asia, and Australia. Had these dwelling places been joined, in all likelihood fewer animals would have evolved to fill this particular niche.

Today, lands separated during the birth of Atlantic have once again touched. The joining of North America to South America, of Africa to Eurasia, has reduced the number of terrestrial mammal provinces from seven to four, the number of mammalian orders from thirty to eighteen. The lower levels of diversity in terrestrial Pangea were reflected in the sea as well. The modern ocean contains as many as thirty-one marine provinces, each with its own distinct community of plants and animals; Pangea's sea may have sheltered as few as fifteen.[3]

Earth's design, a single continent in a single sea, may have left the planet vulnerable. A wide variety of habitats increases the odds that some may be equipped to survive in the event of catastrophe. Large numbers of isolated communities offer the possibility that some will serve as refuges. Compared with other continents in other times, Pangea, with contiguous, interconnected lands surrounded by uninterrupted seas, was a place with little room to hide. As Pangea's tenure came to a close, earth experienced the worst extinction the planet has ever known. The death toll is unparalleled. The more recent, and better-known, demise of the dinosaurs was far less severe. The Great Dying, as this older extinction is known, was twice as devastating, eliminating 85 to 90 percent of species living in the ocean, 70 percent of vertebrates living on land, substantial numbers of plants, and even insects, who usually weather such assaults unscathed. At no time in the planet's long history, either before or since, has life on earth come so close to being extinguished. No one has permanent tenure on earth, but in the fossils of Pangea, the story of extinction is writ large.

No one can fully explain why. Uniformity may have drained earth's only ocean of its vitality. As Pangea drifted north, off the South Pole, glaciers melted and Antarctica warmed; its land held *Glossopteris* forests and coal swamps. In the sea, temperature differences between high and low latitudes were small compared with those of today, slowing down the deep-water circulation. Less oxygen-rich water sank at the Poles to spread along the seafloor and rejuvenate the depths. Lacking fresh air, earth's sea asphyxiated. While such a thing is hard to imagine, a few surviving pieces of Tethys seafloor hold the memory of these lean times. The rock is layered: brick red where skeletons of single-celled dwellers of the surface were rusted by the wash of oxygen, gray and black where no oxygen reached the deep water. When oxygen depletion reached all the way to the surface, killing off entire populations of floating sea life, no minerals accumulated on the bottom.

When the water was refreshed, first at the surface and then in the deep, the animals returned, their remains once again layering the seafloor, first in gray and black and then in red. According to the timetable written into this rock, the sea suffocated for a long time, thirteen million years at the surface, twenty million years at the bottom. Stale water crept in everywhere, at high latitudes and along the equator, leaving its signature in places as far apart as the Dolomites of northern Italy and the fjords of desolate Svalbard. Sluggish circulation exhausted Tethys, rendering her waters inimical to life.[4]

It is possible that carbon dioxide poisoning took the lives of land dwellers as well. Carbon dioxide, once the breath of life, had by this time become toxic to large numbers of earth's inhabitants. As recently as 1986, a bubble of carbon dioxide, inexplicably released from a volcanic lake in Africa, wafted invisibly into the air, killing fifteen hundred people. In this case, an underwater landslide may have uncorked the gas. Scientists cannot yet imagine what might have belched enough carbon dioxide into both air and sea to devastate an entire planet. Crashing comets spew dust and carbon dioxide into the atmosphere. A meteor may have blasted earth, its debris turning day to night, blotting out earth's life.

Gas trapped in rock laid down at this time seems to have an extraterrestrial origin. An Australian crater is a suspect, but its impact date is uncertain. Other evidence—shards of material ejected from the comet, layers of rock shocked by the impact—is still missing.

Raging volcanoes spew carbon dioxide into the air, and there were certainly plenty at the time of the extinction. Eruptions in Siberia flooded the area around Lake Baikal with lava. They never wrenched Pangea apart to form a new ocean, but their volume was immense. It was as if the 1783 eruption of Iceland's Laki volcano, which swallowed villages, poisoned livestock, and led to starvation, had repeated itself year after year, for thousands of years. Similar outpourings have occurred at the time of other mass extinctions. When Atlantic first opened, many reptiles became extinct; when India separated from the Seychelles, the dinosaurs died. Seven out of ten major extinctions are associated with major outpourings of lava. The forces that give birth to an ocean may prune back earth's tree of life. Timing, however, is not everything. Scientists who note the synchronicity between lava floods and extinctions do not all agree that coincidence implies cause. In the case of the Pangean extinction, other factors must figure into the record high level of annihilation.[5]

No single theory fully satisfies, fully accounts for such a great loss of life. It is possible that an event totally unfamiliar to us, one we have never seen before, caused the Great Dying. During our short stay here, it is not likely that we have seen all there is to see, known all there is to know, and our rocks and fossils, our window onto the past, record but a fragment of earth's biography. It is equally likely, though, that an accumulation of stresses strangled earth, that a confluence of circumstances—a large and vulnerable continent, a single ocean filled with warm water, a flood of lava—created an ecological disaster, unhinged the balance of life on the planet. Whatever its cause, the Great Dying was catastrophic, erasing most life on earth within a mere one hundred thousand years. Desolation followed. The recovery, taking five million years, was painstakingly slow.[6]

The Darwinian idea that evolution proceeds in an orderly, predictable way, weeding out the least fit at each round, progressing forward to ever better-adapted species, does not appear to hold up during a mass extinction. Before the Great Dying, no one could have predicted who would live and who would die, who was most and who least fit to survive. The circumstances of mass extinction impose unfamiliar challenges, beyond the demands of day-to-day, year-to-year, or even millennium-to-millennium existence. Reptiles seemed better adapted to the harsh climate of Pangea than amphibians, better adapted to the challenge of living in a dry environment, but in this extinction, three quarters of all reptile families disappeared, and only two thirds of amphibians.

Evolution may indeed select for the fittest, but according to rules that change abruptly, capriciously, unpredictably. What was once flourishing, successful, seemingly well adapted, disappeared—wiped out entirely or reduced to such low numbers that survival was impossible. Widespread colonies of delicate bryozoans, massive meadows of sea lilies, whole reefs of rugose and tabulate corals proliferated in Pangean seas. Today they are unfamiliar relicts of a time gone by, of a past that seems to bear no relation to our present. Their fossils gently warn that under duress, nothing is given, nothing assured, and hegemony is no advantage.

The Great Dying erased whole communities from the sea, wiping the slate almost clean, leaving the waters vacant. Some who survived the wreckage went on to prosper. They grew from the ranks of smaller, marginal forms of life struggling at the edges of once robust communities. The sedentary gave way to the roamers, as sea lilies and lampshells, accustomed to a quiet life on the bottom, yielded to swimmers and burrowers, to fish, squid, and crabs. The passive dwindled; the predators arrived in force. Stationary animals who filtered their food from the water were devastated. Their carbonate shells, slow metabolism, and permeability to carbon dioxide made them more susceptible to the poisonous effects of the gas. Snails and vertebrates better able to adjust to carbon dioxide surges continued their lives more or less uneventfully. No one could have predicted that an ability to regulate carbon dioxide intake would be a saving grace. The Great Dying, by opening vast realms to those whose existence had been more marginal, more tentative, restructured the pattern of life in the ocean and gave the waters their modern countenance.[7]

Pangea itself underwent a transformation as well. Its time on earth drew to a close, and the old continent began to come apart. Plumes of heat born in a mantle still cooling from the planet's fiery birth rose toward earth's surface. The pressure cracked Pangea open, making way for a new sea. Atlantic, earth's second-largest body of water, would grow from a narrow tear in this continent. The opening of an ocean is long and slow; signs of the impending rift are written into the landscape long before the first whiff of salt comes in on a breeze, long before the first water trickles in. Year by year, decade by decade, century by century, millennium by millennium, the firm continent weakens, the resoluteness of hard rock fades. The land swells up, valleys open and fill with lakes, volcanoes form. The land that parted to make way for Atlantic no longer shudders, for this sea is seasoned, but in East Africa, earth's youngest, newest ocean is being born. This nicked continent offers a living, palpable view of how Pangea might have appeared in the early stages of Atlantic's birth. East Africa's singular landscape, reenacting an older dance of continents, breathes life into stone long buried, re-creates the path traveled by Atlantic long ago.

Each year in Tanzania, as the rainy season ends, great herds of zebra and wildebeest migrate through the Serengeti Plains in search of water. Their route takes them along a tear in the earth, a tear that may someday turn into a sea, someday place Nairobi and Addis Ababa on a coast. Tears in the landscape of Africa, visible to astronauts orbiting earth, run from the Zambezi River in Mozambique up through East Africa, through the desert depression of Afar in Ethiopia and into the Red Sea and the Gulf of Aden. In them lie the snowcapped peak of Mount Kilimanjaro, the rich fossil beds of Olduvai Gorge, and a string of sparkling lakes, among the largest and deepest in the world. In them, the land is coming apart.

Forty-five million years ago, when the plains of the Serengeti and the parched shores of Kenya's Lake Turkana were thick tropical rain forest, a plume of heat rose from deep in earth's interior, down near the planet's core. Originating below Atlantic's southeastern basin, the moving African continent dragged it north. Lifted by the plume, Africa rides high above the sea, higher on average than any other continent. Where its pressure cracked the cold, hard crust of the continent, lava flooded out. Thick piles of ash and volcanic rock from the eruptions accumulated in Ethiopia, building a volcanic plateau hundreds of miles wide, high enough to send the Nile River coursing to the Mediterranean. Today lava continues to spread below the surface, undercoating the continent with thick layers of coarse-grained rock. The swollen plume still buoys up East Africa, still stretches and thins the land. Melted rock rising from the plume built Mount Kilimanjaro, Africa's highest mountain. It spewed forth from Ngorongoro, a large volcanic crater just south of the Serengeti.[8]

Today the seething earth finds its expression elsewhere, in other volcanoes, and Ngorongoro is still, its top layers of ash and rock turned to fertile soil, its liquid lava replaced by a freshwater lake. Plants and animals have colonized the once scalding earth and made it home. Two thousand feet (600 meters) below the volcano's rim, flamingos wade in the lake, zebra and wildebeest graze in the grasslands, lions stalk the woods. Closer to the Serengeti is another dormant volcano, Sadiman. Between three and four million years ago, it sputtered ash down upon the nearby savanna. Light rain cemented the ash, hardening the footprints of animals walking by. Before the footprints weathered away, another dusting of ash buried them. They lay preserved, undisturbed, until the heat of the plume raised and stretched the land, once again exposing them to view.

Amidst thousands of animal tracks, tracks of elephant, of giraffe, of mammals long extinct, are the footprints of three hominids. Three animals, no longer fully ape, not quite yet human, walked through the grass in the shadow of the volcano. One walked in the footsteps of the other, the third walked alongside. Perhaps they were a family, perhaps not. Where they were going, and why, we can only imagine. The start of the Laetoli track is eroded away, the end disappears in a nearby gully.

Man's forebears, however they came to walk in the grass, witnessed the very beginnings of a new ocean. They lived within sight of rising volcanoes and walked the restless land. We do not yet know why the first apes came down from the trees and began to walk on two legs. The emergence of the Isthmus of Panama intensified deep-water circulation in Atlantic, chilling Europe, sending down ice that thinned the rain forest, forcing apes out onto the grass. At the same time, tears in East Africa helped break up the thick wooded canopy, and rising volcanoes, blocking rain carried in on westerly winds, helped dry the dense forest and turn it to grassy savanna. The dawn of a new sea not only coincided with the dawn of man, it may have precipitated his arrival.

The continued upheaval in East Africa, the continued shifting of rock, creating valleys, cutting gorges, exposes the record of this distant time, revealing to us the particulars of our own evolution. Not far from the Laetoli footprints is Olduvai Gorge, a cut in the Serengeti made by the stretching earth. Earthquakes have lifted rocks from the shore of an ancient lake into the steep sides of this ravine, exposing the bones and stone tools of *Homo habilis*. Farther north, near the western shore of Kenya's jade-colored Lake Turkana, in a hot, dry stream bed, are stone tools more than two million years old, the oldest to date. The users of these tools lived on an ancient floodplain of a nearby river. Today the land is bare; then it was a grassy shore interspersed with stands of trees, watered by flowing streams. Early man worked by these streams, chipping stone cobbles into flakes. The reassembled cobbles suggest that those who labored here were not randomly smashing rock. Practiced at their trade, they knew the material, understood how hard to hit it and at what angle, and applied the same principles repeatedly and successfully. They honed their skills and fashioned their first tools from clumps of lava, gift of volcanoes created by an emerging sea, another iteration of earth's motion enabling, nurturing life.

Farther north from Lake Turkana, in the valley of the Awash River, in the desert of Ethiopia's Afar, shifting ground has revealed just how these tools were used. As the land is pulled apart, blocks of rock slip down, exposing steep cliffs built from sand and clay deposited in ancient lakes. Buried in the sediment are the bones of antelope and three-toed horses who once lived along the lake and who were butchered with the stone tools of early man. There is a jawbone, cut with a stone flake as the tongue was removed. There is a tibia, cut, chopped, and hammered with a stone tool. There is a thighbone, scarred where the meat was trimmed off, where the bone was hacked to extract the marrow. Whether early man hunted these animals or dismembered them after they had died by other means, we cannot yet discern. It does seem, though, that this shore, unlike the stream bed at Turkana, had no lava cobbles, no ready material for tools. Those presiding at this abbatoir were farsighted enough to bring their own equipment.

The butchered bones of the Awash and the lava tools from ephemeral stream beds west of Lake Turkana are recent finds, uncovered in only the last few years, but the East African Rift is littered with fragments of our evolution. Among the best-known of the fossils released by the stretching land is Lucy, the walking ape, named *Australopithecus afarensis* for the sunken desert of Ethiopia's bleak Afar depression where she was found. Many of the fossils tracing our evolution have been unearthed in the rift valleys of East Africa; where moving earth pulls away before the sea, evidence of our heritage has surfaced. As earth continues to shift and stretch, exposing more and more layers of that distant past, we may come to understand more clearly how our learning to walk, to make tools, and to eat rich, fatty meat and marrow combined to produce a large brain.

Man's early ancestors, aided by the forces of moving rock, evolved at a fast pace. In the last three million years, East Africa has barely ruptured, while we have come out of the forest, brought the land under cultivation, harnessed natural resources to build great cities. Three million years ago, we learned to design basic tools. Today, we are building complex computers. During that same period, Africa has come apart ever so slowly, the two pieces drifting away from each other at the snail-like pace of one quarter inch (six millimeters) each year.

This emerging ocean has preserved and exposed sediment recording the advent of man. Eventually, the slow but inexorable motion of drifting continents will take that record away. As land holding the Laetoli footprints has been stretched, the middle portion has fallen in, in some places by as much as 15.6 inches (40 centimeters). The footprints of Laetoli, the ravine at Olduvai Gorge, the crater of Ngorongoro, the whole of the Serengeti Plains, and the jade waters of Lake Turkana: all may subside beneath an incoming sea, all may become the submerged edges of a rifted continent. The traces of ancient families of hominids who once roamed the plains together, once gathered beside a lake to pound stone into tools or cut apart a carcass, may become separated, to lie buried on opposite shores of a new ocean.[9]

That ocean has not yet emerged. To the north, though, the Gulf of Aden and the Red Sea have separated Africa from the Middle East. The Gulf of Aden came into being ten million years ago, the Red Sea is being born now. The Gulf of Aden is a full-fledged sea; the Red Sea is part continent, part ocean. At its southern end, its floor is oceanic, its rock volcanic basalt. In northern Ethiopia, in the remote region of Afar, where the Red Sea meets the Gulf of Aden, a piece of the Red Sea is exposed. High, steep cliffs bound this desolate area, which encompasses a number of prehistoric sites: Hadar, where anthropologists found the bones of Lucy; Gona, a site containing stone cobbles; and the Awash River's Bouri, where early man began carving meat. Afar, politically part of Africa, geologically belongs to the sea. Were it submerged, the Arabian Peninsula would fit snugly against Africa's high cliffs. It once did, before the land cracked and the Red Sea and the Gulf of Aden split them apart.

In this neat puzzle, only Afar doesn't seem to fit. In this barren place, broad terraces of rock descend toward a depression below sea level. The terraces are faulted and broken, the shifting land has split cones of volcanoes and moved the pieces apart. In the midst of the desert lies one of Africa's most active volcanoes, draped with bands of fresh black lava. The summit of Erta Ale holds a hot lake of liquid rock. The volcano has erupted seven times in the last 125 years, almost continuously since 1967. Its lava is heavy basalt, the rock of the sea, come to fill the space where the continent has thinned and broken apart. Nearby Mount Asmara is made of volcanic glass, of lava chilled by seawater. The land is dry now, but it belongs to the Red Sea. Heat rising beneath Ethiopia raised this piece of land from the water, narrowed the passage between Africa and the Arabian Peninsula to the tight strait of Bab el Mandeb, but seawater once flowed freely here. Dips in the rocky terraces are filled with salt left behind when it evaporated. Seashells and coral line the lava fields.

In the southern part of the Red Sea, magnetic stripes of rock tracking the spread of an ocean have already appeared. Moving north, where the ridge has not yet formed, they grow fainter and disappear. Here the ocean floor gives way to the rock of continents. Just off the coast of Mecca, this rock is cut by deep basins where the ridge will emerge. The deep troughs of the Red Sea are thick with layers of salt, left as the waters of this incipient ocean advanced and then retreated. The salt turns the water to brine, two to six times saltier than the open ocean. Water with such high salinity easily carries high volumes of metal, and the Red Sea's deep basins, hot with the fluids of hydrothermal springs, are laden with millions of tons of copper, iron, and zinc. These metals, carried up from the depths as pieces of continent begin to drift apart, have formed some of the richest deposits in the world.

As the sea grows, the basins will become valleys, and the melting rock now underlying the seafloor will break through and form a ridge of undersea mountains. As the sea widens, the tear has wound its way north, opening the Gulf of Suez and the Gulf of Aqaba, but not yet reaching the Mediterranean. Whatever the ultimate shape of this new ocean, these are its beginnings: a rip in East Africa, the widening waters of the Red Sea. At some point in its youth, Atlantic looked like this, a young and narrow seaway, a small tear in a large continent.[10]

Atlantic still bears the stretch marks of her birth, and from them we can reconstruct her early days. About 200 million years ago, a molten plume of rock, originating deep in earth's interior, ascended through the mantle, carrying heat from fires ignited at the dawn of the solar system. In all likelihood, the plume was also flecked with wisps of primordial helium, a light, evanescent gas that exploded into being in the first minutes of the Big Bang. Though it is the stuff of stars, helium3 is flighty. Most of the He3 once held by earth is gone, having long ago freed itself from the pull of the planet and floated away. What remains is imprisoned down near the core. This vestige of our far-distant past appears occasionally, percolating up through hot springs on the seafloor or rising with deep-mantle plumes. Scientists have found this gas trapped in the solidified rock of plumes tearing other land apart; in all likelihood it also lies within those that opened Atlantic. The blowtorch rising under Pangea melted its way up through the base of the continent, beneath Florida, to the seam stitching North America, South America, and Africa together. Trapped by the continent's hard crust, the head of the long, thin plume mushroomed out, undercoating vast areas of northern Brazil and the eastern United States with a thick layer of lava.

Swollen with heat, the land rose. It stretched, and thinned, and cracked, opening deep valleys. Today, Newark, New Jersey, sits in one of those cracks; the Civil War battlefield at Gettysburg, Pennsylvania, sits in another; the tobacco fields of Connecticut in yet another. These tears in the old continent still rest above the water, while others are drowned in the Bay of Fundy, Georges Bank, and the Grand Banks off Newfoundland. Approximately twenty basins, left from when Pangea first split apart, are strung along the eastern edge of North America. Ten more, once adjacent to these, now lie across the ocean, in Portugal, the French Aquitaine, Morocco, Algeria, and Tunisia. Two hundred million years ago they resembled the chain of lakes lining the rift valley of East Africa today.

As the seasons passed, the valleys slowly filled with sand and silt worn off the high mountains and washed down by streams and rivers. The detritus piled up, layer upon layer, stained to a deep, dark red in the tropical heat. It dried out in the hot sun and cracked, retaining this imprint of aridity when it turned to stone. The rains came and the mudflats were washed in rivers, and after a time the water stayed, pooling in deep lakes. Lake Lockatong, long gone from New Jersey's rift valley, has a modern counterpart in Tanzania's Lake Tanganyika.

Fish lived in the lakes, and vegetation grew along the shore. The fish died and sank to the bottom, along with pollen and spores blown in on the wind. The adjacent land was soggy, and in nearby swamps peat was buried and turned to coal. As time went by, the deep lakes evaporated, reverting to river deltas, and then to playa baking in the dry heat. Returning rains replenished the rivers, refilled the lakes, and each stage—dry mudflat, braided river, shallow pool, deep lake—left its signature as the dust of disintegrating mountains was once again pressed into rock. The cycle, tuning to distant celestial rhythms, repeated itself for hundreds of thousands of years. Like a metronome, the rising and falling lake levels and the swelling and retreating rivers beat out the rhythms of precession and obliquity, the eccentricities of earth's orbit. The record is thick, as much as four miles (seven kilometers) in the Newark Basin. Atlantic now separates many of Pangea's rift valleys, but however scattered they have become and wherever they are, whether flooded by the Bay of Fundy or lifted into the mountains of Morocco, their layers of sediment tell the same story.[11]

Eventually, the hot, molten rock of deep earth melted its way through the continent, releasing floods of lava. Erupting in the rift valleys of Brazil, the eastern seaboard of the United States, Nova Scotia, and West Africa, it spread over 2.7 million square miles (7 million square kilometers), in the most voluminous outpouring of liquid rock recorded on earth. Ultimately, it tore the continent apart, opening a new ocean. The lava floods also coincided with another mass extinction, earth's third largest. The lava floods spewed carbon dioxide, turning the atmosphere into a hothouse. More than half the planet's species were extinguished. In the sea, many bivalves and mollusks disappeared. On land, amphibians and reptiles suffered great losses, opening the way for dinosaurs.

Among the hardest hit were the trees of Pangea's forests, the ginkgoes and palm-like cycads. These trees with large leaves all but disappeared, their demise recorded in the fossil beds of eastern Greenland and southern Sweden. Plants breathe through their leaves, but as carbon dioxide levels rose, the number of leaf pores declined, with deleterious consequences. Summers were already warm in this region of Pangea, at least 86 degrees Fahrenheit (30 degrees Celsius), but as atmospheric levels of carbon dioxide quadrupled, the heat became lethal, the air too hot to breathe. In addition to suffocating, the trees, with fewer leaf pores, couldn't stay cool. They struggled and succumbed. The larger the leaves, the worse the devastation. In this extinction, more than 95 percent of large-leafed species disappeared.

Extinction, closing the door on many species, opened it for others. Flooding lava—pouring into the rift valleys, scorching vegetation, drying lakes—arrived in pulses, often leaving a bed of molten rock hundreds of feet thick. When the burning earth cooled, the rivers returned, the ash-covered basins greened, and sand washed down from the mountains. Then all of it—sand and silt, animal tracks and plant remains, and the marked courses of rivers—was buried in the next volcanic outpouring. The layers of sediment buried in the rift valley running through central Connecticut and Massachusetts contain a chronicle of those who lived and died there. The rock records the departure of reptiles and amphibians whose time on earth had come to an end, and the arrival of others whose tenure was just beginning. It may be that evolution depends on the restless motions of earth, uniting and dividing continents and seas. The volcanic floods that periodically give birth to an ocean may also periodically clean out earth's habitat, emptying the planet's crowded ecological niches, making room for others to prosper.

In the younger sands and muds walked the first dinosaurs, their way made possible by reptiles who had perished. As the stretching earth tipped the valley, tilting the stack of rock on its side and exposing the different layers, thousands of buried dinosaur tracks were returned to the surface. Glaciers ground up the sandstone and streams wore it away, creating the valley's fertile red soil. In 1802, in the Massachusetts portion of the valley, a child plowing his father's field found the first of many rocks covered with dinosaur footprints. Today the softer rock continues to erode, while the more resistant lava stands high above the farmland in a ridge of hills, Mount Tom and the Holyoke Range in Massachusetts, Mount Higby and the Hanging Hills of Meriden, Connecticut. In New Jersey, the steep cliffs of the Palisades, rising from the west bank of the Hudson River, are remnants of lava that opened a continent. From the Palisades, succeeding lava flows are visible in the Watchung Mountains rising to the south and west. Fragments from eruptions that gave birth to Atlantic are scattered far and wide, in the Bay of Fundy, the continental shelf of Morocco, the remote rain forest of northeastern Brazil. It is difficult for scientists to estimate the full magnitude of this massive outpouring, since so much has been buried, worn away, and dispersed to distant places.[12]

The Connecticut valley, the Newark Basin, the Bay of Fundy, and the Mississippi River valley are all tears in Pangea that were never completed. The sites of the final break are buried deep in the continental shelf, well below all the sediment that has since washed in on rivers, and all the plant and animal debris that has drifted from the water's surface in the last 200 million years. The break roughly followed the old suture left from the closing of Iapetus, although some pieces of continent that once drifted on one side of the ancient ocean now are pinned to the opposite shore of Atlantic. Politically and geographically, Florida and South Carolina may belong to North America, but geologically they are African.

The rift valleys severed, North America separating from Africa and South America to open the central Atlantic, the oldest part of the ocean. These continental cracks, which long ago disappeared into the sea, continued to grow wider and wider. Eventually, Tethys trickled in from the east, from Morocco and Gibraltar, a heavy brine seeping over the red sands of Pangea. The water came in fits and starts, spilling in and then evaporating, leaving behind layers of salt hundreds of feet thick. As the mountains of the undersea ridge began to form, newly emerging seafloor filled the space left by separating continents, splitting the salt basin in two. The salt deposits, testament to those early days when Atlantic was a young sea and her waters still tentative, today lie on either side of the ocean, on the Grand Banks of Newfoundland and the continental shelf of Morocco.

Atlantic's basin grew wider and deeper, and the water committed. Salt water, which at first had appeared only intermittently, flooded in and stayed, providing shelter for marine animals. Ammonites, octopus-like animals living in chambered shells, thrived. They grew in many shapes and sizes, as small as coins, as large as serving platters. Their shells were smooth and streamlined, rough and ribbed, spiraled like corkscrews, coiled like rope. Some ammonites were fast and agile swimmers, and others, more clunky, their shells encumbered with knobs and ridges and other ornamentation, lived a more sedentary life on the bottom. Today ammonites have all but disappeared, the nautilus their only living relative, but back then earth's seas were home to at least seventy-five hundred species. Their fossils track Atlantic's early circulation, indicating that the long narrow seaway, resembling the Red Sea, was linked to the world ocean at both ends. Aligned along the equator, young Atlantic received water from both Tethys to the east and what would be known as the Pacific to the west, opening the way for ammonites from both seas to migrate in.[13]

As the ridge grew, buoying the seafloor, water levels rose. Seawater flooded much of northern Europe, and millions upon millions of coccoliths bloomed in the warm, clear water. When continents were drowned, erosion ceased, leaving only the myriad tiny sheaths of these marine organisms to litter the ocean bottom. Their pineapple-ring-shaped plates, visible only with a scanning electron microscope, piled up over millions of years, cementing a pure limestone hundreds of meters thick. When Europe was squeezed to make the Alps, these deposits were gently folded upward, and today a few of them lie exposed, on opposite sides of the English Channel, in the startling white cliffs of Dover and the chalk cliffs of Brittany. Each year the cliffs recede as the soft chalk, fashioned in a warm, tranquil sea, falls away in the rough waves and storms of the English Channel. The cliffs crumble, but coccoliths still bloom in

Atlantic, in the English Channel and the fjords of Norway. There, dense concentrations of coccolith sheaths, one hundred thousand in a few drops of water, mirror sunlight back into the sea, turning it milky turquoise, just as they might have done 100 million years ago.

Wind drove the circulation of young Atlantic, pushing water in from Tethys and out into the Pacific. Rivers washed in nutrients, which welled up at the surface, nourishing an explosive radiation of floating organisms. Diatoms, radiolarians, foraminifera, tiny single-celled dwellers at the surface who would weave the food web of the sea, who would become the source of the sea's bounty, proliferated in the new ocean. Beneath the sunlit waters, though, the deep circulation was sluggish and weak, the bottom poorly ventilated. When plants and animals in the sea died, they sank, and in the oxygen-poor water their remains, rich in carbon, turned to black shale. These shales are buried in the Gulf of Mexico and the continental shelf off Morocco and the Bahamas, windows onto moments in an airless past.[14]

While central Atlantic was growing, another crack opened in Pangea. Lava flooded over southern Brazil, Uruguay and Argentina, and adjacent Namibia. Today, flat reddish brown hills rise from the stark Namib Desert, remnants of the great outpouring of melting rock that opened Atlantic in the south. The lava flood has left its mark in South America as well, in the rich purple soil of the Paraná Plateau, which sustains Brazil's coffee crops, and in the odd drainage of rivers flowing nearby. The plateau, raised by molten rock, has turned its rivers away from the sea. The headwaters of the Iguaçu lie within miles of Atlantic, but the river drains west over the plateau, inland. At the edge of the plateau, it cascades into the Paraná River. Misted by the spray of rushing water, splashed with the light of rainbows, swifts plunge through the falls to nests in this lava cliff built by the architect of continental drift. After leaving the gorge, the river continues through Argentina, detouring south around the plateau before turning back and emptying into the sea at Buenos Aires, its oddly circuitous drainage pattern dictated by a plume of rock ascending from earth's mantle.

The plume of hot rock that wrenched Africa from South America is still active. It now rises below Tristan da Cunha, six small and lonely islands halfway between Buenos Aires and the Cape of Good Hope. The largest island, a volcanic cone rising above the sea, is still racked by eruptions. Although the continents have drifted far from the plume, its shadow still lies beneath Brazil, a narrow cylinder of warmth, 120 miles (200 kilometers) down in the mantle, a quiet relict from a more tumultuous time.

Lava flowed for one million years here, along a long, narrow rift extending up through Brazil and Angola. It tore the rift valleys apart, splitting Pangea, breaking up a chain of volcanoes into what have become the Serra Geral mountains of southern Brazil and the Kaoka volcanoes of Namibia. At the same time, the plume continued to deliver melting rock into the sea, building a high rocky ridge between the drifting lands. The ocean opened from the south, water flowing in from the widening channel between the tips of Argentina and South Africa, but only small amounts trickled north through the ridge. They repeatedly evaporated, leaving thick deposits of salt off the coasts of Brazil and Angola.

The waters of central and southern Atlantic evolved separately, independently. In each instance, a massive outpouring of lava opened a long line of rift valleys, some of which broke apart to create a small ocean basin. Seawater spilled in but then evaporated, spilled in again and then dried up, each time leaving salt to line the basin. The water, when it came to stay, moved sluggishly. In the close, stale sea, oxygen was not replenished, and the remains of those who dwelled there were compressed into layers of black shale. The two basins remained separate until the central Atlantic grew to half its current width. By then the ridge linking Africa and South America had cooled and subsided into the deepening basin, opening the way for water to move north. The two continents separated, and their young seas were joined. Swimming carnivores, huge lizards and jawed fish whose descendants would evolve into piranhas and sharks thrived in this warm tropical ocean. It was a place for predators, for sea monsters, for large and fearsome plesiosaurs and mesosaurs. Coiled oysters and burrowing mollusks hid amidst the long coral reefs and meadows of sea grass, ferreted out by crabs and lobsters who easily crushed their prey and snails capable of drilling through shells to find the meat.[15]

Such prosperity was not to last. Sixty-five million years ago, India and the Seychelles drifted away from Madagascar, releasing another outpouring of lava, another gush of carbon dioxide into the atmosphere. At the same time, a large asteroid, six miles (ten kilometers) across, crashed into Mexico's Yucatán Peninsula. The collision stirred up the sea, flooding the plains at the continents' edge with enormous waves. Tiny shards of glass from the broken asteroid rained down on the planet, and smoke and sulfur darkened the sky for days, for months, perhaps for years. Scientists continue to debate which catastrophic event was more responsible for the mass extinction that ensued.

Dinosaurs who had dominated the continents disappeared, along with large reptiles of the sea. Three quarters of the species that had evolved to thrive in the young ocean became extinct. There seems to be no apparent reason why some survived and others didn't. Mollusks who lived in deep water were no better protected than those who made their homes nearer to shore. Clams burrowing in the sand fared no better than the more exposed scallops and oysters. Many succumbed: those who were large and those who were small, those who filtered their food from the water and those who rooted for sustenance in the detritus left on the bottom. Once again, the tried-and-true strategies for survival that had ensured the succession of generations suddenly became irrelevant under the stress of ecological disaster.

One of the clearest records of the extinction lies in the sea east of Jacksonville, Florida, on the edge of a carbonate platform rising from the seafloor. The platform was built in the young waters of Atlantic after Pangea had thinned, after volcanic eruptions had wrenched it apart. Six miles (ten kilometers) thick, cemented from the skeletal fragments of tiny surface dwellers, it rests above layers of ancient continent and flowing lava. Seventy species of single-celled foraminifera lived in the sea before the extinction. Only three or four survived. Along the Blake Nose, the bank's sloping edge, are places undisturbed by cutting waves, bypassed by eroding currents. Cylinders of mud extracted here illustrate the loss. The gray, older layers, crowded with large ornate fossils of forams, abruptly give way to a green layer, three to eight inches thick, filled with bits of glass rained out of the sky after the comet crashed and exploded. Above the green is a rusted layer of unmelted cometary debris and a concentration of iridium, hard metallic fallout from the comet. The impact, comparable to the detonation of more than one billion nuclear bombs, devastated life in the young sea. For the next five thousand years, the layers of gray mud are virtually devoid of any sign of life.[16]

And yet, earth rebounded from the dark apocalypse as survivors found their niches in impoverished seas and empty continents. In time, the sea would fill again with life, with corals and crabs, sea urchins and sand dollars, walruses and whales. A profusion of flowering plants and grasses would blossom in forest and field, and mammals would replace dinosaurs at the top of the food web. Mammals evolved slowly for ten million years after the extinction, and then suddenly new orders appeared, their arrival assisted perhaps by stirrings in young Atlantic. In the years following the extinction, archaic mammals, who have no living relatives, roamed the basin of Wyoming. Generation after generation, the population changed little, until earth warmed. In the benign climate, crocodiles lived up near the Arctic Circle and broad-leaved evergreens blanketed lands at high latitudes. The warming began gradually, then suddenly accelerated. In only a few thousand years, the sea heated up. Temperatures in the cold depths and polar surface waters surged by an almost unimaginable 10 degrees Fahrenheit (6 degrees Celsius).

A core of mud, laced with methane, partially corroded by carbonic acid, and filled with muddy rubble strewn about in an underwater landslide, identifies the cause. The core also comes from the Blake Nose, where vast reservoirs of methane were imprisoned in ice on the seafloor. As the planet warmed, the ice melted, releasing the gas. One trillion tons of methane burst through the sediment, bubbling up through the water and into the atmosphere. This blast of greenhouse gas released from Atlantic warmed earth even further. In that heat, the progenitors of modern mammals appeared. From them would descend horses and rhinos, cows and camels, antelopes and apes. It is possible that our own primate forebears arrived on earth in this spike of global warming. The new mammals appeared simultaneously in many places—in Europe, in Asia, in North America. Where they originated, no one yet knows, but the warming may have opened overland routes across the Arctic, encouraging their dispersal.

At the peak of the warming, young Atlantic delivered as much greenhouse gas to the atmosphere as humans do today burning fossil fuels. Mud cored from the Blake Nose reveals that global warming in the twentieth century is not unprecedented, that nature herself has nudged the rhythm of carbon dioxide cycling between atmosphere and ocean, altering it as much as we have, and in as short a time. The cores also reveal, though, that earth lived with the consequences of that abrupt, rapid blowout of methane for the next 120,000 years.[17]

Through it all—through earth's collision with a comet, through the sudden release of methane—Atlantic continued to grow. She began to lose her equatorial character when the solid mass of Europe and North America came apart. The same mechanism that split Africa from North and South America, that opened the central and southern basins of the sea, was at work here. Floods of lava bled into the continent as it stretched and tore. Greenland pulled away in two directions, first toward Europe and then toward North America, each time letting in water to separate lands once joined. Both Baffin Bay and the Labrador Sea to the east and the Greenland and Norwegian Seas to the west were born as Pangea unraveled. Undercoating Atlantic in these northern waters is a wedge of lava, five miles (eight kilometers) thick, the mushroom head of a huge plume of rock solidified beneath the seafloor. It stretches between Greenland and Norway, reaching south to the tiny island of Rockall, off Great Britain. It touches the edge of Labrador. It may extend north through the ice; its bounds are not yet marked. Now invisible beneath the waves, the lava erupted in the open air, first bursting through the thinning continent, then filling the rift basins, and finally becoming the floor of a new sea.

Heat determines the quantity of lava, the thickness of the seafloor. Hot mantle rock rising from earth's interior is about 2,440 degrees Fahrenheit (1,340 degrees Celsius). A slight elevation in temperature, an increase of only 10 percent, more than doubles the amount of liquid rock rising up toward a rift. Hot spots, the plumes of rock that break apart the continents, carry that extra heat. A hot spot once rose under what is now São Paolo, Brazil, opening Atlantic in the south and raising the Paraná Plateau, creating the unusual river drainage. Another hot spot opened Atlantic in the north and lifted Iceland out of the sea. The Cape Verde Islands rise from seafloor swollen by a plume whose mushroom head is more than 1,000 miles (1,600 kilometers) in diameter. The plume that built the Azores still makes its presence felt. Although grapevines, hydrangea, and mimosa drape the slopes of dormant volcanoes, heat from the plume still warms the islands' lakes and steaming sulfur springs and keeps the soil hot enough to bake a vegetable and sausage stew.

The Iceland, Cape Verde, and Azores plumes rise from beneath the sea; in the United States a plume rises at the continental divide underneath Yellowstone National Park. The hot spot created a crater more than 50 miles (80 kilometers) wide, and in the past two million years its explosive eruptions have shrouded the western United States in ash. Heat from the melting rock has created the park's geysers, steam vents, and boiling mud pots. It may split North America apart.

Some sixty million years ago, as North America and Europe split open, the landscape of Great Britain may have looked like Yellowstone. Volcanoes spewed lava and ash over Northern Ireland and the Hebrides, covering Skye and Mull and the tiny island of Staffa. Some evidence remains. On the coast of Northern Ireland, across the water from Scotland, is a wide swath of lava left from the eruption. As it cooled, it contracted, and fractured into thousands of flat-topped pillars. Steppingstones to the sea, the Giant's Causeway begins in the misty cliffs and disappears in the water, reappearing on the rainy coast of Staffa, in Fingal's Cave, a recess whose dark and gloomy countenance Felix Mendelssohn captured in his Hebrides Overture. Rising sea levels may have divided Northern Ireland from Great Britain, but geology and myth once joined them. Both lands are of the same continent, and local lore, on both sides of the water, tells of giants building the causeway to link their dens.

As Pangea's northern lands twisted apart, the pieces scattered, some to new homes across the young sea. During the time of Iapetus, Scotland and northern Ireland belonged to North America, their desolate moors and bare rock once an integral part of Greenland and Canada, but as Atlantic opened, they joined Europe. Other pieces of Pangea were lost to the sea. Hidden in waters north of Newfoundland is Orphan Knoll, a shred of North America ripped off, then isolated at sea as Atlantic grew. Between Ireland and Iceland, a piece of granite rises from the water. Rockall Islet, the high point of an underwater plateau about twice the size of Scotland, is built of the rock of continents, wrenched from Pangea when North Atlantic opened, and carried out to sea.[18]

The opening of Atlantic in the north added 3,100 miles (5,000 kilometers) of new ridge to the growing sea. Once it was possible to walk from Newfoundland across Greenland and down through Scotland to Normandy. As the continents drifted apart, Atlantic's waters deepened and widened, and the character of the sea changed, becoming more varied, more complex. As her three basins became one, Atlantic grew from a narrow equatorial seaway to a wide ocean reaching from pole to pole. This rearrangement of separating continents drastically altered the flow of water.

The transition took place slowly, over millions of years. It began in lands far to the south, when Australia broke free and drifted north and when the Drake Passage opened, isolating Antarctica in a ring of cooling water. It continued closer to the equator as Morocco pivoted up toward Europe, restricting the flow of water at Gibraltar. The Tethys spigot was turned off completely farther east as the Arabian corner of Africa slid into Iran, leaving a pocket of seawater that would become the Mediterranean. In the far north, ridges connecting Iceland, Greenland, and the Faeroe Islands sank into the deepening basin, releasing cold Arctic water into Atlantic.

No longer was Atlantic a warm, tropical sea whose bottom waters were subject to long periods of suffocation. Cold, fresh water spilled down along the seafloor from the Arctic, pulling warm equatorial surface water north to replace it. The new circulation, mixing waters deep and shallow, stirred the sea, replenishing the depths with oxygen. Arctic water was cold, but not as cold as the icy currents circling Antarctica. When northern water reached these southern latitudes, it welled to the surface, throwing out heat and vapor to rain ice and snow upon Antarctica. The white reflected the sun's heat, and the temperature dropped further. Thick glaciers grew to cover the once green continent. In the still-widening sea, North and South America shifted position, edging closer to buckle the continental shelf and raise an isthmus that would cut off Atlantic's last link to equatorial waters.[19]

As geographic features go, the Isthmus of Panama is small. Between Panama City and Colón, only forty miles (sixty-five kilometers) of land separate the Pacific from the Caribbean. Yet this seemingly insignificant strip of land profoundly altered Atlantic's circulation and, in turn, earth's climate. The isthmus rerouted warm, salty water flowing across Atlantic from the east, blocking it from the Pacific, propelling it north in the Gulf Stream. Upon reaching the Labrador and Greenland Seas, it lost heat and vapor to the clouds. Colder and heavy with salt, the current sank to the bottom, intensifying the northward flow of the warm Gulf Stream. In the winter, snow from the moisture-laden atmosphere fell down on Greenland, and sea ice clogged coastal waters. In the spring, it all melted, and thawing clumps of snow and ice dumped their loads of stone and grit onto the seafloor. Each year winter melted into spring, until the angle of earth's tilt, which increases and decreases in a regular rhythm, moved to a time of greater obliquity, to a time when the planet felt less heat from the sun. Northern summers cooled. In the colder seasons brought on by variations in earth's orbit, winter snow and ice lingered, accumulating year after year until thick glaciers plunged the planet into an ice age.

The narrow Isthmus of Panama gave Atlantic her modern face, creating the circulation that brought on earth's ice ages. Serving as both well-traveled land bridge and impenetrable barrier, the isthmus changed the course of evolution on land and opened up new habitat in the sea. Before dry land connected the two continents, the mammals of North and South America evolved in isolation, independently. Buried in the Andes, in rock pushed into lofty mountain heights, are fossils of animals who once roamed this lush, grassy island continent. There were sloths, marsupials, and armadillos. There were rodents, whose descendents became guinea pigs and porcupines, and hoofed animals, some as slender as rabbits, others as hefty as hippopotami, and even the skull of a monkey. When the land bridge opened, some animals—parrots and toucans, armadillos and sloths, hedgehogs and opossums—traveled north, but only a few found a safe niche amidst North America's carnivores. Tapir and deer, bear and fox came south. The North American emigrants thrived in the fertile river valleys of their new home, displacing many of those already dwelling there. Today, at least half the mammals in South America are descended from northerners who arrived over the Isthmus of Panama.

The land bridge linked continents, separated seas. As land habitats merged, creating one continent, the ocean realm divided. Animals dwelling in each sea became isolated, and their populations diversified, splitting into different species. The change took place gradually as the isthmus rose higher and higher. Deep-water communities became isolated first, and then those who lived in the shallows, among the mangroves hugging the shore. Snapping shrimp once swam freely between the two seas, once mingled and mated, but as the isthmus rose, they went their separate ways. Caribbean and Pacific populations, once the closest of relatives, have become genetically distinct. DNA sequencing confirms what is apparent in their bellicose behavior. Snapping shrimp ferociously attack all but their own kind, clawing at each other and emitting the loud popping sounds that give them their name. The most aggressive fighting occurs among the most genetically diverse. The Isthmus of Panama has made Pacific and Caribbean shrimp strangers, now more likely to attack each other than to mate.

While the number of mammal families declined as the isthmus joined North and South America, the diversity of marine life increased as the rising strip of land separated seas. The variety of clams and snails in the Caribbean increased by 40 percent. The water, too, became less homogeneous, each sea developing its own distinctive characteristics. Atlantic is saltier, the Pacific fresher. Near the coasts of Central and South America, the shallow waters of Atlantic became depleted while the deeper Pacific grew fertile from nutrients welling up from the bottom. The Pacific's cooler waters came to nourish rich fisheries, and Atlantic's warmth sheltered reefs, nurturing the elkhorn and staghorn corals that evolved after the isthmus rose.[20]

With the elevation of the Panama land bridge, Atlantic became a longitudinal sea, closed off at the equator, open at high latitudes. She became a sea we would recognize, a sea whose currents flow from pole to pole in a long, winding conveyor belt. The flowing water carries heat up from the equator and freshens the depths with oxygen from the surface. Atlantic's eastern and western basins, sculpted by currents, dressed with runoff from rivers, have now developed their own characteristic appearances, with the plains of the abyss larger in some basins, the shelf of the continent broader in others. Yet the foundation below remains the same, a volcanic seafloor built at the ridge and pushed little by little, year by year, out toward the edges of the sea.

Throughout her life, Atlantic, like Iapetus before her, has witnessed the rise and fall of species, the blossoming and extinction of life. Drifting continents and disappearing seas have not merely witnessed and recorded the pulses of life on earth. Restless lands and waters have furthered evolution, periodically separating and mixing species, periodically thinning the ranks to make room for others. The dance of wandering lands and emerging oceans has enabled life to thrive here, recycling nutrients and minerals, rejuvenating waters. Land and water are both alive, evolving. Were their motions to cease, this planet that sustains us might, like the Permian Tethys, stagnate.

Atlantic is now a mature ocean, one of earth's oldest, and like that of the seas that came before, her tenure is coming to a close. In the next few million years, earth's plates will once again shift direction and Atlantic will cease expanding. Her foundation will buckle, rubbing against continents to raise high mountains and descending back into the depths from whence it came. Her waters will drain away. Seen in the larger vistas of time, the span of Atlantic's life, about 200 million years, will be brief. If earth's entire history were collapsed into a day, Atlantic would be present for ten minutes. Like a fish jumping from the water, barely seen, this vast sea has come and will be gone in a flash. While Iapetus graced earth, her basin recorded the evolution of life. When that sea closed and pieces of the basin were welded onto the continents, the record was partially preserved. We can only imagine what Atlantic might preserve of our stay here.

Part 4

FULL CIRCLE

CHAPTER 11

Fraying Edges

I spend as many deck watches as possible out on the bowsprit of the *Cramer,* where I feel closest to the sea. One day, when the sun is bright and high overhead, I catch a glimpse of my shadow, a tiny black dot barely visible in the water, a fleeting nothingness. On time's grandest scales, we humans are evanescent shadows, basking in earth's beneficence for but a moment. It is tempting, each time I find myself alone on a vast body of water, to take comfort in our smallness, our insignificance. Yet there is mounting evidence that we ourselves constitute a great presence, that our passage is hardly as light or clean as that of a passing shadow. Atlantic is not due to begin her full return to earth's depths for another ten or twenty million years, not due to disappear for another 200 million years. Long before her time, however, her edges are beginning to fray. The unraveling begins in the seas and great estuaries along Atlantic's rim, in the Gulf of Mexico and the Caribbean, in Chesapeake Bay and the Baltic, in the Mediterranean and the North Sea.

We *Homo sapiens* have come to exert a persuasive influence here. More than the push and pull of tides, more than the sweep of currents, more than the fiercest northeast storm, our species has become the primary architect of Atlantic's shores and adjacent seas. We are making an indelible mark, which Atlantic will carry until the end of her days. When her waters have drained away to fill other oceans, this mark will remain, written into pieces of her basin strewn ashore. These remnants will record a time in earth's history when a terrestrial species grew strong enough to redesign the web of marine life. From our home on dry land, we have penetrated the many layers of marine ecosystems, restructured the lives of tiny unicellular organisms at the base of marine food webs, and toppled the dominant large predators. In our own image, we have altered entire realms along Atlantic's edges, reshaping mangrove forest, seagrass meadow, and coral reef, altering the temperature and salinity of seawater and even its chemical composition.

We have become a powerful force of nature, at this point perhaps one of the more powerful forces on the planet. We are well on our way toward doubling the amount of carbon dioxide in the atmosphere, and our production of fixed nitrogen already equals that of all other sources combined. We have altered 40 to 50 percent of the ice-free portion of our terrestrial habitat and profoundly transformed our rivers, reigning them in and redirecting their flow for irrigation and flood control, releasing into them billions of tons of industrial, agricultural, and domestic waste. The effects of human enterprise have reached out beyond land into the sea. Our fishing industry has taken out the sea's top predators and substantial numbers of their prey, depleting or exploiting at least half the world's fisheries beyond sustainable limits. Persistent chemical pollutants, disposed of on land, have worked their way into the ocean and up through marine food webs, into the blubber of sperm whales feeding in the deep sea, into the eggs of seagulls living on Norwegian islands north of the Arctic Circle.

Sixty percent of the human population lives within forty miles (sixty kilometers) of the ocean, but all of us live in watersheds draining to the sea. Atlantic's adjacent seas and coastal waters bear the burden. We have added a surfeit of nitrogen, hosts of viruses and bacteria, suites of chemical pollutants, species from faraway places. We have taken out fish, wetland buffers, and shellfish that serve as water filters. The effects accumulate, exceed some threshold, and then begin cascading through entire communities. New symptoms and diseases appear; known diseases suddenly infect new hosts; estuaries suffocate for lack of oxygen. Atlantic's rim is losing its resilience, weakening from what we have taken away and what we have given in return. Already the seabed is recording the changes.

Sometimes the damage occurs slowly, almost—but not quite—imperceptibly. Horseshoe crabs have abandoned the tidal creek near my house. These large, clunky-looking animals have been around for a long time, 200 million years. They survived when others succumbed, weathering the heavy blows of mass extinction, enduring the intense heat of a greenhouse earth, the numbing cold of glacial ice, but now even these stalwarts are disappearing. In only fifteen years, their numbers have dwindled. Each spring, at the highest night tides, they used to come out of the mud to mate and lay their eggs. The creek nursed the young crabs, and they stayed in its embrace all summer. With each molt, they left an unblemished shell on the shore. This year, I found just one shell. Out at the beach, a larger crab, very much alive, overtook my younger daughter one afternoon as she walked in the warm water. It was a cause for celebration.

Gone are the days when an incoming tide would cast a pile of empty crab shells up on the beach, and gone are many of the people who remember. Summer still comes to this tidal creek. The water turns green with blooming plankton, and stripers, following big schools of menhaden, come in to feed. Early in the morning, just as the sun comes up, when wind and tide are just right, the bay roils with fish. It is a beautiful and moving sight, but it still leaves me with a sense of longing. For me, a summer without the horseshoe crabs seems wanting.

Perhaps some natural rhythm in the marsh forced them out, but I don't really think so. More likely, our slovenliness eventually made their home uninhabitable, for their departure coincides with our increasing presence. The decline came ever so slowly, creeping in a little at a time, season by season, year by year. One summer, paralytic shellfish poisoning closed mussel beds. Another summer, green slime crept into the mud near the bridge. Another year, runoff from a summer storm temporarily contaminated the clam flats. A few years later, the shellfish warden closed the flats every time it rained, and a few years after that, he closed them altogether. By the time contamination made the creek unsafe for swimming, horseshoe crabs were long gone.

Little by little, the creek deteriorates. With a change of venue, the decline along Atlantic's edges isn't quite so subtle or gentle. When my husband's parents still lived in Florida, we sometimes ended a visit with an excursion to the Gulf Coast. We'd idle away the last few days of our vacation on Sanibel Island, swimming in the warm water or sitting in the sand, combing through shells. It was on the beach at nearby Captiva that Anne Morrow Lindbergh wrote *Gift from the Sea.* I often wonder whether she could be similarly inspired today.

Our last trip to Sanibel was a sad one; the idyllic vacation we'd planned never happened. We arrived, unpacked, ran out onto the sand, and choked. On this visit, coughing, congestion, sore throats, and teary eyes were our gifts from the sea. Covering our mouths with our T-shirts, we edged our way along the beach, picking our way through dead and dying fish strewn out along the sand. Large mullet with empty eye sockets and baby swordfish littered the beach. Each roll of the tide brought in more dead fish, each break of the waves released more irritants into the air.

Somewhere out in the Gulf of Mexico, currents had swirled some of the sea's smaller inhabitants, tiny single-celled organisms floating on the surface, into a pool of nutrients. They thrived in all the nourishment, multiplying again and again, overpowering their competitors, crowding out their neighbors, blossoming into a swath of chlorophyll hundreds of miles across. As the organisms died, they released a powerful toxin that asphyxiated fish and accumulated in shellfish, rendering them unsafe to eat. Breaking waves lofted minute shreds of these organisms into the wind. An offshore plankton bloom left the beach awash in death and decay. We felt dismay, while others for whom this collapse had become the norm were nonchalant.

All around Atlantic's edges, her waters are changing, her communities giving way to pressures inflicted by the food web's most aggressive and messy predator. A few examples from Atlantic's estuaries, bays, and adjacent seas tell the story. They are illustrative but by no means exhaustive, for losses occur everywhere. Each example emphasizes a different aspect of the problem; the arrival of an introduced species, the burden of untreated waste, the onset of an unknown disease. Together they reveal, sadly, how multiple insults combine to weaken a robust sea.

The Albemarle–Pamlico Estuaries: Hog Manure and Toxic Plankton Blooms

Off the coast of North Carolina, a long chain of barrier islands, the Outer Banks, enclose the shallow waters of Albemarle and Pamlico Sounds. The barrier islands, thin strips of sand, some no more than a few feet above the water, hold the sea at bay, bearing the full brunt of raging ocean storms. Sheltered by the sand dunes, away from the open sea in the calmer waters of the sounds, lie nurseries for the southeast Atlantic's fisheries. Flounder, croaker and menhaden, crab and shrimp all spend crucial parts of their lives in the marshes and estuaries behind the dunes. These days the calm comes at a price, for the Outer Banks are a two-way barrier, keeping out the open sea while holding in animal waste and other pollution washed down from North Carolina's rivers.

A hog produces, on average, four times the waste of a human, a fact that might not seem to matter except that in the last fifteen years, hog production in the United States has increased while the number of hog farms has dropped by two thirds, concentrating production on large industrial farms. Many are located in North Carolina, where today more hogs live than people, and where manure, once a valuable fertilizer, has now become a pollutant. It used to be that a farmer raising a few pigs fertilized his fields with manure. Today's industrial hog farms produce tons of manure, much too much for any farmer to use. Clean water laws require treatment of human waste, in septic systems or sewage treatment plants, but farmers simply hose hog waste, with its equally high concentration of pathogens, into pits and then spray the liquid onto fields, where it drains off into nearby streams. In North Carolina, these cesspits, some larger than a football field, hold millions of gallons of waste. Some rest in the floodplains of rivers, others lie in or near the water table. When the pits leak or spill, or when heavy rains flood the low-lying land, the contaminants wash away, occasionally leaching into drinking wells, more often spilling into rivers flowing to the sound.

Dense schools of menhaden feed in the Albemarle–Pamlico estuaries. In 1991, one million menhaden suddenly died, poisoned by a bloom of toxic plankton. A tiny unicellular organism, *Pfiesteria,* was the agent. *Pfiesteria* takes many different forms over its complex life cycle, developing quickly from cysts lying dormant in the mud to active swimmers that inject lethal toxins into the water when menhaden are passing overhead. The fish, once attacked, grow lethargic and quickly die. For years *Pfiesteria* apparently dwelled harmlessly in the waters of the sound, resting dormant in mud along the bottom or floating through the water, making energy from sunlight with chloroplasts stolen from tiny floating organisms they consumed, a behavior referred to in scientific journals as kleptochloroplastidy. In 1991, scientists first noticed that the organism had become toxic to fish. Since then, it has killed millions of fish in North Carolina estuaries. *Pfiesteria* tends to bloom in slow-moving waters rich in nutrients from sewage or animal waste. In 1995, 25 million gallons (95 million liters) of hog waste accidentally spilled from a North Carolina swine cesspit, and *Pfiesteria* bloomed a week later. Ten thousand fish died, and shellfish beds were closed.

Within a few years, toxic *Pfiesteria* appeared outside the Albemarle and Pamlico estuaries. In 1997, a *Pfiesteria*-related fish kill occurred in Chesapeake Bay's Pocomoke River, in Maryland. The 100 million chickens raised each year in the crowded poultry houses of this watershed produce twenty-five times more waste than their human neighbors. In the last thirty years, the number of broiler farms in the United States dropped by one third while production tripled. The consolidation took place on the eastern shore of Chesapeake Bay. On the peninsula shared by Maryland, Delaware, and Virginia, chickens are big business the way hogs are in North Carolina. Here industrial farmers produce 600 million chickens and one and a half billion pounds of manure each year. Ultimately, it drains into the sea.[1]

Pfiesteria is only one of many toxic organisms blooming in Atlantic's estuaries, their growth fueled by massive amounts of nutrients washed into coastal waters. Red tides, as they are popularly called, have a long history: they gave the Red Sea its name, and, in the story of Exodus, inspired Moses to kill the fish of the Nile by turning its waters to blood. Red tides appeared along coastal Florida long before the Spaniards arrived. In the last twenty years, they have become ubiquitous around Atlantic's edges. Before 1970, they were an occasional presence; now they persist. On Florida's west coast, red tides that once lasted a month or two or three now drag on and on, sometimes for more than a year. Once, red tides were a local phenomenon; now they thrive in distant waters. Blooms once limited to the Gulf of Mexico have spread to North Carolina; blooms of other toxic algae once associated with northern Maine and Canada now appear regularly in Massachusetts and seem to have slowly worked their way down to Long Island and New Jersey. In Europe, harmful blooms, rare before 1980, now appear with increasing frequency in the fjords of Norway, the estuaries of Great Britain and Ireland, the coasts of Brittany, Normandy, and Portugal, and the Côte d'Azur.

When red tides first appeared, they were news, but as the years went by, people came to accept them as part of the ecology, a regular feature of the seascape as spring arrives and the water warms. I barely remember the days when Gloucester had no red tides, had no seasons when sea clams were off-limits for days, sometimes weeks at a time. This disturbance slipped so easily, so quietly into the rhythm of life in the bays and tidal rivers, it somehow seemed normal. But now red tides are growing, in number, in extent, in severity; their increasing presence is a disturbing change in Atlantic's coastal waters. In 1984, scientists had identified twenty-two single-celled dwellers of the sea that cause red tides, but by 2000 the number had quadrupled. It is still increasing. Two new Atlantic species were recently added to the list, one from Chesapeake Bay and the Albemarle–Pamlico estuaries in North Carolina, and the other from Europe. Cultivated mussels from Ireland, England, and Norway now harbor a whole new family of toxins, never observed before, which can persist in shellfish for as long as eight months.

Often the organisms responsible are tiny dinoflagellates, single living cells invisible to the naked eye, equipped with two whiplike tails to propel them through the water. Some dinoflagellates are plantlike, manufacturing energy from sunlight. Others, like *Pfiesteria,* are consumers, engulfing single-celled plants and attacking and preying on fish. Although dinoflagellates once reigned alone in the kingdom of toxic blooms, they now have company. Diatoms, tiny single-celled organisms encased in a shell of silica and lacking the mobility of dinoflagellates, float through coastal waters, blooming profusely in spring and summer, providing nourishment for the sea's grazers. Diatoms were considered benign until 1987, when scientists first detected their toxicity. The toxin accumulated in blue mussels, poisoning one hundred people on Prince Edward Island, four of them fatally.

Harmful blooms have become a public health hazard. Having poisoned the water, we are being poisoned in return. Eating contaminated clams, mussels, oysters, or scallops harboring the toxins can cause a host of frightening diseases, with symptoms that can include dehydration, nausea, dizziness, tingling and numbing of the lips and face, paralysis, spleen and liver damage, and death.

While health boards prohibit human consumption of shellfish from waters contaminated by red tides, the shellfish themselves, and other animals dwelling in coastal waters, are still affected. Shellfish develop lesions in their shells. Blooms inhibit their feeding and reduce their ability to reproduce. A toxic bloom begins with a tiny single-celled organism at the foundation of marine food webs, and its effects travel through to the top predators. One year, toxic blooms off the Florida coast killed over 150 manatees, as well as hundreds of birds, brown pelicans and cormorants that ate fish harboring the poison. Another year, fourteen humpback whales feeding off the coast of Cape Cod suddenly died after ingesting mackerel laden with toxins from a bloom.[2]

A multitude of toxic organisms are emerging in the ocean. Some are truly new. For others, like *Pfiesteria,* only their increased toxic activity is new. A diatom now toxic at certain times of the year in eastern Canada had floated harmlessly in the Gulf of Mexico for at least one hundred years. In 1993, *Cryptosporidium,* a tiny unicellular organism living in rivers and lakes, became sufficiently concentrated to poison Milwaukee's water supply, sickening four hundred thousand people and killing one hundred. Now scientists report finding this organism in the marine environment, where it accumulates in the gills and circulatory systems of Chesapeake Bay oysters, presenting a health risk to consumers.

Cholera lives in the Chesapeake as well. People once believed that sailors and ballast water brought this disease from port to port, but cholera gets about on its own. The pathogens live in the warm waters of coastal bays and estuaries, persisting for many years, hardy enough to travel on currents to distant shores thousands of miles away. When conditions are right, they multiply profusely. One thousand cells make an infectious dose, and as many as ten thousand cells can stick to a single copepod. When cholera is abundant, swallowing only one copepod is all it may take to contract the disease. In 1991, a massive outbreak occurred almost simultaneously along the coast of Peru, quickly spreading through South and Central America.

It was an outbreak nourished by warm waters from the sea and heavy rains washing nitrogen off the land. *Cryptosporidium,* too, thrive in nutrient-rich waters; in the Milwaukee outbreak, runoff from dairy farms may have fueled the population explosion. *Pfiesteria* blooms in the Albemarle–Pamlico estuaries coincide with major increases in swine production. The appearance of toxic diatom blooms near Prince Edward Island was linked to rising water temperatures and nitrogen discharges. Elevated nutrient levels seem to have contributed to the recent spate of harmful blooms in the coastal waters of Norway and Denmark, Brittany, and Great Britain. Every year, Italy's Gulf of Trieste fills with algal blooms triggered by nutrients flowing in from the Po River. On both sides of Atlantic, burgeoning populations of toxic organisms seem to thrive in waters sloshing with nutrients, nutrients provided in abundance by man.[3]

The Gulf of Mexico Dead Zone: Farm Fertilizer Floods the Seafloor

On certain late-summer nights, on the Gulf Coast shores of Alabama and Louisiana, it is possible to go fishing without a boat or a net or even a line. In the hours just before dawn, when the tide is rising, and the sea is slick and calm, and the wind is stirring ever so slightly, droves of flounder, shrimp, and blue crabs leave their homes in the sea and crowd onto the shore, where people grab them by the bucketful. Jubilees, these events are called, but for the bay's marine inhabitants this is no occasion to rejoice. Out in the bay, wind and current are pushing a pool of stale water toward the beach. Fish and crustaceans, trapped as oxygen-poor water moves in upon them, are forced ashore, gasping for breath. Each jubilee lasts only a few hours. When the sun rises and the tide ebbs and the wind shifts, the low-oxygen water drifts back out into the bay.

Further offshore, out in the Gulf of Mexico, a similar event takes place, but on a much larger scale. Each summer, when the wind stills and storms die away, fresh water pouring in from the Mississippi River fails to mix with salty water in the Gulf. Instead, it sits there, a layer of warm, fresh water riding above a cold, dense sea. The enriched river water is manna for the sea's floating plants, fueling bloom after bloom. The blooms harm without being toxic; their frequency and intensity alone impair the health of seafloor inhabitants. Copepods graze the blossoming plants. Their waste, along with detritus from the dying plants, sinks and decomposes, depleting bottom waters of oxygen. Without wind or wave to stir the water and replenish the supply, bottom communities can't breathe. As fish and crabs and shrimp begin to suffocate, they leave. Those who lead sedentary lives—clams, mussels, and starfish—can neither escape nor cope. Their metabolism slows, their activity slows, and soon they cease to feed. Animals who live beneath the seafloor, buried in the mud and sand, leave their burrows. Standing on the tips of their arms and legs or stretching their siphons up into the water, they reach for waters where oxygen is more plentiful. Most often they fail, and, failing, perish.

Each summer, when its waters become layered, a "dead zone" appears in the Gulf of Mexico. Depending on the year, it can stretch from the mouth of the Mississippi River west across the Texas border, covering as much as 7,800 square miles (20,000 square kilometers) of water—an area about the size of New Jersey. Established communities within its boundaries disappear, replaced by black mats of malodorous sulfur-oxidizing bacteria that thrive in the absence of oxygen. Those that live closer to shore are affected as well, particularly young shrimp maturing in the nurseries of coastal marshes. When the call comes to leave, to head for the deeper waters where they spawn, their route may be impassable.

If there were a terrestrial equivalent of the dead zone out in the open air, visible in broad daylight—if each summer, for example, every worm, bug, frog, salamander, and other animal living its life close to home suffocated—those responsible would be immediately identified and called to task. In the Gulf of Mexico, the summer dead zone lies out at sea, out of sight of land dwellers. Identification of the factors that have lead to its existence has been a long time in coming, and in the twenty-five years it has taken to develop recommendations to address the problem, the dead zone has been expanding. In some years, its waters have doubled in size.[4]

The problem lies onshore. In my own neighborhood, it is easy to walk the watershed. The hills are low, flanked by seasonal streams, mere trickles in summer, gurgling brooks in spring. When alewives come in to spawn, I can trace their path in a matter of minutes. From the highest hill, it's a short walk through the woods to the sea. The Mississippi River watershed covers 40 percent of the lower forty-eight states, draining the heart of the nation's farmland, draining corn and soybean fields from Iowa and Illinois, Indiana and Ohio. Its breadth is hard to imagine, but the same principles apply: water runs downhill, rivers flow to the sea. Farming practices out on the plains, though many miles from the ocean, nonetheless affect water quality in the Gulf of Mexico. Agricultural crops absorb less than half the nitrogen they receive. Irrigation and rainwater carry the rest away, into streams and rivers flowing to the sea. Nitrogen washing off agricultural fields in the nation's breadbasket covers great distances, more than 1,500 miles (2,500 kilometers), to reach the Gulf of Mexico. It travels well; much arrives intact.

Once fertilizer is washed into a river, how much arrives in the Gulf depends more on the river's size than on its distance from the sea. In a shallow river, such as the upper reaches of the Missouri, where a greater proportion of water touches the riverbed, plants and animals living on the bottom use more of the available nitrogen. Where river channels deepen, as in the Ohio, Tennessee, Mississippi, and lower Missouri, less water and less nitrogen reach bottom dwellers. Consequently, the nation's mighty rivers carry more of their nitrogen load downstream than smaller rivers much closer to the coast, washing as much as 90 percent into the sea. The distant reaches of the Mississippi watershed contribute heavily to nutrient enrichment in the Gulf of Mexico. Half the fertilizer entering the Gulf originates above the junction of the Ohio and Mississippi Rivers, and the Ohio itself contributes another third. Farmers from the Midwest, tending their fields, are also, whether they know it or not, tending to the lives of shrimp in the Gulf.

More than one and a half million tons of nitrogen pour into the Gulf of Mexico each year from the Mississippi–Atchafalaya River Basin. Half comes from fertilizer washing off fields. Wetlands soak up some nitrogen; they could soften the blow. But swamps and marshes are mostly gone, filled or drained to build cities and farms and highways. Eighty percent of the wetlands in Ohio, Indiana, Illinois, and Iowa have been destroyed, and with them, millions of acres of buffer. Half the coastal wetlands along the Gulf of Mexico have disappeared as well. In Louisiana, the Mississippi strains its banks, but levees hold its waters back, starving coastal marshes of water. Diverted rivers and dammed tributaries, levees, navigation canals, and oil and gas pipelines have all resulted in the loss of Louisiana's wetlands, at a rate that is one of the highest on earth.[5]

As farming practices intensified, the amount of nitrogen flowing into the Gulf of Mexico tripled, creating the dead zone. Sediment cores from the Gulf of Mexico link upland farming practices to dying waters, showing that dramatic increases in phytoplankton remains coincide with dramatic increases in fertilizer use. Today single-celled floating organisms bloom profusely in the plume of the Mississippi River when the river rises and its nutrient load is high. Nourished by fertilizers, carbon production of some species of single-celled organisms has almost doubled. However, the wealth of additional energy does not seem to reach larger, longer-living members of the food web. Instead most of it falls to the bottom, depleting the water of oxygen as it decays, extending the reach of the dead zone.

Copepods do not swarm in to feed on these fertilized meadows. In the midst of burgeoning food supplies, their populations fall, from more than 75 percent of the grazers to less than 30 percent, perhaps because some of the blooms are not particularly palatable. With fewer copepods, the numbers of larval fish also fall. The sea may be green, but neither copepods nor the larger predators, the shrimp and menhaden, seem to benefit. Instead, the arrival of the summer dead zone shortens the rhythm of life within its waters. An ecosystem of small, opportunistic species with shorter life cycles replaces communities of slow-growing, long-lived animals fed by grazing copepods and larval fish. In the dead zone, diversity, abundance, and biomass decline as larger animals disappear, as premature death or loss of habitat inhibits successful reproduction. As summer ends and autumn winds mix the waters, replenishing the oxygen and dissipating the dead zone, the animals return. Communities begin to form once again, but the time between the reprovisioning of autumn and the stagnation of summer may not be sufficient for bottom-dwelling communities to recoup their losses. If low oxygen levels persist, or recur on a regular basis, the structure of the community changes, and established communities disappear.[6]

The Baltic Sea: A Dead Zone Writ Large

The Baltic Sea has had an erratic history. It came into being ten thousand years ago when the ice ages ended, as a salty inlet of Atlantic. As the ice continued to draw back, the continent, relieved of the glacier's heavy weight, rebounded. The land rose, cutting off the Baltic's saltwater connection, turning the ocean inlet into a freshwater lake. The land continued to rise, but unevenly, higher in the north than the south. Seventy-five hundred years ago it tipped the basin of the lake, rejoining it to Atlantic. An uneasy truce between fresh and salt water created earth's largest body of brackish water. Today fresh water pours in from the Oder and Vistula Rivers, and a narrow connection to the North Sea between Sweden and Denmark, the Kattegat, supplies the salt. Baltic waters are sluggish, residing in the sea for decades before they are refreshed with water flowing in from the "cat's throat." The Kattegat has proved a fickle purveyor of seawater. Every few years, under certain weather conditions, it sends a pulse of salty, highly oxygenated seawater along the bottom, renewing the deep Baltic. In recent years, this infusion has not quite sufficed. Initially it failed to penetrate 7,800 square miles (20,000 square kilometers) of deep Baltic water, an area the size of the Gulf of Mexico dead zone.

For a while the periodic surges of salt water did nourish rich communities of life throughout the rest of the sea. Thick growths of bladder wrack, a large seaweed kept afloat by small air bladders attached to slender fronds, covered the rocky floor. Amidst the seaweed grew communities of mussels, snails, and shrimp. The abundance of food nurtured dense schools of herring, cod, and sprat and large numbers of salmon. It was a sea of plenty, sustained by a tenuous but adequate supply of oxygen. Beginning in the 1950s, this precarious balance tipped as industrialization, urbanization, and intensive farming took their toll in the Baltic's watershed. PCBs, DDT, toxic metals, radioactive waste, and nutrients all washed into the sea. While the eighty million people living in the Baltic's drainage area have controlled a number of these pollutants, nitrogen loads from chemical fertilizers, from sewage, and from the burning of fossil fuels have tripled and continue to rise. Each summer, huge meadows of plankton mottle almost the entire sea in an unpleasant wash of brown and green, creating the most persistent, extensive plankton blooms Baltic scientists have ever seen.

Phytoplankton production in the Baltic, fed by a glut of nitrogen, has doubled in the last twenty-five years, doubling the debris sinking to the bottom without a commensurate increase in oxygen to decompose it. Now, all the oxygenated water coming in from the North Sea is used up long before the next pulse arrives, creating about 39,000 square miles (100,000 square kilometers) of low-oxygen water along the bottom of the Baltic, an area equal to one third of the sea. Here, oxygen starvation is a persistent condition, no longer dissipating with the changing seasons, the freshening of wind, or the influx of salt water. Here, massive plankton blooms and a lack of oxygen have impoverished life on the bottom. Fast-growing strands of green and brown algae, fed by the abundance of nutrients, cloud the water, shading the bladder wrack. Unable to capture sunlight in the increasingly dark and turbid water, it can't grow as deep as it once did; today the deepest fronds lie ten feet (three meters) closer to the surface.

As the bladder wrack thins, tiny crustaceans, snails, and shrimp disappear with it. Along the bottom, toxic hydrogen sulfide forms after the last breath of oxygen is taken, and any animals who remain disappear. Animals no longer live in the deeper waters of the central Baltic, and cod no longer reproduce there successfully; their eggs cannot survive in oxygen-starved water. A lack of oxygen chased herring from waters outside Helsinki, cod from the coastal waters of Poland, salmon from rivers draining into the sea. It killed millions of shellfish in the Kattegat, where the occurrence of low-oxygen waters has eliminated bottom-dwelling communities. Norway lobsters, once plentiful, have disappeared from affected areas, their dwellings replaced by mats of sulfur bacteria.[7]

Atlantic's coastal waters suffer from too much of a good thing. On land, crops would not grow so bountifully without nitrogen, essential building block of life, critical element in amino acid and protein synthesis. Often it seemed as if we could never get enough of it, for usable nitrogen was hard to come by. Nitrogen molecules, two atoms held tightly to each other, are quite stable, so although nitrogen fills 78 percent of earth's atmosphere, plants absorb very little of it. What they do assimilate has been broken down by lightning and a few kinds of specialized bacteria. When a plant or animal dies and decomposes, its nitrogen is recycled—some returned to the atmosphere, some turned back into nutrients to nourish a new generation of life. A delicate balance of nutrients, of nitrogen and phosphorus, of silica and iron, sustains and nourishes the food webs of the sea.

Human dwellers in the uplands have introduced a staggering amount of nitrogen into the equation. Earth's population of humans, three billion in 1960, had doubled by the turn of the century. To feed these growing numbers, agricultural production also doubled, and the use of nitrogen fertilizers quintupled. Agriculture is the largest but by no means the only source of additional nitrogen onto the planet. Each year, humans add over 150 million metric tons of fixed, or usable, nitrogen to earth's nitrogen pool. Eighty million metric tons come from fertilizer, 40 million from the production of soybeans and alfalfa, which harbor nitrogen-fixing bacteria in their roots, and 30 million metric tons from fossil-fuel combustion in power plants and motor vehicles. Living our lives in the style to which we have become accustomed, we have doubled the amount of fixed nitrogen produced on earth each year; our contribution now equals that of all other sources combined. We are now the major contributor to earth's nitrogen cycle, adding a potent greenhouse gas to the atmosphere, creating smog and acid rain, thinning earth's protective ozone cover, and now reconfiguring the cycles of life in our estuaries.[8]

The consequences, though unintended, are damaging. Large amounts of anthropogenic nitrogen now leak into Atlantic all around her edges, drastically altering the rhythm of life in coastal waters. Each summer, the symptoms—high levels of chlorophyll, low levels of dissolved oxygen, nuisance or harmful plankton blooms—appear in more than half the estuaries of the lower forty-eight states of the United States. Seasonal bouts of oxygen depletion occur all along the western edge of Atlantic, from Cobscook and Casco Bay high in Maine, down through Long Island's Great South Bay, through the Chesapeake and North Carolina's Albemarle–Pamlico estuaries, around into Florida Bay and over to the Mississippi River Plume, and into the Texas bays of Galveston and Laguna Madre. Scientists predict that by 2020 the problems will worsen in many estuaries.[9]

The eastern edge of Atlantic also gasps for breath, not only in the Baltic and Kattegat but also in the North Sea, in the estuaries of Great Britain, the bays of Ireland, the fjords of Sweden and Norway. There nitrogen pollution is taking a new form as a burgeoning aquaculture industry substitutes for decimated wild fish populations. Between 1993 and 1998, nitrogen discharges from Norwegian fish farms increased by 45 percent; they are now the country's largest nitrogen polluter. The circle goes round and round. Emptying the sea of wild fish, we fill coastal waters with waste from fish we now must grow. Elevated nutrient levels then stimulate more plankton blooms, which release toxins and deplete coastal waters of oxygen, killing salmon in aquaculture pens and entire communities of life on the seafloor, further eroding the richness of the sea.

It is a sorry cycle. Along the coasts of Germany and the Netherlands, where the Rhine, Weser, Ems, and Elbe Rivers empty the waste of Europe into the North Sea, nuisance *Phaeocystis* blooms occur regularly. Few are the grazers who can crop these runaway algae; their thick, foamy masses clog the nets of fishermen and pile up on beaches. Decaying, they consume oxygen, and when it is used up, black, foul-smelling spots appear in the sand. Sea worms die, and birds go elsewhere to feed. In 1996, the black spots covered a full 10 percent of sandy tidal flats in

Germany's portion of the Wadden Sea. Further offshore, *Phaeocystis* blooms can kill cod larvae and force older, larger fish out into deeper waters, away from their feeding grounds.[10]

Atlantic's blooms persist in the summer when seas are sluggish, then dissipate in the fall when rising winds mix the water and disperse the nutrients. Where the sea suffocates, bottom communities die. Where waters are refreshed, communities return. How strong and stable they can be depends on the length and severity of oxygen deprivation. If, year after year, young communities are repeatedly felled, they will never fully recover unless a time comes when they can breathe in peace. That time cannot come until we choose to halt the flow of anthropogenic nutrients into the sea. It has taken us years to sap life from the edges of Atlantic. It will take years to resuscitate her dying waters, but it is possible. When nitrogen enrichment ceased in one Swedish fjord, the number of plankton blooms declined, leaving oxygen in the depths to decompose the mass of organic material accumulated from season after season of intemperate growth at the surface. When, finally, it was gone, broken down and recycled, there was oxygen to support the animals who had once lived there. After eight years, a stable bottom community was reestablished. In other estuaries, restoration will not be so easy or so fast. In estuaries starved for oxygen, entire ecosystems collapse, one loss cascading after another, one decline leading to many others. When the damage is great enough, repairing one strand is insufficient; the whole web must be rewoven.[11]

We don't have forever. Over half the fresh water draining off dry land flows into Atlantic. Compared with freshwater runoff into other seas, this amount is disproportionately large: six times more fresh water empties into Atlantic than into the larger Pacific. Atlantic's well-being is ultimately affected by the quality of water she receives. In the northern portions of Atlantic, the Mississippi and North Sea watersheds have lavished coastal waters with nutrients, upsetting marine food webs, causing plankton blooms, depleting bottom waters of oxygen. In both these cases, small flows of water create large problems: the

Mississippi contributes only 4 percent of the freshwater flow into Atlantic, the North Sea basin 3 percent. The Amazon watershed, on the other hand, contributes 50 percent of the fresh water, but only 1 percent of the population. Every day, however, more rain forest is felled, converted to grazing lands for cattle, to fields for crop cultivation, to highways and airstrips to service Brazil's interior, and to pavement for rapidly growing cities. If Brazilians repeat the mistakes of their sloppy northern neighbors, the potential for trouble is great: the plume of the Amazon reaches far into the sea.[12]

The North Sea: Chemical Contaminants Enter Marine Food Webs

In the North Sea, off the coast of Germany and the Netherlands, lies a long string of barrier islands and sandbars. Sheltered behind them is the Wadden Sea, and its vast expanse of tidal flats and salt marshes. Here, millions of birds, on their way to breeding grounds in Arctic Greenland and Svalbard, stop to feed, fattening up for long, exhausting flights. Between six and nine million birds migrate through each year, pausing on the sands and marshes of the Wadden Sea, the flyway's most important staging area. They come from as far away as Siberia and South Africa. Terns, ducks, oystercatchers, plovers, knots: all depend on the Wadden Sea to replenish their energy reserves. It provides an ample repast. The marshes, the largest expanse of tidal wetland in Europe, are a nursery for herring, plaice, and sole. The sea and its mudflats are, or were, a luxurious home for oysters, cockles, and mussels.

Under the name of reclamation, the Wadden Sea has been whittled down. Half its marshes have been drained to build agricultural fields, shipping channels, and sea barriers. Reclamation is an odd term for this activity, which could more accurately be described as taking, not taking back. The destruction is ongoing, as the Ems, Elbe, and Weser estuaries are diked and deepened. Wetlands disappear, and with them other living resources of the Wadden. Overly enthusiastic fishermen take two thirds of the North Sea fish stocks at unsustainable rates, and they are depleting inshore waters of oysters and mussels as well. This level of plunder has a price. As the riches of the Wadden thin, fishermen turn to cultured mussels, but culturing requires larvae, or spat taken from the wild. Now fishermen empty wild mussel beds not only of mature mussels but of spat as well, jeopardizing the future of their own industry.

Eventually, the sea may reclaim what has been taken, as rising seas and swollen rivers threaten farms and cities built on borrowed land. Over one third of the Netherlands lies below sea level, and the water, held back for so many years, still approaches. Tidal marshes and wetlands could soak up waters spilling over riverbanks, but straightened rivers with paved floodplains leave no room for floodwaters, and dikes and embankments can only be built so high. After hundreds of years of pushing back the sea, of containing the flow of rivers, the Dutch may be ready to return some of what they have taken, to occasionally breach some of their dikes, and to give certain rivers more freedom to wind their way to the sea.[13]

We take, and we also give. While a massive effort to clean up the Rhine has lowered levels of toxic cadmium, mercury, and copper in the North Sea, while nuclear reprocessing plants have dramatically reduced radioactive discharges and cities their dumping of sewage sludge, many problems remain. Nitrogen discharges from agriculture, aquaculture, and motor vehicle emissions continue, but nitrogen is only one of many pollutants. Even though dumping is prohibited, large quantities of plastic still litter the sea. Even though their production is banned, PCBs continue to pollute, along with TBT, a chemical used in antifouling paint on ships. For each chemical whose discharge is regulated, there are many, many others whose effects are still unexamined, whose release is unregulated. DDT is banned, but flame retardants, plasticizers, synthetic musks, industrial detergents, and dioxins flow into the sea instead.[14]

Animals feel the effects. Birds are strangled by plastic sheeting, turtles eat it and die. Dog whelks, usually one sex or the other, become both in the estuaries and coastal waters of the North Sea and the Kattegat, along Portugal and Spain, Great Britain and Ireland, where they are exposed to the endocrine-disrupting effects of TBT. This giving is also a form of taking. Destroying marshes and overfishing takes homes and their inhabitants, but polluting the sea also takes the health and well-being of those who remain. Liver tumors are beginning to appear in North Sea flatfish, and PCBs accumulate in mussels, birds, and marine mammals. Herring from the Wadden and Baltic Seas are laced with PCBs and DDE, a derivative of DDT. Time and time again, in study after study, seals feeding on polluted herring prove less healthy, less robust, than seals eating cleaner food. Pollution from the Rhine flows into the western reaches of the Wadden Sea, where seals, whose bodies contain PCBs, have difficulty reproducing. In a two-year experiment, seals fed polluted herring from the Wadden Sea had significantly lower rates of reproduction than seals fed healthy herring from the northeast Atlantic.

Seals riddled with PCBs also have compromised immune systems. Those feeding for more than two years on polluted Baltic herring have higher white-cell counts and more difficulty fighting infections than those that eat cleaner herring from the open Atlantic. When the threat of infection arises, they are vulnerable. In the past twenty years, viruses have taken record tolls on the lives of marine mammals; hundreds of diseased dolphins have washed ashore along the Mediterranean, the Gulf of Mexico, and the east coast of the United States.

In 1988, distemper killed eighteen thousand seals in the North Sea, a mass dying whose roots lay ashore. The epidemic began in Greenland, with sled dogs. The huskies died, and their bodies were thrown into the sea, where they infected the local harp seals. At the same time, human fishermen were drawing down the fish stocks, forcing the hungry seals south in search of food. En route, they infected their North Sea neighbors, seals whose immunity had already been lowered by their diet of PCB-laced herring. Dead seals carried high levels of PCBs in their bodies. Had they been living in cleaner waters, perhaps they would have been stronger, more able to resist the virus. Had the Greenland harp seals' food stocks been more plentiful, perhaps fewer North Sea seals would have been exposed. Had dead huskies not been tossed into the sea, perhaps the disease would have stayed on land. In each case, human activity induced the stress. One assault might have been manageable; together, the multiple threats proved catastrophic.

Our shared domain extends far beyond the tide line, far beyond the inland reach of waves. Postmortems on porpoises washed ashore in Great Britain since 1990 continue to show a link between organochlorines and death from infectious disease; PCB levels are three times higher in diseased porpoises than in healthy porpoises caught in fishing nets. In 1999, seals experiencing respiratory difficulty stranded on the coast of the Netherlands. They were suffering from influenza, from a strain humans had contracted four years earlier. Before the 1995 human outbreak, there was no evidence of this flu in seal blood samples. Its appearance is one more manifestation of how porous is the boundary between land and sea, how blurry the edge.[15]

The Mediterranean: A Newcomer Moves In

The Mediterranean is a sea of immigrants. The first wave arrived five million years ago when Atlantic spilled over the Strait of Gibraltar. Another more recent wave, this time from the Red Sea, began when the Suez Canal opened. Later, after Egypt built the Aswan Dam, the pace of immigration into the Mediterranean picked up again. Today, very little of the Nile reaches the Mediterranean, rendering its eastern waters a little saltier, a little more attractive to Red Sea inhabitants. Because water levels are higher in the Red Sea, it's easy for shrimp, fish, and mollusks to catch a ride through on the current. Hundreds of Lessepsian migrants, as they are called (after the canal architect who opened the door to the Mediterranean), have successfully taken up residence in their new home. They tend to stay in the eastern waters, along the shores of Israel and Lebanon, where they have become important targets of the trawl fishery.[16]

Sometime in the early 1980s, another immigrant arrived in the Mediterranean, by another route. A seaweed, it took hold quickly and spread rapidly, crowding out existing communities. *Caulerpa taxifolia,* normally a tropical plant, grows in waters off Brazil and Africa, Polynesia and Japan, and in the Caribbean. It is a small, unobtrusive plant, found here and there in the Tropics but not far beyond. It usually doesn't do well in cold water and wouldn't be expected to survive the 27 degree Fahrenheit (15 degree Celsius) temperature swings of the Mediterranean. At some point, no one is quite sure how or when, a cold-water variety did come into existence; it lives in man-made aquariums. In 1984, Mediterranean scientists first noticed a small clump off the coast of Monaco, not far from the Oceanographic Museum. Most likely someone had cleaned out an aquarium and thrown its contents into the sea. By 1997, the lush, bright green plant had spread out along the northern coast to Majorca, to Sicily, and into the Adriatic. Today it lives in ninety-nine different sites, filling 23 square miles (60 square kilometers).

The virtually tideless Mediterranean, this sea "between the lands," is enclosed; water pours in from Atlantic through the narrow Strait of Gibraltar and stays for seventy-five or one hundred years before it leaves. The Mediterranean is, or has been, a sea poor in nutrients, one of the most nutrient-poor seas on earth, but humans have profoundly altered her waters. Humans have exploited the sea's rich fisheries. Rich meadows of the seagrass *Posidonia oceanica* provide spawning and nursery areas for Mediterranean fish, but increasingly, dredging and trawling churn up the seabed, cutting up the meadows. The Mediterranean is a busy sea. Twenty percent of the world's oil shipping moves through her waters, dumping one million tons of crude oil each year. Seventy rivers drain into this sea; many are pipelines for industrial, agricultural, and human waste. The edges of the Mediterranean are crowded. Two hundred million people live in the cities along the coast, and by 2025 that number is projected to double. One of the most heavily visited tourist areas in the world, the Mediterranean received 135 million visitors in 1990. Twice as many are expected by 2025.

All cities on the coast dump their sewage into the sea; half is untreated. Intensive shellfish farming, which has tripled in the last twelve years, adds more waste. The surge of nutrients—from fish farms, from fertilizer, from sewage—has choked the sea's lagoons and shallow bays. Plankton bloom profusely, in numbers far too great for herbivores to graze them down. In the upper reaches of the Adriatic Sea, in the Gulf of Venice and the Gulf of Trieste, in the lagoons of Ravenna and the bays of Croatia, water circulation is reduced, rivers carry nitrogen in from rich farmland, and cities dump untreated sewage into the sea. Here populations of plankton grow and grow. In the past twenty years, the number of blooms has increased, discoloring the water, clogging the nets of fishermen, choking channels and lagoons, leaving clumps of brown and yellow scum along the shore. In Venice, these waters now have a name: *laguna morta,* "the dead lagoon."[17]

The plethora of harmful algal blooms signals distress at the surface. *C. taxifolia* signals distress in the waters below. It first appeared in waters degraded by slaughterhouse effluent and urban waste, where the bottom had been ditched and covered, and where nearby seagrass meadows had been converted into a soccer stadium. Into this wasteland was tossed a clump of *C. taxifolia*. It thrived. *Posidonia* seagrasses spread as much in one hundred years as this more recent arrival expands in a season. *C. taxifolia* is indiscriminate, growing well in mud, sand, or rock. Vigorous, highly invasive, this fast-growing seaweed is overwhelming almost everything in its path. Despite its size, it is a single cell. Its mats are thick: hundreds of feet of stolons (stems), bearing thousands of fronds (leaves), anchored by tens of thousands of rhizoids (roots), crowd into eleven square feet (one square meter) of seafloor.

Where *C. taxifolia* arrives, others soon depart. It shades out *Posidonia* seagrasses and drives away the crabs, sponges, coral, and lobsters living in their midst. A feathery carpet of green has taken over an entire cliff, once home to red and yellow coral, to hundreds of species of algae and invertebrates, to lobsters and moray eels. Even those who merely pass through are vulnerable. An octopus swimming through this meadow can't hide. Fluorescent green is not on its palette of camouflage colors. *C. taxifolia* is a repellent. Local animals don't eat it, and sea urchins, placed in tanks with only this seaweed for food, will starve. Fish sharing a tank with *C. taxifolia,* or consuming mussels who have filtered in bits and pieces of it, have trouble ridding their bodies of its toxins.

C. taxifolia that is spreading rapidly through the Mediterranean is genetically identical to the aquarium variety. Each piece is a clone of the same individual. Humans delivered it to the Mediterranean, created an environment in which it could thrive, and then helped it spread. Storms break up the seaweed and disperse it along the seafloor. Fishing gear and anchors from pleasure boats do so as well. So far, nothing has stopped its advance; it has no predators.

Across Atlantic, another species of *Caulerpa* grows at the edges of the sea, in the nutrient-rich waters of Florida's mangrove swamps. In the past five years it began to grow in deeper waters offshore, north of Palm Beach. It smothered a coral reef in a thick carpet of seaweed, replacing—as it has in the Mediterranean—a diverse community with a monoculture. The mat is six miles (ten kilometers) long and spreading rapidly. A number of scientists believe that *C. taxifolia* in the Mediterranean owes its existence to the careless disposal of aquarium wastewater, although a few still blame the Suez Canal. In either case, it thrives in a sea filled with agricultural runoff and sewage. On the western edge of Atlantic, a nitrogen-loving plant simply extended its range as we made the sea more fertile. These *Caulerpa* species, in their current aggressive, insatiable forms, are our own creation.[18]

The Chesapeake Bay: An Oyster Water Filter

To the unknowing, appearances deceive. I visited Chesapeake Bay for the first time more than twenty-five years ago. Vacationing at the home of a college friend, I saw a quiet, idyllic backwater. We spent our days lazing about in the inlets of Broad Creek and the Choptank River, and our evenings eating. We'd hang a chicken neck in the water, and a little later haul up a bone draped with crabs, enough for dinner and dessert. Our little lures held more crabs than I'd ever seen, and more than I would ever see in the colder marshes where I live now. The element of surprise added a nice twist to our nightly anticipation of dinner. We never knew exactly how large our catch was until we pulled it up. Leaning over the edge of the dock, peering into the water, dreaming of crabs, or sailing in the creeks, I never could see very far into the water, but that's how I thought the bay was supposed to be, murky. Its waters were so beautiful, it never occurred to me that something was terribly wrong, that I should have been able to see crabs clinging to chicken necks, and seagrasses waving in the shallows beneath our boat. I didn't know any better because I didn't know the bay, and I didn't know its history.

Once Chesapeake Bay was thick with oysters. They grew one upon the other, crowded together on great reefs more than a mile long, piled so high they were exposed at low tide. When they spawned, sometime in the summer, there was plenty of hard substrate for the young larvae to settle on. The newly attached spat, filtering their food from water swirled in by the currents, grew fat and juicy. They cleaned the bay as they fed, their population large enough to filter a volume of water equivalent to the entire bay in a few days. Vast meadows of underwater seagrasses grew in the clean water, providing 600,000 acres (2,400 square kilometers) of food and shelter for the Chesapeake's famous blue crab. The oyster reef itself teemed with life, with anemones, barnacles, sponges, and mussels all nestled within its nooks and crannies, offering a veritable feast for blue crabs and stripers coming by to feed.

The name Chesapeake derives from the Native American "Tschiswapeki," meaning "great shellfish bay." It is a name few would consider apt today. For a number of years the Chesapeake oyster has been on the verge of disappearing. At one time, successful navigation of the bay required knowing the oyster reefs, but oyster dredges have flattened them and picked them clean. Between 1870 and 1920, watermen scoured the bay of oysters, taking 20 million bushels a year, dredging them from sailing skipjacks or tonging them by hand. This was taking that could not last. From the 1920s into the 1970s, they took 3 million bushels a year, still too many. Now, with one of earth's greatest oyster beds diminished to a mere 2 percent of its former abundance, the take hovers at a paltry 200,000 bushels. The oysters that remain can't filter the bay in less than a year.

Decreases in the number of oysters and increases in pollution have led to further losses. The bay has become cloudy. Uneaten plankton, left to bloom unchecked, flourish in the wash of nitrogen from sewage, agricultural fields, chicken farms, power plant and motor vehicle exhaust. Eelgrass has withered in the shade. Unable to photosynthesize in waters dimmed by sediment and plankton blooms, the meadows have shrunk to 40,000 acres (160 square kilometers). Eelgrass once grew routinely at a depth of nine feet, but rarely does anymore; the water is too turbid.

Oysters are what they eat. They live in a bath of fertilizer and sewage, of toxic chemicals, of silt and sand washed down from developed uplands. Their home is murky with algal blooms and low in oxygen. It is perhaps no wonder that young oysters, overcome by disease, fail to reach full size. In the past twenty years, millions of oysters have been ravaged. Fishermen take the healthy ones, leaving few who have developed resistance. When waters are warm and salinity levels high, mortality rates are high. The Chesapeake oyster, victim of overfishing, pollution, and disease, represents an ecosystem unraveled. Years of abuse, of pollution, of overdevelopment and overfishing ate away at the bay, degraded the United States's largest estuary to barely 25 percent of its historical abundance and fertility. Half the bay's wetlands and forest buffers have been destroyed. Nutrient loading into the bay is seven times what it once was. Forty-five percent of the nitrogen entering Chesapeake Bay comes from agriculture, 24 percent from wastewater treatment plants, and 21 percent from power plants and automobiles. As a result, bay waters are no longer clear, and many areas are depleted of oxygen.

Over the years, from all directions, the Chesapeake has been besieged: by fertilizers washing in from as far away as New York and Pennsylvania, by manure from poultry farms, by urban sprawl, by the depletion of precious fisheries, by industrial discharges of toxic chemicals, by waste generated by millions of people living in the watershed. A massive effort is under way to restore the bay, to repair, strand by strand, the damaged web. Every year, the Chesapeake Bay Foundation evaluates the bay, comparing its current status with that of a healthy bay. A healthy bay receives a rating of 100. In 1983, the state of the Chesapeake was rated at 23. By 2000, the rating had gone up to 28. It took many years to weaken the resilience of the Chesapeake, to undermine its vitality; it will take many more years to return the bay to health. In June 2000, the states in the Chesapeake Bay watershed—Virginia, Maryland, Pennsylvania, as well as the District of Columbia—renewed their commitment to restore the bay. They pledged, by 2010, to restore thousands of acres of seagrasses, riparian forests, and wetlands, to stem the discharge of nutrients, sediments, and chemicals, to rebuild oyster reefs, to manage the blue crab fishery sustainably, to preserve permanently from development 20 percent of the land area in the watershed, and to reduce the rate of sprawl by 30 percent.

It is an ambitious plan. Its beauty is its vision of the bay as a whole, living entity, its recognition that all is intertwined, that farming practices in the Susquehanna River valley affect seagrass meadows several hundred miles downstream, that reseeding oyster reefs can succeed only if plankton blooms decline. The plan represents a commitment to restoring the riches we have taken and to withholding what is toxic and deleterious. It acknowledges that merely treating part of the problem is to hold the health of the bay hostage. If urban sprawl is not contained, the benefits of controlling runoff from animal manure may be neutralized. If the signatories honor their commitments, the effects should reverberate throughout the bay. Oysters resistant to disease can spawn in reef sanctuaries, increasing healthy stock, building the bay's filter, and improving water clarity. Seagrasses can blossom in clear water, creating more habitat for blue crabs and fish. As nitrogen discharges are reduced, plankton blooms can decrease, raising oxygen levels, allowing the bay to breathe once again. If action follows rhetoric, this is a plan full of promise.

The Chesapeake Bay Foundation's annual State of the Bay Report is a yardstick by which it is possible to measure, year by year, how far we have come in restoring the bay, and how far we have to go. The report emphasizes results, not merely effort. The 100 rating holds the memory of abundance. The high standard is essential, for as time goes by and each passing generation sees a different bay, people too easily forget what has been and what could be, and too easily come to accept what is.[19]

The Caribbean and the Gulf of Mexico: Corals Lose Their Resilience

I felt keenly the limits of my imagination upon seeing a coral reef for the first time; its vitality and opulence exceeded my wildest fantasy. The fish are flashy, their colors of blood red, brilliant yellow, and electric blue combined in bright, zany patterns. Black-and-white-striped sergeant majors, blue tangs, and foureye butterfly fish dart over the reef. The variety astounds. The fish move around a dazzling array of sedentary sponges that more nearly resemble pieces of rope and tubing, barrels, collections of vases. Larger, more somber-looking fish, grouper and barracuda, swim by. Mantis shrimp burrow into the reef's crevices, and clawless spiny lobster crawl along its surface. Here, in this kaleidoscope of color and flurry of activity, even the stone breathes. Delicate sea fans, massive brain corals, branching elkhorn and staghorn coral, motionless amidst the bustle of the reef, still as stone, are nonetheless alive—quietly, unobtrusively building what is perhaps earth's largest living structure. By day, tiny single-celled algae living in the guts of coral use energy from the sun to make carbon. The zooxanthellae, as they are called, give coral both their color and the wherewithal to build the limestone reef. They are essential. Without their algal tenants, Atlantic reefs would cease to grow.

By night, the rhythm on the reef changes, but it still hums with life. Parrot fish, tired after a day spent crunching on coral, spin a clear, thin, predator-proof mucus cocoon around themselves. They rest, while octopus, spider crab, and red shrimp come out to feed and moray eels slide from their holes. Brittle stars, a skinny variety of starfish, flashing green to fend off predators, emerge from tube sponges or empty coral where they hide in the daylight. The coral awaken, extending their tentacles to gather food in the watery darkness. Once a year, on a night sometime after the full August moon, they spawn, releasing millions of tiny bundles of egg and sperm into the water. Buoyed by drops of nourishing fat, the bundles float like bubbles to the surface, where they are divided and rejoined by waves. In a few days, the survivors settle on the bottom to form a new colony. Each spawning, lasting only a few hours, is precisely timed; along a reef, each species moves in synchrony, releasing its future offspring together.

The reef is beautiful, and violent. Sea worms and sponges bore their way into coral, dissolving or scraping out the skeleton to make a home. Corals stun their prey with barbs and stinging cells attached to their tentacles, turn their enemies to mush with chemical poisons. Competing for space on the crowded reef, corals turn these deadly weapons on each other. Competition is fierce, but each contestant has strengths. Slower-growing corals hold their own in the fray, making up in aggression what they lack in size, creating a balance that has sustained the reef for thousands of years.

A multitude of species, anywhere between one and nine million, representing thirty-two of thirty-three phyla found on the planet, live on coral reefs, a diverse gathering that rivals the richness of species in a rain forest. Ironically, only a thin sea can support such largesse. Few coral reefs live across the Isthmus of Panama, where deep water facing an open sea, strong tides, and cold, nutrient-rich water welling up from the depths make a fertile ocean. Here, on Atlantic's side, the waters of the Caribbean and Florida Bay are warm, calm, shallow. Few winds pull nourishment up from the deep. Around the edges of the Caribbean and along the Florida Keys, reefs break the rolling waves, creating sheltered lagoons where thick carpets of turtle and manatee grass line the sandy bottom and forests of mangrove line the shore.

The three communities nurture and need each other. Corals front the sea, providing calm and tranquillity for mangroves gnarled and twisted into an impenetrable forest. Red mangroves, with their stiltlike roots arching out of the water, and black mangroves, surrounded by breathing tubes sticking out of the thick, close mud, trap sediment and drink in nutrients coming off the land, leaving clear, lean water for coral and seagrasses. Where the water is truly clear, turtle grasses can grow as deep as one hundred feet (thirty meters). Each shelters its own community of animals—crabs in the mangroves, conch in the seagrass, coral on the reef—but many move among the different habitats. Sea urchins, commuters, live on the reef by day and in the grass by night. Other fish and crustaceans such as spiny lobsters spend their early days in the seagrass and then move onto the reef when they grow older.

In waters where nutrients are scarce, corals are both plant and animal, producer and consumer. Photosynthesizing zooxanthellae and coral polyps are consummate recyclers, thriving on each other's wastes. Zooxanthellae use the polyps' carbon dioxide and nitrogen wastes to make carbon, which in turn nourishes coral animals as they secrete their limy skeleton. It is a delicate balance, this symbiosis, predicated upon an abundance of predators and a dearth of nutrients. It is a magnificent existence in a sea of paucity.

Multiple redundancies give the reef resiliency. A coral harbors many species of zooxanthellae. If one disappears, another can substitute. If three quarters of the zooxanthellae are expelled from the coral, it lives on, relying on reserves and night feeding until symbiosis can be restored.

Plankton blooms and seaweed can overwhelm a coral reef. Nutrient-poor water is one defense; grazers are another. A large number and variety of grazers keep a reef strong. Hurricanes strike, reducing the reef to rubble, but in the aftermath, fish and sea turtles nibble at encroaching seaweed, leaving corals room and time to rejuvenate.[20]

Not all reefs lie beneath the sea. Born in warm, shallow, sun-drenched waters, coral reefs may, as continents move or oceans recede, rise above the waves to become dry land. The Atlantic edges of Central America, the coasts of Caribbean islands, and all the Florida Keys were once underwater, teeming with lobster and crab, tarpon and snook, and living coral. Coastal Puerto Rico and Jamaica are dry reefs, pushed from the water as the Caribbean and North American plates grind past each other. The

Florida Keys were exposed one hundred thousand years ago, during the last ice age, when advancing glaciers grew fat with water from the sea.

A coral reef may seem a place apart, a place not of our world, but on the Florida Keys, both dry land and underwater reef are made of limestone fashioned by living coral. Today, seventy-eight thousand people have built full-time residences on these old reefs, and another twenty-five thousand move in for the winter. Each year several million more come to visit. They come for warm weather, clear water, great fishing, vibrant reefs. To support so many people, huge chunks of the intertwined community of mangrove forest, seagrass meadow, and coral reef were dismantled. People dredged more than two hundred canals, taking the spoils to fill mangrove swamps and seagrass beds, to make real estate.

Humans generate a lot of wastewater, 45 gallons (170 liters) per day per person, full of viruses and bacteria, nitrogen and phosphorus. On the Florida Keys, human wastewater is dumped into 4,000 cesspools, 20,000 septic tanks, and 750 groundwater injection wells. The limestone, porous as Swiss cheese, can't hold it all, and the reef releases its burden to the sea. Within hours, viruses and bacteria discharged into septic tanks and injection wells appear in the groundwater and canals. Onshore beaches become too polluted for swimming, and fecal coliform bacteria show up several miles offshore. Large quantities of nitrogen—from sewage, from fertilizer, from grass clippings—deplete many canals of oxygen, killing fish and bottom communities. In some canals, the sheer volume of decaying matter precludes the return of bottom communities, even when autumn winds and cold mix the water. Where nitrogen discharges are high, seagrasses disappear from the canals. Some canal entrances also show signs of nitrogen overload: water cloudy with chlorophyll, thick mats of seaweed growing in the grass beds, seagrass leaves encrusted with algae.[21]

Nitrogen-enriched water swirls toward living coral, not only from the heavily populated keys but from farther away as well. Four thousand years ago, when a rising sea inundated low-lying parts of the exposed reef, creating tidal passes between the keys, water began flowing south from the Everglades. Florida Bay was once washed in water trickling from Lake Okeechobee down through the Everglades. In the past fifty years, half the Everglades has been destroyed to make room for sugar plantations and sprawl from Miami and Ft. Lauderdale. Hundreds of billions of gallons of water from the Everglades have been diverted to irrigate farms and to prevent flooding in "reclaimed" land. Productivity in Florida's vegetable and sugarcane fields depends on the heavy use of fertilizer. Water managers pumped agricultural runoff saturated with fertilizer into Lake Okeechobee and nearly killed it. The lake, losing oxygen, overcome with algal blooms, began to suffocate.

To save the lake, managers rerouted the dirty water through the Everglades into Florida Bay. Along the way, it lost phosphorus but retained much of its nitrogen. It is this nitrogen, some scientists believe, which now triggers plankton blooms in Florida Bay, sending cloudy water down toward the reefs. Water managers plan to replumb the Everglades, returning some of the water that has been taken for farm fields and flood control. However, additional fertilizer-laden water emptying into Florida Bay may intensify the problem of noxious algal blooms. Restoring a watery lifeline to the Everglades is key to its survival. It should not be too much to ask of agriculture that this water be clean; the health of downstream coral reefs may be at stake.[22]

Man's presence permeates the sea, and coral reefs feel the effect. Corals on the Florida Keys and in the Caribbean are dying in unprecedented numbers, from fatal fungal infections, from tissue-eating bacteria, from strange lesions disintegrating their skeletons. Appearing in the last twenty years, the diseases—black, red, or white band disease, white pox, coral plague, rapid-wasting syndrome—are new, unfamiliar, and virulent. In 1995, a new disease attacked Florida star corals, killing their tissue within three days. It spread rapidly, from Key West to the Dry Tortugas and then up to Miami, infecting sixteen other species of coral. The pathogen, a new species of bacteria, is of unknown origin. The extensive outbreaks of coral diseases are troubling. Over the past fifteen years, *Aspergillus* has infected 40 percent of Caribbean sea fans. The fungus is ubiquitous onshore, in moldy grains and vegetables, but until this outbreak it hadn't been considered toxic to marine life. Of the many people exposed to this fungus, usually only those with compromised immune systems become ill.

Perhaps corals, too, are growing weak. The disease outbreaks are adding up, pointing to a contagion widespread and growing worse. The news from a monitoring system at 160 stations along 40 reefs in the Florida Keys National Marine Sanctuary is dispiriting. Between 1996 and 1998, the number of stations with diseased coral increased from 16 to 82 percent and the number of affected species increased from 27 to 85 percent. Many corals haven't recovered. At 88 percent of the sites, species were lost. Since 1975, coral cover on Carysfort Reef, the largest and most diverse reef on the keys, has declined 90 percent.

Carysfort is one reef, but the devastation occurs throughout Atlantic coral. From Belize to the Bahamas, from Florida to the West Indies, large, treelike elkhorn and staghorn corals, 10 feet (3 meters) high and 60 inches (152 centimeters) wide, once dominated. In the 1980s, many were felled, by a still unidentified disease. They have yet to recover, these hardy corals that have thrived on Caribbean reefs for several thousand years. At no time in their past does the fossil record describe a similar catastrophe. For the corals of the Caribbean, today's crisis is unique.[23]

Overfishing has further weakened coral reefs, further lowered their resilience. Algae grow on healthy reefs, but they don't tend to dominate: lack of nitrogen and an abundance of grazing animals limit their growth. Herbivorous fish, sea turtles, and sea urchins all clean a reef of algae. We took out the turtles long ago, and the herbivorous fish, leaving long-spined sea urchins to scrape the reefs. With only one strong defense, they became vulnerable. In 1980, Hurricane Allen swept through the Caribbean, breaking up the reefs of Jamaica, killing coral. Algae moved onto the ruins, but sea urchins nibbled it away, and the coral began its slow resurgence.

In 1983, Caribbean sea urchins began to die, shedding their spines in response to unexplained illness. In Jamaica, the urchin population fell by

99 percent. On the reef, where overfishing had already removed other grazers, sea urchin losses ended the coral recovery. With no grazers left to hold off the seaweed, it moved in unimpeded, covering the reefs in thick, fleshy mats. Reefs grow slowly, a few feet every thousand years. Young colonies of coral had no time to become established. The seaweed moved in first. Large brain corals, surviving the hurricane, continued to grow until 1989, when Hurricane Gilbert tore into this usually storm-resistant coral. The slow-growing brain coral stood no chance against the unchecked advance of seaweed; where the coral was destroyed, it could not begin to recover. Today, on a reef dominated by seaweed, it has all but disappeared. In the past thirty years, coral cover on Jamaican reefs has dropped from 52 percent to 3 percent.

No one knows what killed the sea urchins in 1983, but in 1997 more died, this time in the Netherlands Antilles. The plague began in the harbor on Curaçao and followed the current downstream. This time scientists isolated the cause, a potent toxin not previously known to kill marine animals. Restoring predators to a reef can only strengthen its resistance to assault. In 1997, managers at the Florida Keys National Marine Sanctuary prohibited fishing in a 12 square mile (30 square kilometer) area. Within a year, spiny lobsters had returned to the reef in the daytime. Gray snapper, hogfish, yellowtail snapper, and grouper returned, in numbers larger than anyone had seen in twenty years.

Lovers of clear, impoverished waters, coral are stressed by an accumulation of insult: by storm and wastewater running off land, by losses of vegetarian grazers, and by seas cloudy with clumps of seagrass churned by boat propellers, and silt that mangroves, besieged themselves, release to the reefs. The loss of predators has accelerated the pace of life on the reef, tipping it toward the pulse of plankton, whose numbers swell when nutrients are lush and whose lives are measured in hours and days, and away from the slower, more steady pace of corals, whose time ticks in years, centuries, millennia. The World Resources Institute now estimates that almost two thirds of Caribbean reefs are at risk from overfishing, pollution, sedimentation, agricultural practices, and coastal development.[24]

Corals are already plagued by mysterious and fatal diseases, by water clouded with sediment and enriched with nitrogen. The additional burden of global warming may prove too much for them to bear. A coral reef is always in a war against attrition; rising carbon dioxide levels may determine the outcome. As coral builds up a reef, other forces tear it down. Passing waves and storms erode the limestone, sea urchins and parrot fish scrape it away, seawater dissolves it. In order for production to trump destruction, coral requires a sea drenched in carbonate. As carbon dioxide levels rise in the atmosphere, they rise in the sea, increasing its acidity, decreasing its concentration of carbonate. Scientists believe that limestone production on coral reefs may have dropped 10 percent between 1880 and 1990, and they predict it could drop another 20 percent in the next hundred years, more than enough for reefs to cease growing, for disintegration to exceed growth, for loss to exceed gain.

The risk is not limited to coral. The white cliffs of Dover are built from tiny single-celled marine dwellers that secrete a limestone shell. Their growth rate, like that of coral, is sensitive to rising and falling levels of carbon dioxide in the water. They, along with coral and other coralline algae, produce well over half the limestone that appears on earth. As earth and her seas warm, limestone production may decrease, altering ocean chemistry in ways we have yet to realize.

Rising water temperatures harm corals as well, but by different means. Corals living in the Caribbean today have a cold-water history; their ancestors date back to the waning of the ice ages. Today's unusually warm water tests their tolerance for heat; hot water forces them to evict their tiny, photosynthetic tenants, without whom they can't grow. In the summer of 1983, the Caribbean heated up, and corals from Panama and the Florida Keys expelled their zooxanthellae, losing their color and turning a ghostly white. The damage was severe, but when the temperature dropped, the algae returned and the coral recovered. Since then the water has continued to warm, and bleaching has increased in frequency and severity, affecting corals in the Cayman Islands, Jamaica, Cuba, the Bahamas, Mexico, and Belize. Only a small rise in temperature, 2 degrees Fahrenheit (1 degree Celsius) above normal, triggers bleaching.

The summer of 1998 was also a hot one in the Caribbean. Water in the central lagoon behind Belize's barrier reef, rarely warmer than 85 degrees Fahrenheit (30 degrees Celsius), heated to a record high of 90 degrees Fahrenheit (32 degrees Celsius). The heat lingered. Cooling rains from Hurricane Mitch delivered a slight respite, but water temperatures didn't drop until November. For many corals growing in the central lagoon, the heat was unbearable. They bleached in October, and sixteen months later they still showed no sign of recovery. This is the first time scientists monitoring reefs have recorded the death of a bleached reef. The devastation covered a 145 square mile (375 square kilometer) area. For this catastrophe, there is no known precedent; three-thousand-year-old cores taken from the reef show no evidence of fatal bleaching.

It is likely, though, that massive coral fatalities, unknown since the glaciers receded, will be common in our future. Whether the cause be disease, changing water chemistry, or warming seas, it is quite possible that most of earth's reefs will be dead by the year 2100. Looking back beyond the ice ages, back to Atlantic's youth, and then further back, to the time of Iapetus, there is a long record of the rise and fall of reefs. Coral reefs have come and gone as seas open and close, as earth's orbit around the sun changes shape, as species rise to prominence and then give way to others. The disappearance of reefs is not new. The difference is that now, we are the driving force of nature, and we work at a fast and furious pace.[25]

CHAPTER 12

Landfall

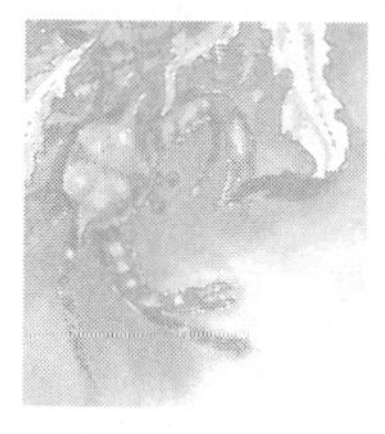

WHERE the *Cramer* is becalmed, out in the Sargasso, the pace is anything but hurried. We spend day after listless day floating in the glassy water, putting up the sails and taking them down. We don't make much headway: it seems as if each time we set a sail, the teasing breezes die. We try to dodge squalls that suddenly blacken the sky, drench us in rain, and then, just as suddenly, dissipate. Occasionally a waterspout whirls at the horizon. In the afternoons, when we're not on watch, we tie knots. One of the tasks is to tie nine or ten knots in less than a minute each. With practice, this is possible. I'm glad I can tie bowlines and clove hitches, sheet bends and rolling hitches, but my favorite is one we're not required to learn, the Turk's head. This decorative knot is a braid woven in a circle, often around a stanchion. The beauty of the knot lies in its unending plait. Tracing the lines of a well-tied Turk's head, it's impossible to know where the braid first began, or where its weaver finished. After many false starts, I manage to tie a Turk's head around my wrist. Its many strands, all formed from a single line, intertwine along a circle that has no discernible start or finish, a circle that is, like the flowing waters of Atlantic, seamless.

I am happy learning to tie knots and splice line in the many moments of calm, but the mates are edgy. After waiting and waiting for a wind that doesn't blow, the captain gives up and turns on the motor. We head south, hoping to catch the trades. They come on quietly. I first notice a shift in the middle of the night. To escape the hot, close air in my cabin, I'd taken to sleeping out on the deck, under the stars. Exhausted, I'd usually sink into a deep sleep, oblivious to the sounds of people tending the sails and sorting through clumps of *Sargassum*. One night, though, I wake early, shivering ever so slightly. The deck is bathed in moonlight. The ship, no longer chugging through the water, rides the waves. The droning motor has been silenced, the sails raised once again. Water slaps the bow. The air is light, the wind gentle, but steady from the northeast. Over the next twelve hours, it grows stronger. When I stand at the wheel, back to the wind, I hear it in both ears. At long last, we have caught the trades. The ship runs before them; we raise every sail. Birds appear. Bermuda longtails soar overhead, flying fish skim over the waves. The birds and fish are local, but high above us, the wind carries dust from the Sahara. The science tow brings up locusts blown in from Africa.

These locusts are quite dead. At some point, they wearied and fell into the sea. A transatlantic voyage on a stiff wind was not for them. For others, though, the passage across Atlantic has become quite easy. Riding on ship's hulls, floating in ballast water, increasing numbers of animals are making the crossing in a matter of weeks. Today earth's continents are scattered, separated by deep and wide seas, but we are overcoming the watery barriers and forging an essentially continuous landscape. Bridging our seas with a vast array of ships and airplanes, engaging in global trade and commerce, we are building another Pangea, linking communities of plants and animals who at one time evolved independently. By our own hand, we are making our lands one, short-circuiting the cycle of moving rock and drifting continents by 200 million years. No longer can the breadth of the ocean guard earth's diversity.

We have opened Atlantic's borders, pushed through her watery mass. Atlantic has not physically closed, but the easy passage from shore to shore makes it seem as if the sea has already narrowed, and her coastal waters already begun to mix. Many newcomers die in their new homes, but enough survive to make a difference. Zebra mussels, come from across the sea, clog American rivers and lakes. A runaway population of American comb jellies, with their insatiable appetite for fish eggs and larvae, saturates the Black Sea. These recent arrivals constitute an alteration to our planet as significant as the injection of carbon dioxide into the atmosphere, as significant as the depletion of earth's vast reserves of wild animals.

The redistribution of species began years ago, almost the moment man took to the sea. Merchant vessels brought the green crab from Europe to the shores of America, and Viking longboats took the American clam back to the Baltic. The creeks near my home are filled with newcomers: European green crabs, periwinkles, and more recent arrivals—Pacific crabs inadvertently shipped through the Panama Canal and European oysters accidentally cut loose from aquaculture trials in Maine. Their arrival has profoundly altered the life and look of American estuaries. The eating habits of the common periwinkle, for example, have given the shores of New England and maritime Canada their rocky character. Periwinkles arrived in eastern North America around 1860 and flourished, satisfying their hunger by scraping vast quantities of algae off the rocky shore. In places where the industry of periwinkles is not yet felt, thick mats of algae grow in over the rock, trapping sediment, making mud. The hard, dark edges of the shore disappear, and soft, green meadow of marsh grass takes root.

Green crabs arrived a little earlier than periwinkles, and over the years they have made their presence known. Their appetite for shellfish is just as strong as ours. In New England green crabs plow through entire flats, devouring both baby and mature clams. So far, nothing has stopped them, although a potential competitor has recently arrived. The Pacific shore crab came to Cape May, New Jersey, in 1988 by way of ballast water. In the intervening years, it has spread along the shore between Massachusetts and North Carolina. Like the green crab, this crab also has a prodigious appetite, gorging on oysters and clams and seaweeds. In a few locations, it has usurped the green crab's reign.

When nonindigenous species lack predators, they dominate. American comb jellies have contributed to the demise of the anchovy fishery in the Black Sea, and European green crabs are helping clammers clean out New England shellfish beds. Farther south, the veined rapa whelk, hailing from the Sea of Japan, has made a recent appearance in Chesapeake Bay. This large marine snail, no stranger to overseas travel, took up residence in the Black Sea, and within twenty years consumed almost the entire oyster population. A ride in ballast water expanded its range to the Chesapeake, where it now threatens depleted oyster stocks.[1]

For centuries, boats plying oceanic waters have introduced species to distant places, giving them homes they might never have known otherwise. As global commerce has increased, nonindigenous species have become commonplace. Cargo ships routinely unload ballast water when they come into port, mixing water from both sides of Atlantic. Ships entering U.S. ports unload two and a half million gallons (nine and a half million liters) every hour. The number of stowaways is staggering. A liter of ballast water carried by ships entering German ports may contain 4,000 specimens of fish and zooplankton, and 110,000 specimens of smaller, unicellular organisms. If only a few of the newcomers live and rise to prominence, the complexion of entire ecosystems may change.

Of approximately 200 newcomers in the Chesapeake Bay, at least 36 have altered the quality and character of life in the bay. Purple loosestrife has appeared in wet meadows and freshwater marshes at the edge of the bay, crowding out native grasses and sedges. In the freshwater recesses of the bay itself, dense mats of water chestnuts, hailing from abroad, outcompete native plants that are the preferred food of waterfowl. Left uneaten, the mats decay, depleting the water of oxygen. Chesapeake Bay oysters have been a frail lot, succumbing in large numbers to bacterial and parasitic diseases. At least one of these infections was carried in on oysters from the Pacific. If oysters can survive these assaults and develop resistance, the rapa whelk may be waiting.

Ships breach the protective barrier of the sea, easily transporting diseases from one shore to another. Foreign ships docking in England and Wales unload ballast water and sediment full of organisms that cause red tides and shellfish poisoning. Ballast water entering Chesapeake Bay carries vast numbers of potential pathogens: each quart contains as many as one billion bacteria, ten billion viruses. American aquaculture farms and processing plants import millions of pounds of shrimp each year. When the shrimp are diseased, the diseases come in as well. In the 1990s, Taura syndrome, white spot syndrome, and a host of other virulent diseases once known only in foreign stocks began wreaking havoc in Texas and South Carolina shrimp farms, decreasing production by more than two thirds. The highly infectious viruses spread rapidly among hatcheries, farms, and processing plants. Now these diseases have begun to appear in wild shrimp and crab, possibly transmitted when wastewater from shrimp farms was released into the coastal nurseries of wild populations.[2]

We are becoming a world without borders. Losing our regional distinctions, we are growing more homogeneous. If the green crab successfully conquers the bay by my home, I shall miss the soft-shell clams that once crowded the sandbars and mudflats. Already, as purple loosestrife takes over the meadow outside my window, I miss the grasses and wildflowers forced to yield to its advances and wonder about the wildlife whose lives have depended on them. Until this perennial herb was imported for use in American gardens, Atlantic's wide expanse held loosestrife at bay; its seeds cannot withstand a long, watery, transoceanic voyage. My sadness is more than simple melancholy over the loss of what is familiar. I know that sooner or later forest will claim the meadow and river silt bury the clam flats, but I mourn the loss of species passing before their time.

We cannot stop what we have begun, nor, I suspect, would we want to, but it is possible to slow the mingling of waters, the homogenization of communities. Rapidly shrinking the grand array of earth's dwellers may reduce our resilience in the face of unforeseen circumstances, our ability to weather unpredicted storms. In coastal waters, bottom-dwelling communities with more species of sedentary, filter-feeding animals show greater resistance to pressure from introduced species than sparer assemblages. Strong are communities that are rich in species. Their members' lives unfold to different rhythms; some grow while others die, keeping all the space filled. We have done much to impoverish this richness. Altering habitats, by removing key predators from coral reefs or inundating estuaries with fertilizer and manure, eliminates species. Thinning a community creates open niches for introduced species, and introduced species further thin a community. Loss begets loss.

Large numbers of losses lead to extinction. Approximately 4,200 species of mammal live on earth today. Scientists calculate that if earth were reassembled as it was in the time of Pangea, if the planet once again held a single continent, that continent could support only 2,000 mammal species. The existence of large numbers of different kinds of plants and animals filling similar niches, doing similar jobs, is a form of insurance. As species inevitably come to the end of their terms, or natural disasters ravage a community, others will be left to carry on. Closing oceans, crossing seas, eliminating redundancies may threaten our long-term well-being, and that of our earth. The empty planet left after the Pangean extinction is a bleak reminder of the dangers of uniformity.[3]

Eventually, my time at sea draws to a close. Once the *Cramer* picks up the trades, the ship skims through the waves, heading toward land. An hour in each of my remaining deck watches puts me out on the bowsprit. While checking for cruise ships and freighters, I see dolphins playing in the bow wake and waters awash in bioluminescence. Plankton shine as waves break, stars fill the sky. In their midst, another light appears, blinking on as the bow rises on a wave, blinking off as the ship drops and dips toward the water. The appearance of the Barbados Light means that my sail is ending. For twenty-one days and nights I have been lucky to be with Atlantic, to experience the dark and stormy Gulf of Maine, the palpable heat and deep, deep blue of the Gulf Stream, the lassitude of the Sargasso. When the ship lands at Bridgetown, I disembark, completing a passage of more than two thousand miles.

Before I fly home, I spend a day in Bridgetown, strolling through the markets, walking out to a beautiful sandy beach with one of the ship's scientists. I have just spent an intense, concentrated time with thirty-five people in very close quarters. For a brief period, we were a world unto ourselves. Relaxing now in the warm water, I think about what this experience has to say about our future relationship to Atlantic. After all, this boat is full of college students, the new generation of Atlantic's stewards.

Perhaps someday, as these kids make their way in a world where wealth is defined by consumption, a few will remember the shards of plastic accumulating in the wide Sargasso and make their purchases with caution. Perhaps, as pollution continues to creep into coastal waters, they will remember when dirty dishwater, released into the sea at an inopportune time, contaminated an entire set of samples. Perhaps, as

Caribbean Sea

nitrogen levels increase in many of our estuaries, they will remember the time when someone, in blatant disregard of the signs posted in the bathrooms, threw something down the toilet, requiring us to spend an entire day unclogging the plumbing. Someday memories of these thoughtless acts might call some of this student crew to work to protect the sea. Or perhaps, if they ever stand on a shore where breakers are crashing, they might recall the enormous swells in the Gulf of Maine and remember that water gracing the eastern seaboard of North America may have rolled in from Africa, that the water they touch will rush on to someone's shore. Or maybe one night, gazing up at a starry sky, they might remember that our place on earth is always relative.

Barbados is the end of my Atlantic crossing. It is the end of another as well. The very existence of this island means that Atlantic too is closing. Lonely outpost at the edge of the sea, Barbados is the easternmost expression of destruction in the deep water below, easternmost sign that Atlantic has begun her descent back into the depths. Above the water, fields of pale green sugarcane grow in soil made from pieces of Atlantic seafloor, sent back to the surface as the deeper basin returns to its source. Barbados records in broad daylight the unseen history written in the water below. The islands of the Caribbean are unlike others rising above Atlantic's waves. Iceland burst forth from the ridge. Other plumes of heat lifted the Canaries and the Azores. Great Britain and Ireland belong to the continents; only a rising sea isolates them from the mainland. The islands of the Caribbean are different. Born where a sea closes, they foretell the future of the rest of Atlantic.

The edges of Atlantic, almost 200 million years old, are cold and heavy. Where they meet the floor of the Caribbean, Atlantic buckles under, furrowing a deep, narrow cut in the seafloor. The trench curves around the Windward Islands and slices through the seabed north of Puerto Rico at a depth of 27,200 feet (8,300 meters), the deepest part of Atlantic. Farther south, it shallows, filling with sand and mud washed in by the Orinoco and Amazon Rivers. In this deep cut, Atlantic seafloor, created long ago in the mountains of the ridge, sinks back into the mantle. One sea does not slide gently, smoothly beneath another; departing seafloor resists its destiny. The heavy volcanic foundation disappears, but lighter layers, laid down as shells and skeletons of dead animals sank to the seafloor, are scraped off and piled high in the trench. As more Atlantic seafloor sinks beneath the Caribbean, more layers are scraped off and the pile grows higher and higher. The stack eventually surfaces—in this case, as the island of Barbados.

A closing sea is a noisy, messy affair. The history of the Caribbean is violent, geologically. Thousands of Europeans retraced the course of Christopher Columbus, seeking the fabled riches of the West Indies. They squabbled over the islands, competing for possession, rushing in to make their claims, oblivious to signs of volatility in the landscape, to denuded mountain peaks rising from lush rain forest, to steaming, sulfurous hot springs, to boiling lakes.

A departing sea is indifferent to the desires of men. In 1655, French colonists arriving on Martinique built a city at the mouth of the Roxelane River, a short distance from a mountain they named Pelée, for its baldness. St. Pierre grew and prospered, exporting rum, coffee, and sugar, using its wealth to build a theater, cathedral, and hospital. When dormant Pelée reawakened, the "Paris of the West Indies" lost its shine. Warnings of imminent catastrophe went unheeded. In February 1902, toxic sulfur fumes rising from the volcano began killing birds and tarnishing silver. By the end of April, the crater's steaming lake had boiled, the mountain had rumbled, and a rain of ash had fallen nearby. Insects fleeing the hills invaded St. Pierre's sugar mill and bit farmers' livestock. Residents were forced to cover their mouths and noses with wet handkerchiefs to ward off the stench of sulfur. Horses suffocated and dropped dead in the streets. By early May, poisonous snakes, routed from the forests by ash and fumes, escaped to the city, where they attacked pigs and chickens. Ash fell like snow, muffling the sounds of carriage wheels, withering sugarcane and field crops. Cattle starved. The boiling lake overflowed, and a torrent of hot mud flowed down the mountain, burying the sugar mill and killing twenty-three people.

The end, when it came, was swift. On May 7, explosions racked the volcano and black clouds rose from the summit. The sun never rose over St. Pierre on May 8, Ascension Day. Dawn came, but the sky remained pitch-black. At 7:52 A.M., the hands on the clock of the military hospital froze and telegraph service was abruptly cut off. A glowing, incandescent cloud of ash and gas, blasted from the throat of Pelée, roared down the mountainside and engulfed the entire city in sulfur and flames. Within minutes, twenty-eight thousand people died. A murderer, jailed in a windowless cell, survived to recount the horror. Pelée exploded throughout the summer, further devastating the countryside as more *nuées ardentes*, "burning clouds," tore from the mountain and swept to the sea.

Pelée is hardly unique. Many a bare mountain rises from many a Caribbean island. Their names may differ, but their source is the same. Les Pitons rise from the sea at St. Lucia, where water heated by volcanoes supplies the island with energy. On Guadeloupe, St. Vincent, and Montserrat, the volcanoes are called Soufrière, the "sulfur mine." A boiling, sulfurous river pours out of Guadeloupe's La Soufrière, a slumbering volcano racked by thousands of earthquakes. Rising above sugarcane fields and coconut groves is St. Vincent's La Soufrière, whose eruptions have spewed ash across the Caribbean Sea onto Haiti and forced thousands of people to abandon their homes. The last eruption, in 1979, destroyed the volcano's lava dome, but a new one has risen. It sits, waiting for the next surge to rise from the darkness and turmoil below. Dominica's Soufrière seethes below the surface, its warmth denuding the landscape in the nearby Valley of Desolation, heating one of the world's largest boiling lakes.

Destruction, when it comes, is total. Guidebooks once touted the possibility of frying eggs on the heated rocks of Montserrat, but few tourists do so today. Once lush and green, Montserrat is now covered by a veil of ash. *Nuées ardentes* blasting from its Soufrière Hills in 1997 engulfed a village and ravaged the capital. The southern half of Montserrat, burned out, buried in ash and lava, is uninhabitable; more than half the residents have emigrated. The eruption lasted for more than three years, releasing ash fine enough to permanently scar the lungs of people who breathed it.

Atlantic's seafloor sinks down toward the mantle, in a volatile farewell that can—in a matter of days, hours, minutes—destroy entire landscapes, obliterate cities.[4]

Salt water makes this finish so explosive. Drifting seafloor, on its journey from the ridge, soaks up water. When the seafloor slides down into the warm mantle, this water is heated and released into the overlying rock, which begins to melt. Thick, viscous magma, full of water vapor, carbon dioxide, and sulfur, rises toward the surface. The pressure drops, blasting the trapped gas from the volcano, just as when a cork is popped from a bottle of champagne. During the explosion, incandescent clouds of burning ash and gas hurtle from the crater at 62 miles (100 kilometers) per hour.[5]

Between eruptions, the volcanic islands of the Caribbean are lush and green. Hills pushed up by rising magma catch moisture from the air, washing the islands in life-giving rain. Where there are hills, there are rivers; more than two hundred flow from the hills of Dominica. Water sustains thick rain forests in Dominica, Montserrat, and St. Lucia, banana plantations in Basse-Terre, Guadeloupe, spice fields in Grenada, pineapple and sugarcane fields in Martinique. As long as the bottom of Atlantic pushes against the Caribbean basin, with one sinking beneath the other, these islands will be volatile. Many a Pelée or Soufrière will rise from the sea, providing soil and catching rain to nourish lush forests and rich plantations of tropical fruit, only to devastate, in one explosive eruption, all that has been given. Then the ash will turn to fertile soil and the land will turn green again, until the next eruption. The cycle, tuned to rhythms in the deep water, will continue until the sea finally closes and its opposite shores join, squeezing the islands onto the land.

The islands of the Caribbean curve into an arc, not unlike the arcs that formed long ago in waters off the ancient coast of North America when Iapetus closed. Remnants of this older arc have been swept ashore, into rounded hills, scattered along the Connecticut River, extending up into New Brunswick and Gaspé, then broadening to make up interior Newfoundland. At one time, these tatters might have resembled

Dominica, St. Lucia, Montserrat. When Atlantic finally closes, these newer islands will come to resemble the older ones, chunks of rock crushed between closing continents.

The Great Arc of the Caribbean, as it is called, rings the sea, encompassing islands in both the Greater and Lesser Antilles, reaching through Cuba and Hispaniola to the north, Aruba and Blanquilla to the south. The arc and its wandering islands hold a rich history; they have traveled far to reach their current location. Although the Caribbean is considered part of Atlantic—their waters mingle freely, and Atlantic's currents swirl through the Caribbean before turning north—the floor of this smaller adjacent sea is of another place, originating far to the west, in the Pacific. In the hills of Caribbean islands, in bits and pieces of rock exposed in mine shafts and old quarries and along road cuts, riverbanks, and railroad beds, are signs of this history. It is long and convoluted.

Two hundred million years ago, as the great continent of Pangea broke apart, North and South America separated, opening the Gulf of Mexico and another anonymous sea, a nameless ancestor of the Caribbean. Like Atlantic and her predecessor Iapetus, this sea rested in a basin of volcanic rock created at a ridge of underwater mountains. Once a few hundred miles wide, it is for the most part long gone. Rocky outcrops in the Dominican Republic, Venezuela, and the tiny island of La Désirade record its birth. Like Atlantic, it had a shallow youth. Its waters rolled in between drifting landmasses and evaporated, rolled in and evaporated again, until eventually the sea deepened and stayed. A piece of the Pacific drifted into the western end of this sea and subsided beneath it. Where the two plates collided, the Great Arc of the Caribbean, a chain of islands 1,500 miles (2,500 kilometers) long, rose from the water. It included what would become the Greater and Lesser Antilles. Even farther west, where other pieces of the Pacific collided, another chain of islands arose, producing what would become Panama and Costa Rica. Approximately eighty million years ago, lava from a rising plume of hot rock flooded the seafloor between the two arcs, building a light, buoyant plateau 8 miles (13 kilometers) thick.

This plateau, born in the Pacific, would become the light, shallow seafloor of today's Caribbean, but eighty million years ago it was far west of its current home. As different pieces of the Pacific shifted position, it began drifting east, pushing one island arc ahead of it, dragging the other behind. Twice as thick as most ocean bottom, it rammed against seafloor from the older Caribbean and forced it to descend. Today, almost none remains. Oil wells drilled deep into the Bahamas have penetrated what is left of this ancient ocean basin. The Caribbean plate still moves east; the seafloor descending beneath it now belongs to Atlantic.

The Great Arc and the plateau squeezed through the gap between North and South America, but it was a tight fit. Not all the pieces slipped through, and the arc fell apart. The southern islands grazed South America. Some were thrust up into the Guajira Peninsula of Colombia and the Caribbean Mountains of Venezuela, and others—Aruba, Curaçao, Bonaire, La Orchila, La Blanquilla—are strung out along the continental shelf of South America, still in the water but firmly wedged into the continent. Tobago, a volcanic island born in the Pacific, leads the line, perched incongruously on the sands of South America, snagged by the shelf of a drifting continent.

The Great Arc's northern islands met a similar fate. As they passed through the narrow gateway between the continents, the islands of the Greater Antilles—Cuba, Hispaniola, and Puerto Rico—were pressed into the shelf of North America. In Haiti, where the Caribbean and North American plates were squeezed together, coral reefs were lifted high above the waterline. Just west of Puerto Rico, a sliver of limestone was raised to form Mona Island. The topography of Cuba contains the whole history of the Caribbean: seafloor formed during the initial separation of North and South America, pieces of the lava plateau built in the Pacific, shreds of volcanoes erupted when Cuba was still an active island in the arc, limestone from the North American shelf.

If the Caribbean were drained, and all the sand and shells that have settled on the bottom removed, the track left in the wake of the Great Arc on its eastward journey would be revealed for all to see. The signage is not purely geological. A colonial kitchen, apothecary, and carpenter shop from the Jamaican city of Port Royal are preserved today not on the city's original site, but under 55 feet (17 meters) of water off the island's coast. On June 7, 1692, this city, best known for the pirates who lived there and the saloons they frequented, slid into the sea, nudged by an earthquake released as the Caribbean and North American plates bumped. Two hundred years later, on the same site, fires from another earthquake burned the city of Kingston. The seam between the two plates is uneven. In Jamaica, they press into each other; elsewhere, they pull apart. South of Cuba, the two blocks separate as they pull past each other, opening the Cayman Trough, a sea within a sea, a younger, smaller version of the mountainous ridge opening Atlantic.

Volcanoes continue to erupt on the Lesser Antilles as seafloor, first from the ancient Caribbean and now from Atlantic, continues its descent into earth's mantle. Subtle motions on the subsiding seafloor, subtle shifts in the farewell dance of Atlantic's basin, are written into these islands. Seething volcanoes and high moist hills once rose from the islands of Antigua, Anguilla, and Barbuda, until hills on Atlantic's seafloor arrived at the edge of the Caribbean. More buoyant than the flat plains of the abyss, they resisted, but ultimately could not withstand, the pull of the depths. Their resistance softened the incline of descending seafloor, shifting volcanic eruptions west. Eruptions ceased on Antigua, Anguilla, and Barbuda, the volcanoes eroded away, leaving the limestone-covered islands flat and dry while fiery magma created the hilly islands of Montserrat, Nevis, and St. Kitts.

Molten rock rising from this more gentle slope doubled the size of Guadeloupe, fashioning its butterfly shape. The eastern island of Grande-Terre, formed before the Barracuda Ridge began its descent, is now flat, covered with limestone, its white sand beaches built from bits of broken coral. Across a narrow river rises the younger Basse-Terre, a craggy island containing boiling waterfalls, thick rain forest, and a smoldering volcano that has erupted at least eight times since Columbus first arrived there. Farther south, on Martinique, the effects of the gently sloping seafloor are less pronounced. The active volcanoes, once sited on the east side of the island, have merely shifted to the west. The Great Arc of the Caribbean—built as one sea, and then another, makes her farewell—may well be one of the oldest active island arcs in the world. The explosive eruptions that created the arc will continue, building islands and then destroying them, until the last piece of seafloor has made its exit.[6]

Departing seafloor is noisy and showy, but only a little of the descending rock creates the show. Most of the exiting seafloor slips back into the mantle, but even these pieces are not necessarily gone for good. The large, cold slabs come to rest down near earth's core. A piece of Atlantic sits at the bottom of the mantle, below the Caribbean. A piece of ancient Tethys lies below the Mediterranean. Mountains are worn down and washed into the sea, ground into minerals to become the shells and skeletons of animals who live and die, whose remains are built into seafloor and raised again into mountains. So, too, earth's interior melts and rises to form the foundation of the sea, then returns to its place of origin, only to rise again in another era, to build another sea. Lava welling up in the mountain valleys at the Mid-Atlantic Ridge, building new seafloor, carries the chemical signature of a recycled ocean basin. And so, it seems, neither earth's water nor its rock is truly new. Our living planet has recycled its resources, moving water from the sea to the atmosphere and back again, moving rock from the heights of mountains to the bottom of the sea and back again, from the mantle to the bottom of the sea and back again, sustaining itself for over four billion years.[7]

Across Atlantic, a plume of melting rock rises under Europe. Perhaps it contains traces of Iapetus or Tethys returning to the surface. It's too early to say, for this plume has yet to split a continent or open a sea. It raised France's Massif Central and carved out the valleys of the Rhine and the Saône Rivers. The plume still nudges the land. As recently as 1992, an earthquake, the worst northwest Europe has experienced in two hundred years, shook the Dutch city of Roermond, toppling hundreds of brick chimneys, shaking the ground in London and Berlin. Yet the cracked land has not come apart: the press of another continent, closing another piece of Atlantic, holds Europe together. Since Atlantic opened, Africa and Europe have engaged in a wobbly dance, veering together, coming apart, veering together again. Their faltering steps have built mountains and reduced ancient seas to puddles, creating Europe's landscape. One crash raised Iran's Zagros Mountains and reduced the vast Tethys to puddles, to the Black, Aral, and Caspian Seas. A separation opened the Mediterranean.[8]

Another nudge brought Africa against Spain, building the Pyrenees. A reversal sent Africa south, carrying Italy. The return shoved Italy against Europe, pushing up the Alps, folding layer upon layer of continental shelf into high mountains holding fossils of seashells. It also lifted the sea's volcanic plumbing out into the air. In the Troodos Mountains of Cyprus, it is possible to walk amidst coarse-grained rocks from an underwater magma chamber, piping that carried liquid rock into the sea and erupted pillow lavas. Pieces of "black smoker" chimneys lie on the ground, the remains of a once prolific field of deep-sea hot springs. The innards of the Troodos Mountains, once seafloor, have been mined since 2000 B.C. From Cyprus, island fashioned from the sea, comes the word copper.[9]

The two continents continue to bear down upon each other, with Turkey caught in between, shaking with earthquakes. Other sites along the Mediterranean manifest the strain. Three hundred years after Cleopatra's armies were defeated, her royal quarters in Alexandria fell into the sea, jolted by moving continents. Today, statues and sphinxes, pedestals and pottery, pieces of temples and pavement from streets all rest at the bottom of Alexandria's harbor, caked with sediment. Hidden in water murky with millions of gallons of untreated sewage lies Cleopatra's palace. When the Mediterranean finally closes and her seafloor is cast up to become dry land, it will contain a record not only of animals who lived in the sea but of humans who lived along the shore. Today, giant religious statues, temples, and houses from the ancient city of Herakleion rest in waters just off Egypt. This city, kept alive in the writings of Seneca and

Aeschylus, slid into the sea following an earthquake. When the Mediterranean finally closes, it too will return to land.[10]

In her youth, Atlantic grew from a necklace of lakes into a narrow, salty sea. As her waters drain away, she will retrace her steps, shrinking to a narrow seaway like today's Mediterranean. The Mediterranean, as it closes, has on occasion dried out. In closing, Atlantic may as well. Six million years ago, Africa and Europe shifted position, turning off the spigot filling the Mediterranean. With no water flowing in from Atlantic, the Mediterranean evaporated. Mammals walked from Spain to Morocco. The climate, tuned to variations in earth's orbit, alternated between wet and dry. In dry years, the Mediterranean was a shallow, briny lake lined with thick layers of salt. In wet years, flowing rivers emptied into the lake, capping it with fresh water. After half a million years, the sea level rose. Atlantic swelled, spilling over the Strait of Gibraltar into the Mediterranean. It was not a trickle but a deluge, the evidence being deep-water sediment lying directly above the older, shallow salt deposits. Basins brim with water and dry out, only to refill and empty again: these cycles, part of Atlantic's youth, may characterize her old age as well.[11]

Today, Africa and Europe continue drifting toward each other, squeezing the Mediterranean to fit its name, a sea between lands. As this sea closes, descending seafloor has pushed up arcs of islands, delivered explosive volcanoes to the edges of continents. It has built the Aegean Islands of Greece, the Aeolian Islands of Italy, carved Etna and Vesuvius into the Italian landscape. The continents approach ever so slowly, at a rate of two-tenths of an inch (four millimeters) each year, but the volcanoes explode violently. About thirty-five hundred years ago, the Greek Cycladic island of Thera blew up. Today, only a few bare rocks at the throat of the volcano and a piece of the rim lie above the water. Deep in the ash and lava are streets and houses from an ancient Minoan city. Collapsed rooms, beautifully detailed wall frescoes, painted jugs, vases, and cooking pots are the remains of a thriving community entombed by the march of continents. Ash from the eruption was lofted as far as

Egypt, where it is possible that the memory of this apocalypse was written into the biblical story of Exodus. It may be no coincidence that the plagues visited upon Pharaoh—soot and dust, vermin and insects, animals dying in the fields, and the descent of darkness—all accompany cataclysmic volcanic eruptions.

Where continents converge, earth is scorched. The Greeks named the Phlegraean Fields, near the Bay of Naples, from their word for "burn." Boiling muds and steaming hot springs characterize this place of unrest where Virgil's Aeneas descended into the Underworld and where Dante may have imagined his Inferno. For hundreds of years the area has swelled and subsided, the height of land governed by pulses of magma rising and falling in the ground below. More than once, water has lapped at the base of Roman ruins as the magma retreated. Today, the crater is lush with vegetation, but the volcano below, dormant since 1538, stirs. The ground has not yet bulged, but temperatures are rising. When it will erupt, no one can say.

Nearby, Vesuvius, the Pelée of the Mediterranean, rests. In 79 A.D., burning clouds of gas exploded from the volcano, and hurtled down its flanks, suffocating the residents of Pompeii and Herculaneum, burying the cities in ash and lava. Ash turned to fertile soil, and the catastrophe was forgotten. New cities arose from the plains. When Vesuvius last erupted, in 1944, lava moved slowly down the mountain, engulfing the village of San Sebastiano, killing twenty-seven people. Vesuvius has been dormant since then; temperatures in the crater's steaming vent have stayed steady, no increases in sulfur dioxide emissions have been detected. The interior often shudders with earthquakes, but between 1995 and 1999, two swarms were particularly strong, the strongest in fifty years.

Whether these earthquakes augur an eruption, no one can say. Solidified lava plugs the throat of Vesuvius. Like a cork on a bottle of spumante, the plug seals the volcano, keeping volatile gasses inside. If it pops, the volcano will explode, releasing clouds of burning ash and gas, the same kind of incandescent clouds that killed twenty thousand people in Pompeii, twenty-eight thousand in St. Pierre. The difference today is that three million people now live in Naples and its surroundings, almost one million of them on the slopes of Vesuvius. As Africa and Europe draw toward each other, closing the Mediterranean, another outpouring from Vesuvius may be inevitable. It is only a matter of time.

Eventually, all things come to an end. What opens, closes. What emerges, recedes. Earth's oceans come and go as drifting continents break apart and come together. While Atlantic's mountainous ridge is still warm and buoyant, her edges are thick and old. In the next ten or twenty million years, this sea, now in her prime, will begin to close. First the Mediterranean will disappear, seaming Africa and Europe with a chain of high mountains. Then the two existing tears in Atlantic's basin, one at the edge of the Caribbean and another south of Argentina, may lengthen, creating deep trenches all along the eastern coasts of North and South America. Strings of explosive volcanic islands will chart Atlantic's descent.

Atlantic took 200 million years to reach her current width. She will take another 200 million to depart. Just as pieces of ancient Iapetus thrust up into the continents contain the history of those who dwelled in her waters, so too will Atlantic seafloor record the comings and goings of her inhabitants. This time, man will figure prominently in the record. Pieces of a stromatolite reef scattered in the woods of Saratoga Springs, New York, describe an ancient ocean where tiny unicellular organisms evolved to breathe oxygen into the atmosphere. Pieces of Atlantic coral reefs strewn on some future landscape may describe a deterioration in diversity brought on by global warming, pollution, and the removal of key predators.

Other pieces of Atlantic seafloor may hold more direct indicators of our presence. When the Mediterranean closes, pieces of Cleopatra's palace and statues from ancient Herakleion may be lifted high into new mountains rising between Africa and Europe. In a few hundred years, portions of Atlantic's edges may be flooded by rising sea levels. When Atlantic closes, pieces of New Orleans and Miami may become the stuff of mountaintops. Whoever might be picking through the rock of these mountains looking for Atlantic's history will find quite a story.[12]

Researching this book, I began to understand how Atlantic's waters move, how her webs of life are woven, how her basin holds her history. My inquiry, leading me from Atlantic's coastal waters to the far reaches of the Sargasso, from tiny food web to top predator, from surface to seafloor, from newly forming ridge to ancient foundation, revealed a sea whose generosity nurtures us, sustains us, and brought us here. A deep-sea hot spring, spewing from the floor of one of Atlantic's primordial ancestors, may have nurtured earth's earliest life. The emergence of the Isthmus of Panama helped initiate an ice age, sending our ancestors down from the trees. Cold, salty water sinking off Greenland and Labrador drags the heated Gulf Stream up from the Tropics, blessing harsh lands with warmth. Pulled by currents, nutrient-rich waters well up from the depths, sustaining what were, and still could be, great fisheries. Each exploration revealed the complexity and mystery of a sea still largely unfathomed.

In the twenty-five years since I was in college, our insight into Atlantic has deepened profoundly, overturning our assumptions, restructuring our understanding. Until we had access to sophisticated microscopes, we didn't realize that the most abundant photosynthesizing organisms in the ocean, *Synechococcus* and *Prochlorococcus,* were ones we can't see. Turning light into carbohydrate in the smallest of spaces, they work on the grandest of scales, building the base of entire marine food webs. We have found, in the black, scalding waters of hot springs, tiny bacteria living without light, redefining for us the realm of life. We have found that currents flowing along the seafloor of Atlantic, rejuvenating bottom communities with oxygen, can pull our climate into and out of a deep freeze. In the past twenty-five years, science has made explicit what, at some level, we already knew but could not explain: we are embraced by the sea.

As vast as Atlantic seems, this sea can be a small, intimate place.

Bluefin tuna feed in New England waters but may spawn out near the Azores. A loggerhead sea turtle born on the beaches of Florida may spend its youth in the Mediterranean. Running water joins faraway shores, carrying dust from the Sahara into the Caribbean, salt from the Mediterranean into Arctic seas. Fast-moving currents wash water away but then bring it back.

In these boundaried, finite waters, where the flow of wind and wave affect the well-being of bacteria and bluefin tuna, human endeavor has made an impression. My initial interest in Atlantic was to make an acquaintance with the sea. Everywhere I looked, I found man. Tiny floating plants, visible only to those with sophisticated fluorescing microscopes, live out their lives according to rhythms determined by the synthetic nitrogen we introduce into the water. High in the Arctic, halibut are laced with pesiticides. Sperm whales feeding in depths few men have plumbed, eating giant squid no man has seen in its natural habitat, nonetheless accumulate toxic chemicals in their blubber. We live ashore and can only visit the sea, and then usually only on its surface, but nonetheless we reach far. At the beginning of the twenty-first century, we no longer stand outside the marine realm but, rather, are fully of it. Atlantic can no longer be understood without considering the work of man.

We hold these life-giving waters in our hands. At the dawn of a new millennium, we are perceiving the fingerprint of a global warming brought on by our actions. The temperature of Atlantic is rising. Greenland glaciers are melting, Arctic sea ice is retreating, and the Northwest Passage is opening. We barely understand the implications, but if Atlantic's history has anything to share, they may be monumental. Discharging nutrients and chemicals into our coastal waters, we have toppled marine food webs and restructured entire communities of marine life, with no real understanding of the effects. The edges of the sea have trouble breathing, while we test the resilience of open waters with massive amounts of fertilizer. Depleting our fisheries, we seem ready to replace our wild fish populations with marine farms. My children's children may never see a wild salmon slapping against a waterfall as it finds its way home. In our own time, we who pride ourselves on our commitment to endangered species are driving the right whale to extinction.

Our own departure is inevitable. The history of Atlantic and her predecessors suggests that our hold on earth is temporary. Motions of drifting plates led to our walking upright, provided raw material for our first tools, but like that of so many who have dwelled in Atlantic, and like that of the sea herself and the seas that came before her, our time has come and will be gone. Like the community of plants and animals crowded on a piece of *Sargassum* that sinks when the seaweed grows too old and tired to keep afloat, we may also sink of our own accumulated weight when earth can no longer sustain us. A difference is that now we may be accelerating our own leave-taking, as the unsustainable practices that brought down ancient human civilizations now threaten the entire planet.

Twenty-five years of scientific sounding of the sea have revealed that in ways too numerous to count, too profound to measure fully, Atlantic's great waters are life-giving. For we who have finished the work of Adam and Eve, who have toiled in the wilderness and reclaimed it, who have subdued earth and come to rule over the fish and fowl, it is time to recognize this gift for the bounty it brings and to seek, in our thoughts, words, and deeds, to return it in full measure.

Notes

PART I: MOVING OFFSHORE

Chapter 1: Waters of Life

1. Oceans constitute 99.5 percent of earth's living space by volume in G. Richard Harbison, "The Gelatinous Inhabitants of the Ocean Interior," *Oceanus,* Fall 1992, 18–24.

2. Floating plants produce as much carbon as land plants with 0.2 percent of the biomass in Sallie W. Chisholm, "What Limits Phytoplankton Growth?," in P. G. Falkowski, R. T. Barber, and V. Smetacek, "Biogeochemical Controls and Feedbacks on Ocean Primary Production," *Science* 281 (5374), July 10, 1998: 200–206, and in C. B. Field et al., "Primary Production of the Biosphere: Integrating Terrestrial and Oceanic Components," *Science* 281 (5374), July 10, 1998: 237–41. One million progeny in three weeks in Henry S. Parker, *Exploring the Oceans* (Englewood Cliffs, N.J.: Prentice-Hall, 1985), 211.

3. *Prochlorococcus* in Conrad Mullineaux, "The Plankton and the Planet," *Science* 283 (5403), February 5, 1999: 801–2. Bacterial and microbial loop in Farooq Azam, "Microbial Control of Oceanic Flux: The Plot Thickens," *Science* 280 (5364), May 1, 1998: 694–96. Bacteria and viruses in Jed A. Fuhrman, "Marine Viruses and Their Biogeochemical and Ecological Effects," *Nature* 399 (6736), June 10, 1999: 541–47.

4. Deep algae in Bayard Webster, "Plant Is Discovered at Record Ocean Depth," *New York Times,* December 28, 1984, A1, A13. Sulfur bacteria in H. N. Schulz et al., "Dense Populations of a Giant Sulfur Bacterium in Namibian Shelf Sediments," *Science* 284 (5413), April 16, 1999: 493–95, and in Bernice Wuethrich, "Giant Sulfur-Eating Microbe Found," *Science* 284 (5413), April 16, 1999: 415.

5. Photosynthesizing animals, consuming plants, and increased efficiency of marine food webs in Diane K. Stoecker, "Animals That Photosynthesize and Plants That Eat," *Oceanus,* Fall 1992, 24–28. *Mesodinium rubrum* in D. E. Gustafson et al., "Cryptophyte Algae Are Robbed of Their Organelles by the Marine Ciliate *Mesodinium rubrum,*" *Nature* 405 (6790), June 29, 2000: 1049–52. Bacteria converting light into energy in Z. S. Kolber et al., "Bacterial Photosynthesis in Surface Waters of the Open Ocean," *Nature* 407 (6801), September 4, 2000: 177–79, in Oded Béjà et al., "Bacterial Rhodopsin: Evidence for a New Type of Phototrophy in the Sea," *Science* 289 (5486), September 15, 2000: 1902–06, and in Elizabeth Pennisi, "High-Tech Lures Hook into New Marine Microbes," *Science* 289 (5486), September 15, 2000: 1869.

6. Copepods in James W. Nybakken, *Marine Biology: An Ecological Approach,* 3d ed. (New York: HarperCollins, 1993), 42, 68, 75–77, in Henry B. Bigelow and William C. Schroeder, *Fishes of the Gulf of Maine* (Washington, D.C.: U.S. Fish and Wildlife Service, 1953), 320, and in A. D. Davis et al., "Myelin-like Sheaths in Copepod Axons," *Nature* 389 (6728), April 15, 1998: 571. *Oikopleura* in James L. Sumich, *An Introduction to the Biology of Marine Life* (Dubuque, Iowa: William C. Brown, 1984), 240, and in Sir Alister Hardy, *The Open Sea: Its Natural History, Part I: The World of Plankton* (Boston: Houghton Mifflin, 1970), 152. Video plankton recorders in Marguerite Holloway, "The Plankton Stalkers," *Scientific American,* April 1992, 20.

7. Phytoplankton blooms in Robert Kunzig, "Invisible Garden," *Discover,* April 1990, 67–74, in "Atlantic Algae Form Bloom Bigger Than Britain," *New Scientist* 131 (1777), July 13, 1991: 22, in Bob Holmes, "Fish TV Presents . . . New Instruments to Track Plankton," *New Scientist* 152 (2054), November 2, 1996: S21–23, and in Sumich, *Biology of Marine Life,* 208.

8. Copepod populations varying with blooms in Nybakken, *Marine Biology,* 70, in J.-M. Fromentin and B. Planque, "*Calanus* and Environment in the Eastern North Atlantic. II. Influence of the North Atlantic Oscillation on *C. finmarchicus* and *C. nelgolandicus,*" *Marine Ecology Progress Series* 134, April 25, 1996: 111–18, in Sumich, *Biology of Marine Life,* 238–39. Plaice in J. Metcalfe, G. Arnold, and I. E. Priede, "Time and Tide Wait for No Plaice," *New Scientist* 125 (1698), January 6, 1990: 52–56.

9. Microbial loop and viruses and bacteria in Azam, "Microbial Control," in David L. Kirchman, "Phytoplankton Death in the Sea," *Nature* 398 (6725), March 25, 1999: 293–94, and in Fuhrman, "Marine Viruses," 541–47. How bacteria compete successfully in J. S. Martinez et al., "Self-Assembling Amphiphilic Siderophores from Marine Bacteria," *Science* 287 (5456), February 18, 2000: 1245–47, in D. A. Hutchins et al., "Competition Among Marine Phytoplankton for Different Chelated Iron Species," *Nature* 400 (6747), August 26, 1999: 858–60; and in David L. Kirchman, "Microbial Ferrous Wheel," *Nature* 383 (6598), August 26, 1996: 303–4. Bacteria and phytoplankton in the Mediterranean in Nigel Williams, "The Mediterranean Beckons to Europe's Oceanographers," *Science* 279 (5350), January 23, 1998: 483–84. Bacteria and Grand Banks fishery in L. R. Pomeroy and Don Diebel, "Temperature Regulation of Bacterial Activity During the Spring Bloom in Newfoundland Coastal Waters," *Science* 233, 1986: 359–62. Resource competition models in Juf Huisman and Franz J. Weissing, "Biodiversity of Plankton by Species Oscillations and Chaos," *Nature* 402 (6770), November 25, 1999: 407–10, and in Ulrich Sommer, "Competition and Coexistence," *Nature* 402 (6770), November 25, 1999: 366–67. Functional redun-

dancies and strong phytoplankton web in Falkowski, Barber, and Smetacek, "Biogeochemical Controls."

10. Extent of ozone loss in Fred Pearce, "Algal Gloom," *New Scientist* 159 (2146), August 8, 1998: 24, and in Richard Kerr, "Deep Chill Triggers Record Ozone Hole," *Science* 282 (5388), October 16, 1998: 391. Continuing chemical pollution of stratosphere in S. A. Montzka et al., "Present and Future Trends in the Atmosphere Burden of Ozone-Depleting Halogens," *Nature* 398 (6729), April 22, 1999: 690–94, and in P. J. Fraser and M. J. Butler, "Uncertain Road to Ozone Recovery," *Nature* 398 (6729), April 1999: 663–64. Ozone loss and global warming in Peter Aldhous, "Global Warming Could Be Bad News for Arctic Ozone Layer," *Nature* 404 (6778), April 6, 2000: 531–33, and in A. W. Waibel et al., "Arctic Ozone Loss Due to Denitrification," *Science* 283 (5410), March 26, 1999: 2064–69. Ozone hole and phytoplankton productivity in R. C. Smith et al., "Ozone Depletion: Ultraviolet Radiation and Phytoplankton Biology in Antarctic Waters," *Science* 255 (5047), February 21, 1992: 952–60, in P. J. Neale, R. F. Davis, and J. J. Cullen in "Interactive Effects of Ozone Depletion and Vertical Mixing on Photosynthesis of Antarctic Plankton," *Nature* 392 (6676), April 9, 1998: 585–92, and in Pearce, "Algal Gloom," 24.

11. Sulfur, phytoplankton, viruses, bacteria, clouds, and climate in Timothy M. Lenton, "Gaia and Natural Selection," *Nature* 394 (6692), July 30, 1998: 439–47. Sulfur and climate in Gillian Malin, "Sulphur, Climate and the Microbial Maze," *Nature* 387 (6636), June 26, 1997: 857–58. Sulfur and shipping in K. Capaldo et al., "Effects of Ship Emissions on Sulphur Cycling and Radiative Climate Forcing over the Ocean," *Nature* 400 (6746), August 19, 1999: 743–46 and Barry J. Huebert, "Sulphur Emissions from Ships," *Nature* 400 (6746), August 19, 1999: 713–14.

12. Carbon dioxide will triple in William K. Stevens, "Experts Doubt Rise of Greenhouse Gas Will Be Curtailed," *New York Times,* November 3, 1997, A1, A12, and in John Houghton, *Global Warming: The Complete Briefing* (Oxford: Lion Publishing, 1994), 37. Phytoplankton production and carbon export in Falkowski, Barber, and Smetacek, "Biogeochemical Controls." Iron enrichment and phytoplankton productivity in K. N. Coale et al., "A Massive Phytoplankton Bloom Induced by an Ecosystem-Scale Iron Fertilization Experiment in the Equatorial Pacific Ocean," *Nature* 383 (6600), October 10, 1996: 495–501, in Bruce W. Frost, "Phytoplankton Bloom on Iron Rations," *Nature* 383 (6600), October 10, 1996: 475–76, in Philip W. Boyd et al., "A Mesoscale Phytoplankton Bloom in the Polar Southern Ocean Stimulated by Iron Fertilization," *Nature* 407 (6605), October 12, 2000: 695–702, and in Sallie W. Chisholm, "Stirring Times in the Southern Ocean," *Nature* 407 (6605), October 12, 2000: 685–87. Entrepreneurs planning to sow iron in open ocean in Gretel Schueller, "Testing the Waters," *New Scientist* 164 (2206), October 2, 1999: 34–36, and in Chisholm, "Stirring Times." Increase in production does not necessarily mean increased export, Adina Paytan, "Iron Uncertainty," *Nature* 406 (6795), August 3, 2000: 468–69, in Boyd et al., "Mesoscale Phytoplankton Bloom," and in Chisholm, "Stirring Times." Greenland Sea returning carbon dioxide in L. A. Miller et al., "Seasonal Dissolved Inorganic Carbon Variations in the Greenland Sea and Implications for Atmospheric CO_2 Exchange," *Deep-Sea Research, Part II—Topical Studies in Oceanography* 46 (6–7), 1999: 1473–96. Carbon dioxide leaking back to atmosphere in M. E. Huntley, M. D. Lopez, and D. M. Karl, "Top Predators in Southern Ocean: A Major Leak in the Biological Carbon Pump," *Science* 253 (5015), July 5, 1991: 64–66. Diatoms up over *Prochlorococcus* in Gretel Schueller,

"Testing the Waters," *New Scientist* 164 (2206), October 2, 1999: 34–36. Thinner diatoms with iron, more dissolution by bacteria in Ed Boyle, "Pumping Iron Makes Thinner Diatoms," *Nature* 393 (6687), June 25, 1998: 733–34, and in Victor Smetacek, "Bacteria and Silica Cycling," *Nature* 397 (6719), February 11, 1999: 476–77. Poor link between phytoplankton and fish biomass in Fiorenza Michele, "Eutrophication, Fisheries, and Consumer-Resource Dynamics in Marine Pelagic Ecosystems," *Science* 285 (5432), August 27, 1999: 1396–98. Effect of diatoms on copepods in A. Miralto, G. Barane, and G. Romano, "The Insidious Effect of Diatoms on Copepod Reproduction," *Nature* 402 (6758), November 11, 1999: 173–76. Depleted water into Mediterranean in Williams, "The Mediterranean Beckons." Blooms off Iberia in Debora MacKenzie, "Algae Stole Our Carbon," *New Scientist* 162 (2185), May 8, 1999: 15. Possible dead zone in Yvonne Baskin, "Can Iron Supplements Make the Equatorial Pacific Bloom?," *Bioscience* 45 (5), May 1995: 314–17. Methane as side effect in Richard Monastersky, "Iron versus the Greenhouse," *Science News* 148 (14), September 30, 1995: 220–22. Sulfur hexafluoride outgassing in D. J. Cooper, A. J. Watson, and P. D. Nightengale, "Large Decrease in Ocean-Surface CO_2 Fugacity in Response to In Situ Iron Fertilization," *Nature* 383 (6600), October 10, 1996: 511–13. Estimate of carbon dioxide drawdown in Antarctica in T. H. Peng and W. S. Broecker, "Dynamical Limitations on the Antarctic Iron Fertilization Strategy," *Nature* 349, January 17, 1991: 227–28, in Boyd, "Mesoscale Phytoplankton Bloom," in Chisholm, "Stirring Times," and in A. J. Watson et al., "Effect of Iron Supply on Southern Ocean CO_2 Uptake and Implications for Glacial Atmosphere CO_2," *Nature* 407 (6605), October 12, 2000: 730–33.

13. Ocean warming in "Tropical Waters in Northern Hemisphere Heating at an Accelerated Rate, National Oceanic and Atmospheric Administration (NOAA) Reports," NOAA press release, July 28, 2000, in R. Monastersky, "Temperatures on the Rise in Deep Atlantic," *Science News* 145 (19), May 7, 1994: 295, in Alain Poisson, "Carbon Dioxide Uptake at Sea," *Nature* 396 (6711), December 10, 1998: 521–22, in Williams, "The Mediterranean Beckons," and in S. Levitus et al., "Warming of the World Ocean," *Science* 287 (5461), March 24, 2000: 2225–29. Warm water absorbs less carbon dioxide in R. Monastersky, "Carbon Dioxide Marches to an Uneven Beat," *Science News* 147 (25), June 24, 1995: 390–91. Effects of global warming on blooming and carbon dioxide removal and export in K. R. Arrigo et al., "Phytoplankton Community Structure and the Drawdown of Nutrients and CO_2 in the Southern Ocean," *Science* 283 (5400), July 15, 1999: 365–67, and G. R. DiTullio et al., "Rapid and Early Export of *Phaeocystis antarctica* Blooms in the Ross Sea, Antarctica," *Nature* 404 (6788), April 6, 2000: 595–98. Shutting off carbon export via phytoplankton triples atmospheric carbon dioxide levels in Chisholm, "What Limits Phytoplankton Growth?" Warming effect on food webs in Dean Roemmich and John McGowan, "Climate Warming and the Decline of Zooplankton in the California Current," *Science* 257 (5202), March 3, 1995: 1324–27; in J. Kaiser, "Vanishing Zooplankton," *Science News* 146 (10), March 11, 1995: 151, and in J. A. McGowan, D. R. Cayan, and L. M. Dorman, "Climate-Ocean Variability and Ecosystem Response in the Northeast Pacific," *Science* 281 (5374), July 10, 1998: 210–17.

Chapter 2: A Diminished Thing

1. First animals on dry land in C. Raymo and M. E. Raymo, *Written in Stone: A Geological and Natural History of the Northeastern United States* (Chester, Conn.:

Globe Pequot, 1989), 77. Whale blubber in James W. Nybakken, *Marine Biology: An Ecological Approach,* 3d ed. (New York: HarperCollins, 1993), 112. Mackerel in H. B. Bigelow and W. C. Schroeder, *Fishes of the Gulf of Maine* (Washington, D.C.: U.S. Fish and Wildlife Service, 1953), 318–19, and in James L. Sumich, *An Introduction to the Biology of Marine Life,* 3d ed. (Dubuque, Iowa: William C. Brown, 1984), 245, 256, 259. Pressure and diving in Sylvia Earle, *Sea Change* (New York: G. P. Putnam's, 1995), 54, 81, 101, 127.

2. Shelf and rise in M. Grant Gross, *Oceanography: A View of the Earth,* 3d ed. (Englewood Cliffs, N.J.: Prentice-Hall, 1982), 37, in K. O. Emery and E. Uchupi, *The Geology of the Atlantic Ocean* (New York: Springer-Verlag, 1984), 33, 46, 57, and in James P. Kennett, *Marine Geology* (Englewood Cliffs, N.J.: Prentice-Hall, 1982), 308–9. Characteristics of Georges Bank in Richard Backus, "Geology," in Richard Backus, "Does Georges Shoal Every Dry?," and in Kenneth Emery, "Georges Cape, Georges Island, Georges Bank," from Richard Backus, ed., *Georges Bank* (Cambridge: MIT Press, 1987), 22–24, 38–39, 139. Succession of climate on Georges in "Prehistoric Man on Martha's Vineyard," *Oceanus,* Spring 1985, 36–37.

3. Copepods waiting out the winter in Nybakken, *Marine Biology,* 70–71. Percentage of primary productivity supporting fisheries in L. W. Botsford, J. C. Castilla, and C. H. Peterson, "The Management of Fisheries and Marine Ecosystems," *Science* 277 (5325), July 25, 1997: 509–15. Trophic levels in marine food webs in A. C. Duxbury, A. B. Duxbury, and K. A. Sverdrup, *An Introduction to the World's Oceans,* 6th ed. (Boston: McGraw-Hill, 2000), 389. Copepods and mackerel in Bigelow and Schroeder, *Fishes of the Gulf of Maine,* 320.

4. Characteristics of cod in Bigelow and Schroeder, *Fishes of the Gulf of Maine,* 182–96, and in Sumich, *Biology of Marine Life,* 246, 268–69. Cod schools in George A. Rose, "Cod Spawning on a Migration Highway in the North-West Atlantic," *Nature* 366 (6454), December 2, 1993: 458–62.

5. Cod declines in Colin Nickerson, "Off Canada, Diminished Fishing Stocks Cause a Crisis," *Boston Globe,* November 26, 1993, 1, 36, in Marcia Barinaga, "New Study Provides Some Good News for Fisheries," *Science* 269 (5227), August 25, 1995: 1043, and in Vaughan Anthony, "Special Advisory Groundfish Status on Georges Bank," presented to New England Fishery Management Council, August 9–10, 1994. Mature fish fifteen inches long in Rose, "Cod Spawning on a Migration Highway." Most Newfoundland cod catch immature in Alan Harman, "Newfoundland Cod Fishery Closed for Two Years," *National Fisherman,* September 1992, 19. No signs of recovery yet in E. L. Dalley and J. T. Anderson, "DFO Canadian Stock Assessment Secretariat Research Document—98/121," Canadian Department of Fisheries and Oceans, 1998. Status of Atlantic fisheries in UN Food and Agricultural Organization (FAO), "The State of World Fisheries 1998," part 1, World Review of Fisheries and Aquaculture.

6. Most stocks fished unsustainably in European Environment Agency, "Environment in the European Union at the Turn of the Century, 1999," 361–62. North Sea cod in David Symes, "UK Demersal Fisheries and the North Sea: Problems in Renewable Resource Management," *Geography,* 76 (331, part 2), April 1991: 1131–42, in R. M. Cook, A. Sinclair, and G. Stefansson, "Potential Collapse of North Sea Cod Stocks," *Nature* 385 (6616), February 6, 1997: 521–22, in C. M. O'Brien et al., "Climate

Variability and North Sea Cod," *Nature* 404 (6774), March 9, 2000: 142, and in Advisory Committee on Fishery Management, "ACFM Report 1999," International Council for the Exploration of the Sea, 234–35.

7. Halibut, pp. 249–58, and redfish, pp. 430–37, in Bigelow and Schroeder, *Fishes in the Gulf of Maine.* Exploitation rate in the Gulf of Maine in R. Mayo and L. O'Brien, "Status of Fishery Resources of the Northeastern United States in 1998," National Oceanic and Atmospheric Administration (NOAA) technical memorandum NMFS-NE-115, revised January 2000. Catch double the target landings in recent years in "Gulf of Maine Cod Near Landings Target for First Time," National Marine Fisheries Service (NMFS) Northeast Region press release, January 18, 2000. Seventy percent of New England groundfish overfished in "Report to Congress: Status of Fisheries of the United States," NMFS, October 1999, and "Status of Fishery Resources of the Northeastern United States in 1998," NOAA technical memorandum NMFS-NE-115, 19. Chances of stock recovery in Conservation Law Foundation et al., "In the U.S. District Court for the District of Columbia: Complaint for Declaratory, Mandatory and Injunctive Relief," May 2000. Long-term cod and haddock collapse in Jeffrey A. Hutchings, "Collapse and Recovery of Marine Fishes," *Nature* 406 (6798), August 24, 2000: 882–85.

8. Challenges of fisheries science in T. Lauck et al., "Implementing the Precautionary Principle in Fisheries Management Through Marine Reserves," *Ecological Applications* 8 (1), supplement, February 1998: S72–78, and in Callum M. Roberts, "Ecological Advice for the Global Fisheries Crisis," *Trends in Ecology and Evolution* 12 (1), 1997: 35–38. Egg-hatching percentages from personal communication, Steve Murawski, Northeast Fisheries Science Center, August 2000. Environmental variations and Atlantic fisheries in E. E. Hofmann, and T. M. Powell, "Environmental Variability Effects on Marine Fisheries: Four Case Histories," *Ecological Applications,* 8 (1), supplement, February 1998: S23–32. European fishery in "Ministers Agree to Deep Cuts in EU Fish Catches," Environmental News Service Daily (ENDS) press release, December 17, 1999.

9. Developing nations and debt in "Our Plundered Seas," *World Press Review* 42 (6), June 1995: 36 (2 pp.). Argentinian hake in "Argentinean Fisheries Crisis Highlights Need for WTO Action," WWF International press release, November 29, 1999. Morocco fisheries in "More than Fish," *Economist* 336 (7931), September 9, 1995: 47, and in "Morocco's Fish Stock: Dangerous Decline," *North Africa Journal and Mahgreb Weekly Monitor,* no. 28, May 16, 1998. Mauritania and Senegal, pp. 20–37, and Reykjanes Ridge, pp. 47–64, in R. Bonfil et al., Worldwide Fund for Nature International, "The Footprint of Distant Water Fleets on World Fisheries," 1998. Protein from fish and European catch from abroad in E. Kemf, M. Sutton, and A. Wilson, "Marine Fishes in the Wild: Special Status Report," World Wildlife Fund, 1996. Northwest Africa overfished in FAO, "State of World Fisheries and Aquaculture."

10. Redfish in Bonfil et al., "Footprint of Distant Water Fleets," 47–64. Toothfish in Fred Pearce, "The Ones That Got Away," *New Scientist* 149 (2012), January 13, 1996: 14, and in "NOAA Proposes New Conservation Measures to Protect Toothfish," NOAA press release 00-R108, March 13, 2000. Spain on the Grand Banks in Carl Safina, "The World's Imperiled Fish," *Scientific American,* November 1995, 46 (8 pp.), and in "Our Plundered Seas."

11. Monkfish in Northeast Fisheries Science Center in "Status of Fishery Resources," NMFS-NE-115, 88. Spiny dogfish in "Spiny Dogfish Plan Implemented," NOAA press release 00-Rlll, April 5, 2000. Fishing in deep waters in Michael Rivlin, "Deep New Sea," *National Fisherman,* November 1996, 20, in "Adding Depth to Ireland's Fish Menu," Ireland Department of Marine and Natural Resources press release, December 8, 1999, and in "Commissioner Fischler Calls for Measures to Conserve Deep-Water Species," European Commission Division of Fisheries press release, July 6, 2000.

12. Fishing down the food chain in D. Pauly et al., "Fishing Down Marine Food Webs," *Science* 279 (5352), February 6, 1998: 860–61. Shetland Island sand eels and seabirds in Mark Avery and Rhys Green, "Not Enough Fish in the Sea," *New Scientist* 123 (1674), July 22, 1989: 28–29, and in "Disaster Strikes Breeding Colonies in Shetland," *New Scientist* 119 (1625), August 11, 1988: 23. Capelin, herring, and Barents Sea cod, seabirds, mammals, and industrial fishing in Phil Aikman, "Industrial 'Hoover' Fishing: A Policy Vacuum," Greenpeace International, January 1997, and in "Scientists Recommend 20,000 Square Kilometer Closure of Sand Eel Fishery," Royal Society for the Preservation of Birds press release, December 14, 1999. Industrial fishing and seabirds in OSPAR, "OSPAR Quality Status Report 2000," chap. 6, Overall Assessment. Grand Banks seals in Debora MacKenzie, "Seals to the Slaughter," *New Scientist* 149 (2021), March 16, 1996: 34.

13. Swim bladders, air pressure, and rapid ascents in Sumich, *Biology of Marine Life,* 252–57, in Brad Warren, "Major Study Points to Many Problems but Few Answers," *National Fisherman,* May 1994, 17, and in Eric Denton, "The Buoyancy of Marine Animals," *Scientific American,* July 1960, reprinted in *Life in the Sea: Readings from Scientific American* (San Francisco: W. H. Freeman), 49–56. Georges Bank discard rate in Northeast Fisheries Science Center, "Status of Fishery Resources of the Northeastern United States for 1994," NOAA technical memorandum NE-108, 37. Whiting fishery in Symes, "Demersal Fisheries and the North Sea," and in Phil Aikman, "Industrial 'Hoover' Fishing." Cod thrown overboard in New England in Conservation Law Foundation et al., "Complaint for Declaratory, Mandatory and Injunctive Relief." Barn-door skates in Jill M. Casey and Ransom A. Myers, "Near Extinction of a Large, Widely Distributed Fish," *Science* 281 (5377), July 31, 1998: 690–92, and Janet Raloff, "Skating to Extinction?," *Science News* 155 (18), May 1, 1999: 280–83.

14. Effects of trawling in Michel J. Kaiser, "Significance of Bottom-Fishing Disturbance," *Conservation Biology* 12 (6), December 1998: 1230–35, in L. Watling and E. A. Norse, "Disturbance of the Seabed by Mobile Fishing Gear," *Conservation Biology* 12 (6), December 1998: 1180–97, in P. Schwinghamer et al., "Effects of Experimental Otter Trawling on Surficial Sediment Properties of a Sandy-Bottom Ecosystem on the Grand Banks of Newfoundland," *Conservation Biology* 12 (6), December 1998: 1215–22, in J. Prena et al., "Experimental Otter Trawling on a Sandy Beach Ecosystem of the Grand Banks of Newfoundland: Analysis of Trawl Bycatch and Effects on Epifauna," *Marine Ecology Progress Series* 181, 1999: 107–24, in Jeremy Collie, "Studies in New England of Fishing Gear Impacts on the Sea Floor," pp. 53–62, in Les Watling, "Benthic Fauna of Soft Substrates in the Gulf of Maine," pp. 20–29, in Fred Bennet, "Changes to the Sea Floor in the Chatham Area," pp. 115–16, and in Richard Langton, "Bottom Habitat Requirements of Groundfish," pp. 38–43, in E. Dorsey and J. Pederson, eds., *Effects of Fishing Gear on the Seafloor*

of New England (Boston: Conservation Law Foundation, 1998). Switch from long-living to short-living species in P. J. Auster and R. Langton, "Effects of Fishing on Fish Habitat," American Fisheries Society Symposium 22, 1999: 150–87. Survival of cod larvae in R. S. Gregory and J. T. Anderson, "Substrate Selection and Use of Protective Cover by Juvenile Atlantic Cod, *Gadus morhua,* in Inshore Waters off Newfoundland," *Marine Ecology Progress Series* 146, 1997: 9–20. Effects of trawling on Georges Bank in J. S. Collie, G. A. Escanero, and P. C. Valentine, "Photographic Evaluation of the Impacts of Bottom Fishing on Benthic Epifauna," *ICES Journal of Marine Science* 57, 2000: 987–1001.

15. Need to reduce mortality in New England in M. J. Fogarty and S. A. Murawski, "Large-Scale Disturbance and the Structure of Marine Systems: Fishery Impacts on Georges Bank," *Ecological Applications* 8 (1), supplement, February 1998: S6–22. Europe fleet in European Environment Agency, "Environment in the European Union," 357–75, in "Argentinian Fisheries Crisis Highlights Need for WTO Action," WWF International press release, November 29, 1999, in "WWF Welcomes Decision on European Fishing Quotas," WWF International press release, December 17, 1999, and in David Schoor, "Towards Rational Disciplines on Subsidies to the Fishing Sector," pp. 146–72 in WWF International, "The Footprint of Distant Water Fisheries on World Fisheries," 1998. Subsidies in "Coalition of Fishermen, Conservationists and Scientists Releases Tally of Fisheries Mismanagement Costs, Urges Congress to Pass Fisheries Recovery Act," Marine Fish Conservation Network press release, March 3, 2000. Marine sanctuaries in Tatiana Brailovskaya, "Obstacles to Protecting Marine Biodiversity Through Marine Wilderness Preservation: Examples from the New England Region," *Conservation Biology* 12 (6), December 1998: 1236–40, in Karin Jegalian, "Plan Would Protect New England Coast," *Science* 284 (5412), April 9, 1999: 237, in "21 Scientists Map Top Priority Areas for Protection in the Gulf of Maine and Georges Bank," Marine Conservation Biology Institute press release, April 2, 1999, in G. W. Alison, J. Lubchenco, and M. H. Carr, "Marine Reserves Are Necessary but Not Sufficient for Marine Conservation," *Ecological Applications* 8 (1), supplement, February 1998: S79–92, in T. Lauk, C. W. Clark, M. Mangel, and G. R. Munro, "Implementing the Precautionary Principle in Fisheries Management Through Marine Reserves," *Ecological Applications* 8 (1), supplement, February 1998: S72–78, and on barn-door skate in Casey and Myers, "Near Extinction." Barn-door skate in Bigelow and Schroeder, *Fishes in the Gulf of Maine,* 60–63.
16. Endangered species listing in David Malakoff, "Atlantic Salmon Spawn Fight over Species Protection," *Science* 279 (5352), February 6, 1998: 777, in "Federal Government Proposes Endangered Species Listing for Atlantic Salmon in Maine," NOAA press release, November 17, 1999, in Robert Braile, "Proposal to Protect Salmon Criticized," *Boston Globe,* October 15, 1999, B3, in National Marine Fisheries Service and Fish and Wildlife Service, "Guide to the Listing of a Distinct Population Segment of Atlantic Salmon as Endangered," November 2000, in Carey Goldberg, "U.S. Classifies Wild Salmon in Maine Rivers as Threatened," *New York Times,* November 14, 2000, A12, and in Beth Daley, "Maine Worries as US Declares Salmon Endangered," *Boston Globe,* November 14, 2000, A1, B14. Sea cucumber in Henry S. Parker, *Exploring the Oceans* (Englewood Cliffs, N.J.: Prentice-Hall, 1984), 204. Salmon and salt in Sumich, *Biology of Marine Life,* 35–36. Salmon finding their way home in Marcia Barinaga, "Salmon Follow Watery Odors Home," *Science* 286

(5440), October 22, 1999: 705–6. Dams and salmon in Bigelow and Schroeder, *Fishes in the Gulf of Maine,* 121–31. Salmon statistics in William K. Stevens, "As a Species Vanishes, No One Can Say Why," *New York Times,* September 14, 1999, D1, D4. Habitats of salmon in U.S. Fish and Wildlife Service, "Report on the Status of Atlantic Salmon," 1999.

17. Challenges for salmon in U.S. Fish and Wildlife Service, "Report on the Status of Atlantic Salmon," 1999, in J. Raloff, "Pollutant Waits to Smite Salmon at Sea," *Science News* 155 (19), May 8, 1999: 293, and in Debora MacKenzie, "River of No Return," *New Scientist* 162 (2189), June 5, 1999: 12.

18. Salmon farms in Robert Braile, "US Officials Sound Warning on Wild Atlantic Salmon," *Boston Globe,* October 9, 1999, in Carey Goldberg, "Stake High in Maine Fight over Fish Farming," *New York Times,* August 28, 1999, A1, A8, U.S. Fish and Wildlife Service, "Report on the Status of Atlantic Salmon," 1999, in "Norwegian Fish Farms Now Major Polluters," *ENDS Daily,* April 11, 2000, in R. L. Naylor et al., "Nature's Subsidies to Shrimp and Salmon Farming," *Science* 282 (5390), October 30, 1998: 883–84, in Carol Kaesuk Yoon, "Altered Salmon Leading Way to Dinner Plates but Rules Lag," *New York Times,* May 1, 2000, A1, A20, and in "WWF International Calls for Action to Save the Wild Salmon," WWF International press release, May 31, 2000. Ivermectin in Rob Edwards, "No Free Lunch," *New Scientist* 157 (2120), February 7, 1998: 23. Scotland sea trout infected by farm salmon in Rob Edwards, "Infested Waters: Sea Lice from Salmon Farms Threaten Scotland's Sea Trout," *New Scientist* 159 (2141), July 4, 1998: 23. Transgenic salmon in Tony Reichardt, "Will Souped-up Salmon Sink or Swim?," *Nature* 406 (6791), July 6, 2000: 10–12. Trojan gene in W. M. Muir and R. D. Howard, "Possible Ecological Risks of Transgenic Organism Release When Transgenes Affect Mating Success: Sexual Selection and the Trojan Gene Hypothesis," *Proceedings of the National Academy of Sciences,* 96 (24), November 23, 1999: 13853–56. Percent of salmon food consisting of fish meal and fish oil, 2.8:1 ratio of food in and pounds of salmon produced in Naylor et al., "Nature's Subsidies." Percent capture fish used for fish meal, fish oil, and aquaculture, and rapid growth of aquaculture in R. L. Naylor et al., "Effect of Aquaculture on World Fish Supplies," *Nature* 495 (6790), June 29, 2000: 1017–24.

19. Language of exploitation and harvest in Northeast Fisheries Science Center, "Status of Fishery Resources," NMFS-NE-115. Language of depleted fisheries and fisheries production in FAO, "World Review of Fisheries and Aquaculture." Terms seriously and dramatically depleted in "Cod Near Landings Target," National Marine Fisheries Service Northeast Region press release, January 2000.

20. Changes in the colonial landscape in William Cronon, "Commodities of the Hunt" and "Taking the Forest," in *Changes in the Land: Indians, Colonists, and the Ecology of New England* (New York: Hill & Wang, 1983). Robert Alter, *Genesis: Translation and Commentary* (New York: W. W. Norton, 1996). Transition from the Middle Ages to the Age of Reason in Susan Griffin, *The Eros of Everyday Life: Essays on Ecology, Gender and Society* (New York: Doubleday, 1995). Theodore Roszak, ed., *Ecopsychology: Restoring the Earth, Healing the Mind* (San Francisco: Sierra Club, 1995). Economics and ecology in Robert Costanza et al., "The Value of the World's Ecosystem Services and Natural Capital," *Nature* 387 (6630), May 15, 1997: 253–61. Walt Whitman, "As I Ebb'd with the Ocean of Life," in *Leaves*

of Grass. Herman Melville, "Loomings," in *Moby Dick* (New York: Bobbs-Merrill), 1964.

Chapter 3: Along the Slope

1. Continental shelf and slope in Henry S. Parker, *Exploring the Oceans* (Englewood Cliffs, N.J.: Prentice-Hall, 1985), 47–49. Florida escarpment in L. F. Pratson and W. F. Haxby, "Panoramas of the Seafloor," *Scientific American,* June 1997, 82–87.

2. Submarine canyons and turbidity currents in K. O. Emery and E. Uchupi, *The Geology of the Atlantic Ocean* (New York: Springer-Verlag, 1984), 80–86, and in James Kennett, *Marine Geology* (Englewood Cliffs, N.J.: Prentice-Hall, 1982), 412–19. Turbidity currents on the Grand Banks, in the Congo Canyon, and the Magdalena River in B. C. Heezen and C. D. Hollister, *The Face of the Deep* (New York: Oxford University Press, 1971), 297–305. Turbidity deposits in the Mediterranean in R. G. Rothwell, J. Thomson, and G. Kahier, "Low-Sea-Level Emplacement of a Very Large Late Pleistocene 'Megaturbidite' in the Western Mediterranean Sea," *Nature* 392 (6674), March 26, 1998: 377–80, and in R. Monastersky, "Giant Seabed Slides May Have Climate Link," *Science News* 153 (13), March 28, 1998: 198. Turbidity current off Venezuela in R. Thunell, E. Tappa, and R. Varela, "Increased Marine Sediment Suspension and Fluxes Following an Earthquake," *Nature* 398 (6724), March 18, 1999: 233–36.

3. Marine diversity in Paul V. R. Snelgrove, "Getting to the Bottom of Marine Biodiversity Sedimentary Habitats," *Bioscience* 49 (2), February 1999: 129 (10 pp.), and in P. V. R. Snelgrove and J. F. Grassle, "The Deep Sea: Desert AND Rainforest, Debunking the Desert Analogy," *Oceanus,* Fall–Winter 1995, 25–29. Rockall Trough in Fred Pearce, "Rockall Mud Richer Than Rainforest," *New Scientist* 147 (1995), September 16, 1995: 8, and in Stephanie Pain, "Mud, Glorious Mud," *New Scientist* 152 (2024), November 2, 1996: S4 (4 pp.).

PART II: THE OPEN SEA

Chapter 4: The Rush of Water

1. *Velella velella* in S. B., "The By-the-Wind Sailor: A Masterpiece of Natural Design," *Discover,* August 1987, 46, in John Dillon, "*Velella velella*," *Oceans,* November 1982, 3, and in Sir Alister Hardy, *The Open Sea: Its Natural History, Part I, The World of Plankton* (Boston: Houghton Mifflin, 1970), 112–14.

2. Genesis 1:6–9, Job 38:8–11.

3. Gulf Stream in "A Primer on Ocean Currents," *Oceanus,* Spring 1994, 3–5.

4. Ponce de León, Cortés, and sea beans in William H. MacLeish, *The Gulf Stream* (Boston: Houghton Mifflin, 1989). Path of Atlantic currents in William J. Schmitz and Michael S. McCartney, "On the North Atlantic Circulation," *Reviews of Geophysics* 31 (1), February 1993: 29–49. British mail boats in Philip L. Richardson, "Benjamin Franklin and Timothy Folger's First Printed Chart of the Gulf Stream," *Science* 207, February 8, 1980: 643–45. Sea beans and sea hearts in Wayne Armstrong, "Floaters," *Sea Frontiers,* May–June 1994, 24–30. Message from Plum Island in "12 Years Later, Message in a Bottle Found," *Gloucester Daily Times,* February 11, 1994, A7.

5. Mississippi River water in "River in a Stream," *Sea Frontiers,* January–February 1994, 11, and in David L. Chandler, "Floodwaters Gone? Check the Gulf Stream," *Boston Globe,* October 23, 1993, 1. Norway in Jerome Namais, "Seasonal Resistance and Recurrence of European Blocking During 1958–1960," *Tellus* 16 (3), 1964: 394. Salinity and temperature anomalies in D. V. Hansen and H. F. Bezdek, "On the Nature of Decadal Anomalies in North Atlantic Sea Surface Temperature," *Journal of Geophysical Research* 101 (C4), April 15, 1996: 8749–58, and in Richard Monastersky, "Oceanography's New Catch: Roving Blobs," *Science News* 149 (18), May 4, 1996: 276.

6. Heat, wind, and currents in Henry S. Parker, *Exploring the Oceans* (Englewood Cliffs, N.J.: Prentice-Hall, 1985), 108–25. Sahara sand in "Sahara Dust Blows over United States," *Science News* 148 (26 and 27), December 23 and 30, 1995: 431, and in J. M. Prospero and T. N. Carlson, "Vertical and Areal Distribution of Saharan Dust over the Western Equatorial North Atlantic Ocean," *Journal of Geophysical Research* 77 (7), September 20, 1972: 5255–65. Gulf Stream and Europe's climate in Raymond Schmitt, "Mysteries of Planetary Plumbing," *Oceanus,* Summer 1992, 38–46, and in Michael S. McCartney, "Towards a Model of Atlantic Ocean Circulation," *Oceanus,* Spring 1994, 5–8. Ice in Hudson Bay and Norway in M. Tomczak and J. S. Godfrey, *Regional Oceanography* (Oxford: Pergamon, 1994), 96, 299.

7. Upwelling off northwest Africa, in Tomczak and Godfrey, *Regional Oceanography,* 274–75. The *Wolston* in Philip L. Richardson, "Derelicts and Drifters," *Natural History,* June 1985, 43–48. Brazil retroflection eddies in P. L. Richardson, G. E. Hufford, and R. Limeburner, "North Brazil Current Retroflection Eddies," *Journal of Geophysical Research* 99 (C3), March 15, 1994: 5081–93, and in Philip Richardson, "Tracking Ocean Eddies," *American Scientist* 81 (3), May–June 1993, 261–71. Routes of the currents in Schmitz and McCartney, "On the North Atlantic Circulation."

8. Atlantic eddies in Richardson, "Tracking Ocean Eddies." Gulf Stream eddies in Peter H. Wiebe, "Rings of the Gulf Stream," *Scientific American,* March 1982, 60–69, and in the Ring Group, "Gulf Stream Cold-Core Rings: Their Physics, Chemistry and Biology," *Science* 212, June 5, 1981: 1091–1100. Eddies in New England in Simon Perkins, "Cruising on the Gulf Stream," *Sanctuary,* January–February 1993, 27, and Sy Montgomery, "Fish out of Waters," *Boston Globe,* June 28, 1999, C1, C4.

9. Links between surface and deep currents, and path of Antarctic bottom water in Schmitz and McCartney, "On the North Atlantic Circulation." Origin of deep water in James F. Price, "Overflows: The Source of New Abyssal Waters," *Oceanus,* Summer 1992, 28–35. Salt- and freshwater influences and Atlantic precipitation in Schmitt, "Mysteries of Planetary Plumbing." Salt and precipitation, pp. 279–80, and Antarctic bottom water, p. 254, in Tomczak and Godfrey, *Regional Oceanography*. Mediterranean eddies in J. F. Price et al., "Mediterranean Outflow Mixing and Dynamics," *Science* 259 (5099), February 26, 1993: 1277–83, in James F. Price, "Dynamics and Modeling of Marginal Sea Outflows," *Oceanus,* Spring 1994, 9–12, in Amy Bower, "Meddies, Eddies, Floats and Boats," *Oceanus,* Spring 1994, 12–15, and in Richardson, "Tracking Ocean Eddies."

10. Greenland Sea ice in P. Wadhams, "Sea Ice—Ocean Interactions in the Greenland Sea," European Subpolar Ocean Program, contract MAS2-CT93-0057, pp. 559–78, and in P. Wadhams et al., "The Development of the Odden Ice Tongue in the

Greenland Sea During the Winter 1993 from Remote Sensing and Field Observations," *Journal of Geophysical Research* 101 (C8), August 15, 1996: 18213–235. Atlantic deep-water formation in Robert R. Dickson and Juan Brown, "The Production of North Atlantic Deepwater: Sources, Rates, and Pathways," *Journal of Geophysical Research* 99 (C6), 1994: 12319–341, and in William J. Schmitz Jr., "On the Interbasin-Scale Thermohaline Circulation," *Review of Geophysics* 33 (2), May 1995: 151–73. Atlantic deep-water formation in Tomczak and Godfrey, *Regional Oceanography,* 89–110, 352, in Price, "Overflows," in John A. Whitehead, "Giant Ocean Cataracts," *Scientific American,* February 1989, 50–67, in " Primer on Ocean Currents," in Schmitz and McCartney, "On the North Atlantic Circulation," and in McCartney, "Atlantic Ocean Circulation."

11. Tritium tracers in H. G. Ostlund, "Tritium and Radiocarbon in the North Atlantic Ocean: An Overview," *Rit Fiskideilar* 9, 1985: 13–19, and Claes H. Rooth, "GEOSECS and Tritium Tracers," *Oceanus,* Summer 1977, 53–57. Tritium and CFC tracers in W. J. Jenkins and W. M. Smethie Jr., "Transient Tracers Track Ocean Climate Signals," *Oceanus,* Fall–Winter 1996, 29–33. CFC tracers in the Gulf Stream and bottom current in R. S. Pickart and W. M. Smethie, "How Does the Deep Western Boundary Current Cross the Gulf Stream?," *Journal of Physical Oceanography* 23, December 1993: 2602–16, and in Robert S. Pickart, "Where Currents Cross," *Oceanus,* Spring 1994, 16–18. Merging and dividing Atlantic currents in Schmitz and McCartney, "On the North Atlantic Circulation," and in M. Susan Lozier, "Evidence for Large-Scale Eddy-Driven Gyres in the North Atlantic," *Science* 277 (5324), July 18, 1997: 361–63. Drifting buoys in Monika Rhein, "Drifters Reveal Deep Circulation," *Nature* 407 (6800), September 7, 2000: 30–31. Diffusion in J. R. Ledwell, A. J. Watson, and C. S. Law, "Evidence for Slow Mixing Across the Pynocline from an Open-Ocean Tracer-Release Experiment," *Nature* 365 (6439), August 19, 1993: 701–3, in Chris Garrrett, "A Stirring Tale of Mixing," *Nature* 364 (6439), August 19, 1993: 670, in J. R. Ledwell et al., "Evidence for Enhanced Mixing over Rough Topography in the Abyssal Ocean," *Nature* 403 (6766), January 13, 2000: 179–82, in Peter Killworth, "Oceanography: Something Stirs in the Deep," *Nature* 396 (6713), December 24, 1998: 720–21, and in Carl Wunsch, "Moon, Tides and Climate," *Nature* 405 (6788), June 15, 2000: 743–44.

12. Antarctic bottom water in James Kennett, *Marine Geology* (Englewood Cliffs, N.J.: Prentice-Hall, 1982), 251, and in Nelson Hogg, "The Deep Basin Experiment," *Oceanus,* Spring 1994, 22–25. World circulation in Schmitz and McCartney, "On the North Atlantic Circulation," and in Schmitz, "Interbasin-Scale Thermohaline Circulation." Drifts in K. O. Emery and E. Uchupi, *The Geology of the Atlantic Ocean* (New York: Springer-Verlag, 1984), 104, 106, and in Kennett, *Marine Geology,* 534. Benthic storms in C. Hollister, A. R. M. Nowell, and P. A. Jumars, "The Dynamic Abyss," *Scientific American,* March 1984, 42–53, in Richard Kerr, "A New Kind of Storm Beneath the Sea," *Science* 208, May 2, 1980: 484–86, and in Victoria Kaharl, "These Deep Waters Don't Run Still," *Technology Review,* February–March 1985, 70–75.

13. Bipolar foraminifera in K. F. Darling, C. M. Wade, and I. A. Stewart, "Molecular Evidence for Genetic Mixing of Arctic and Antarctic Subpolar Populations of Planktonic Foraminifers," *Nature* 405 (6782), May 4, 2000: 43–47, and in R. D. Norris and C. de Vargas, "Evolution All at Sea," *Nature* 405 (6782), May 4, 2000: 23–24. Casting sins into the sea, Micah 7:19. Pesticides and PCBs in the Arctic in J.

U. Skaare et al., "Organochlorines in Top Predators at Svalbard—Occurrence, Levels and Effects," *Toxicology Letters* 112–13, March 15, 2000: 103–9, in "Organochlorines Lace Inuit Breast Milk," *Science News* 145 (7), February 12, 1994: 111, in Janet Raloff, "The Pesticide Shuffle," *Science News* 149 (11), March 16, 1996: 174–75, in Frank Wania and Donald MacKay, "Tracking the Distribution of Persistent Organic Pollutants," *Environmental Science and Technology* 30 (9), 1996: 390A–96A, and in Rob Edwards, "Unfit to Eat," *New Scientist* 159 (2153), September 26, 1993: 13. Pesticides, PCBs, and radioactivity in the Arctic in R. W. MacDonald and J. M. Brewers, "Contaminants in the Arctic Marine Environment: Priorities for Protection," *ICES Journal of Marine Science* 53, 1996: 537–63, and in Arctic Monitoring and Assessment Program, "Arctic Pollution Issues: A State of the Arctic Environment Report," 1997. Radioactivity and European reprocessing plants in Rob Edwards, "Sea Change," *New Scientist* 155 (2098), September 6, 1997: 10, in Fred Pearce, "Sellafield Leaves Its Mark on the Frozen North," *New Scientist* 154 (2081), May 10, 1997: 14, in Rob Edwards, "Now You See It," *New Scientist* 162 (2183), April 24, 1999: 17, in Rob Edwards, "Set in a Nuclear Sea," *New Scientist* 161 (2175), February 27, 1999, in OSPAR Commission for the Protection of the Marine Environment of the North-East Atlantic, "OSPAR Quality Status Report 2000," chap. 6, Overall Assessment, June 2000, and in BirdLife International et al., "Joint Response to the OSPAR Quality Status Report 2000," July 2000. New water in James Trefil, *A Scientist at the Seashore* (New York: Scribner's, 1984), 5.

Chapter 5: Climate and Atlantic

1. Tigris–Euphrates civilizations in H. M. Cullen et al., "Climate Change and the Collapse of the Akkadian Empire: Evidence from the Deep Sea," *Geology* 28 (4), April 2000: 379–82, in R. Monasterksy, "The Case of the Global Jitters," *Science News* 149 (9), March 2, 1996: 140–41, in Arie S. Issar, "Climate Change and the History of the Middle East," *American Scientist* 83, July–August 1995: 350–55, in Clive Ponting, *A Green History of the World: The Environment and the Collapse of Great Civilizations* (New York: St. Martin's Press, 1991), 68–73, in William K. Stevens, "Study of Ocean Currents Offers Clues to Global Climate Shifts," *New York Times,* March 18, 1997, C1, and in H. M. Cullen and P. H. deMenocal, "North Atlantic Influence on Tigris–Euphrates Streamflow," *International Journal of Climatology* 28 (8), 2000: 853–63.

2. Sahel in Sharon E. Nicholson, "Sahel, West Africa," *Encyclopedia of Environmental Biology* 3, 261–75, in M. Hulme and M. Kelly, "Exploring the Links Between Desertification and Climate Change," *Environment,* July–August 1993, 4 (15 pp.), in Norimitsu Onishi, "Where Dwelling Is Kept from Sand Dune, One Scoop at a Time," *New York Times,* January 14, 2000, A1, A8, in Richard A. Kerr, "The Sahara Is Not Marching Southward," *Science* 281 (5377), July 31, 1998: 633–34, in S. E. Nicholson, C. J. Tucker, and M. B. Ba, "Desertification, Drought and Surface Vegetation: An Example from the West African Sahel," *Bulletin of the American Meteorological Society* 79, 1998: 815, and in P. J. Lamb and R. A. Peppler, "Further Case Studies of Tropical Atlantic Surface Temperature Patterns Associated with Subsaharan Drought," *Journal of Climate* 5, May 1992: 476–88. Sand encroachment on mosques in 14th session of World Heritage Committee, "List of the World Heritage Sites in Danger," 1990, and in Joshua Hammer, "Timbuktu Postcard," *New Republic,* November 13, 1995, 12. Sahel in Peter Lamb, "Waiting for Rain," *The*

Sciences, May–June, 1986, 30–35, and in P. J. Lamb and R. A. Peppler, "West Africa," chap. 5 in M. H. Glantz et al., eds., *Teleconnections Linking Worldwide Climate Anomalies: Scientific Basis and Societal Impact* (Cambridge: Cambridge University Press, 1991), 121–89. Rain forest and drought in Fred Pearce, "Lost Forests Leave West Africa Dry," *New Scientist* 153 (2065), January 18, 1997: 15. Sahel in F. A. Street-Perrott and R. A. Perrott, "Abrupt Climate Fluctuations in the Tropics," *Nature* 343, February 15, 1990: 607–12. North Atlantic Oscillation in A. L. Gordon, S. E. Zebiak, and K. Bryan, *EOS* 73 (15), April 15, 1992: 161–76, in J. W. Hurrell and H. VanLoon, "Decadal Variations in Climate Associated with the North Atlantic Oscillation," *Climate Change* 36, 1997, in R. Dickson et al., "Long-term Coordinated Changes in the Convective Activity of the North Atlantic," *Progressive Oceanography* 38, 1996: 241–92, and in Michael S. McCartney, "North Atlantic Oscillation," *Oceanus,* Fall–Winter 1996, 13.

3. Great Salinity Anomaly in R. A. K., "Did the Great Salinity Anomaly Cool the Atlantic?," *Science* 255, March 20, 1992: 1509, and in P. Schlosser et al., "Reduction of Deepwater Formation in Greenland Sea During the 1980s: Evidence from Tracer Data," *Science* 251, March 1, 1991: 1054–56. North Atlantic Oscillation and Labrador Sea in J. F. Read and W. J. Gould, "Cooling and Freshening of the Subpolar North Atlantic Ocean Since the 1960s," *Nature* 360, November 5, 1992: 55–57. Atlantic freshwater pools and the Sahel in Street-Perrott and Perrott, "Abrupt Climate Fluctuations." R. R. Dickson et al., "The Great Salinity Anomaly in the Northern North Atlantic 1968–1972," *Progress in Oceanography* 20, 1988: 103–51. Pools in R. Monastersky, "Oceanography's New Catch: Roving Blobs," *Science News* 149 (18), May 4, 1996: 276, and D. V. Hansen and H. F. Bezdek, "On the Nature of Decadal Anomalies in North Atlantic Sea Surface Temperature," *Journal of Geophysical Research* 101 (C4), April 15, 1996: 8749–58. Warming and North Atlantic Oscillation in James W. Hurrell, "Influence of Variations in Extratropical Wintertime Teleconnections on Northern Hemisphere Temperature," *Geophysical Research Letters* 23 (6), March 15, 1996: 665–68. Labrador Sea cascade in A. Sy et al., "Surprisingly Rapid Spreading of Newly Formed Intermediate Waters Across the North Atlantic Ocean," *Nature* 386 (6626), April 17, 1997: 675–79. Labrador Sea and Greenland Sea sinking in Bob Dickson, "From the Labrador Sea to Global Change," *Nature* 386 (6626), April 17, 1997: 649–50. Odden ice in P. Wadhams et al., "The Development of the Odden Ice Tongue in the Winter 1993 from Remote Sensing and Field Observations," *Journal of Geophysical Research* 101 (C8), August 15, 1996: 18213–35, and in personal communication, Leif Toudal Pedersen, Danish Center for Remote Sensing, September 2000. Whalers in Fred Pearce, "Is Broken Ocean Pump a Global Warning?," *New Scientist* 141 (1917), March 19, 1994: 4. Cold winters in Europe in Richard A. Kerr, "Climate: A New Driver for the Atlantic's Moods and Europe's Weather?," *Science* 275 (5301), February 7, 1997: 754–55. Ice in Holland in "With Thick Ice, a Hot Event for the Dutch," *New York Times,* June 5, 1997, 8. North Atlantic Oscillation and sea surface temperature in Yochanan Kushnir, "Europe's Winter Prospects," *Nature* 398 (6725), March 25, 1999: 289–91, and in M. J. Rodwell, D. P. Rowell and C. K. Folland, "Oceanic Forcing of the Wintertime North Atlantic Oscillation and European Climate," *Nature* 398 (6725), March 25, 1999: 320–23.

4. Fifteen-hundred-year cycles in G. Bond et al., "A Pervasive Millennial-Scale Cycle in North Atlantic Holocene and Glacial Climates," *Science* 278 (5341), November 14,

1997: 1257–65, and P. deMenocal et al., "Coherent High- and Low-Latitude Climate Variability During the Holocene Warm Period," *Science* 288 (5474), June 23, 2000: 2198–2202. Medieval warm period in Lloyd D. Keigwin, "The Little Ice Age and Medieval Warm Period in the Sargasso Sea," *Science* 274 (5292), November 29, 1996: 1503–8, in S. H. Schneider and R. Londer, *The Coevolution of Climate and Life* (San Francisco: Sierra Club, 1984), 111–14, in Ponting, *Green History of the World,* 100–106, in Marlise Simons, "Glaciers Are Star Witnesses to Earth's Warming," *New York Times,* December 19, 1995, C4, and in J. Gribbin and M. Gribbin, "Climate and History: The Westvikings' Saga," *New Scientist* 125 (1700), January 20, 1990: 52–54. The Little Ice Age in Jean M. Grove, *The Little Ice Age* (New York: Routledge, 1991), 2–21, 112–43, 308, 391–414, and in Josef Eisinger, "Sweet Poison," *Natural History,* July 1996, 51.

5. Earth's orbit and climate in John Houghton, *Global Warming: The Complete Briefing* (Oxford: Lion, 1994), 24–25, 53–69. Orbit and glaciers in W. S. Broecker and G. H. Denton, "What Drives Glacial Cycles?," *Scientific American,* January 1990, 49–55.

6. J. Kutzbach et al., "Vegetation and Soil Feedbacks on the Response of the African Monsoon to Orbital Forcing in the Early to Middle Holocene," *Nature* 384 (6610), December 19 and 26, 1996: 623–26, in J. E. Kutzbach and Z. Liu, "Response of the African Monsoon to Orbital Forcing and Ocean Feedbacks in the Middle Holocene," *Science* 278 (5337), December 17, 1997: 440–43, and in deMenocal et al., "Coherent High- and Low-Latitude Climate Variability During the Holocene Warm Period."

7. Sea cores and ice age cycles in Daniel P. Schrag, "Of Ice and Elephants," *Nature* 404 (6773), March 2, 2000: 23–24, in G. M. Henderson and N. C. Slowey, "Evidence from U-Th Dating Against Northern Hemisphere Forcing of the Penultimate Deglaciation," *Nature* 404 (6773), March 2, 2000: 61-66, and in D. B. Karner and R. A. Milner, "A Causality Problem for Milankovitch," *Science* 288 (5474), June 23, 2000: 2143–44. Antarctic deep-water circulation slowing in W. S. Broecker, S. Sutherland, and T.-H. Peng, "A Possible 20th Century Slowdown of Southern Ocean Deep Water Formation," *Science* 286 (5442), November 5, 1999: 1132–35, and in Richard A. Kerr, "Has a Great River in the Sea Slowed Down?," *Science* 286 (5442), November 5, 1999: 1061–62. Southern and northern current balance in Thomas F. Stocker, "The Seesaw Effect," *Science* 282 (5386), October 2, 1998: 61–62, and in R. Monastersky, "Rhythm of the Ice Age: North Versus South," *Science News* 154 (8), August 22, 1998: 119.

8. Vema Channel in K. O. Emery and E. Uchupi, *The Geology of the Atlantic Ocean* (New York: Springer-Verlag, 1984), 587–88. Sea animals measure pace of deep-water circulation in Broecker and Denton, "What Drives Glacial Cycles?," in E. A. Boyle and L. D. Keigwin, "Deep Circulation of the North Atlantic over the Last 200,000 Years: Geochemical Evidence," *Science* 218, November 19, 1982: 784–86. Orbital changes insufficient explanation for deep-water circulation in L. D. Keigwin et al., "Deglacial Meltwater Discharge, North Atlantic Deep Circulation, and Abrupt Climate Change," *Journal of Geophysical Research* 96 (C9), September 15, 1991: 16811–26. Rare-earth indicators in R. L. Rutberg, S. R. Hemming, and S. L. Goldstein, "Reduced North Atlantic Deep Water Flux to the Glacial Southern Ocean Inferred from Neodynium Isotope Ratios," *Nature* 405 (6789), June 22, 2000: 935–38. Surface waters in S. J. Lehman and L. D. Keigwin, "Sudden Changes in

North Atlantic Circulation During the Last Deglaciation," *Nature* 356 (6372), April 30, 1992: 757–62. Prehistoric paintings in Marlise Simons, "Stone Age Art Shows Penguins at Mediterranean," *New York Times,* October 20, 1990: C1, C10. Gulf Stream in Jean-Claude Duplessy, "Climate and the Gulf Stream," *Nature* 402 (6762), December 9, 1999: 593–95, and in J. Lynch-Stieglitz, W. B. Curry, and N. Slowey, "Weaker Gulf Stream in the Florida Straits During the Last Glacial Maximum," *Nature* 402 (6772), December 9, 1999: 644–48. Cooling in the Tropics in L. G. Thompson et al, "Late Glacial Stage and Holocene Tropical Ice Core Records from Huascarán, Peru," *Science* 269 (5290), July 7, 1995: 46–50, in M. Stute et al., "Cooling of Tropical Brazil (5°C) During the Last Glacial Maximum," *Science* 269 (5222), July 21, 1995: 379–82, in Wallace Broecker, "Glacial Climate in the Tropics," *Science* 272 (5270), June 28, 1996: 1902–4, in G. S. Dwyer, T. M. Cronin, and P. A. Baker, "North Atlantic Deepwater Temperature Change During Late Pliocene and Late Quaternary Climatic Cycles," *Science* 270 (5240), November 24, 1995: 1347–51, and in W. Aeschbach-Hertig et al., "Palaeotemperature Reconstruction from Noble Gases in Ground Water Taking into Account Equilibration with Entrapped Air," *Nature* 405 (6790), June 29, 2000: 1040–44.

9. Sudden climate changes in Stefan Rahmstorf, "Rapid Climate Transitions in a Coupled Ocean-Atmosphere Model," *Nature* 372 (6501), November 3, 1994: 82–85, in Stefan Rahmstorf, "Ice-Cold in Paris," *New Scientist* 153 (2068), February 8, 1997: 26, in Wallace S. Broecker, "Chaotic Climate," *Scientific American,* November 1995: 62–68, in Lehman and Keigwin, "Sudden Changes in North Atlantic Circulation," in J. E. Smith et al., "Rapid Climate Change in the North Atlantic During the Younger Dryas Recorded by Deep-Sea Corals," *Nature* 386 (6627), April 24, 1997: 818–21, in J. F. Adkins, H. Cheng, and E. A. Boyle, "Deep-Sea Evidence for Rapid Change in Ventilation of the Deep North Atlantic 15,400 Years Ago," *Science* 280 (5364), May 1, 1998: 725–28, and in Richard A. Kerr, "Deep-Sea Coral Records Quick Response to Climate," *Science* 280 (5364), May 1, 1998: 679.

10. Broecker and Denton, "What Drives Glacial Cycles?," and Broecker, "Chaotic Climate." D. C. Barber, A. Dyke, and C. Hillaire-Marcel, "Forcing of the Cold Event of 8,200 Years Ago by Catastrophic Drainage of Laurentide Lakes," *Nature* 400 (6742), July 22, 1999: 344–48. K. A. Hughen, J. T. Overpeck, and S. J. Lehman, "Deglacial Changes in Ocean Circulation from an Extended Radiocarbon Calibration," *Nature* 391 (6662), January 1, 1998: 65–69.

11. Greenland ice 120,000 years ago in Christine Schott Hvidberg, "When Greenland Ice Melts," *Nature* 404 (6778), April 6, 2000: 551–52, in K. M. Cuffey and S. J. Marshall, "Substantial Contribution to Sea-Level Rise During the Last Interglacial from the Greenland Ice Sheet," *Nature* 404 (6778), April 6, 2000: 591–94, in W. Krabill et al., "Rapid Thinning of Parts of the Southern Greenland Ice Sheet," *Science* 283 (5407), March 5, 1999: 1522–24, in W. Krabill et al., "Greenland Ice Sheet: High-Elevation Balance and Peripheral Thinning," *Science* 289 (5478), July 21, 2000: 428–30.

12. Fresh water and the Bering Sea in G. Schaffer and J. Bendtsen, "Role of the Bering Strait in Controlling North Atlantic Ocean Circulation and Climate," *Nature* 367 (6461), January 27, 1994: 354–57, and in E. Cortijo et al., "Eemian Cooling in the Norwegian Sea and North Atlantic Ocean Preceding Continental Ice-Sheet Growth," *Nature* 372 (6505), December 1, 1994: 446–49. The *Jeanette* in M. Tomczak and J.S.

Godfrey, *Regional Oceanography* (Oxford: Pergamon, 1994), 84–95. Bath toys in Judy Mathewson, "Make Way for Ducklings—They May Be Headed Here," *Salem Evening News,* December 11, 1995, A1, A8.

13. Opening of the Northwest Passage in Colin Nickerson, "On Thin Ice," *Boston Globe,* March 27, 2000, E6, and John Noble Wilford, "Ages-Old Icecap at North Pole Is Now Liquid, Scientists Find," *New York Times,* August 19, 2000, A1. Arctic ice declines in O. M. Johannessen, E. V. Shalina, and M. W. Miles, "Satellite Evidence for an Arctic Sea Ice Cover in Transformation," *Science* 286 (5446), December 3, 1999: 1937–39, and Richard A. Kerr, "Will the Arctic Ocean Lose All Its Ice?," *Science* 286 (5446), December 3, 1999: 1828. Decline is unnatural variation in K. Y. Vinnikow et al., "Global Warming and Northern Hemisphere Sea Ice Extent," *Science* 286 (5446), December 3, 1999: 1934–37. Predictions for liquid North Pole in Ola M. Johannessen and Martin W. Miles, "Arctic Sea Ice and Climate Change—Will the Ice Disappear in this Century?," *Science Progress,* in press.

14. Arctic warmer than in the last four hundred years in J. Overpeck et al., "Arctic Environmental Changes of the Last Four Centuries," *Science* 278 (5341), November 14, 1997: 1251–56. Arctic warming and circulation in E. C. Carmack et al., "Evidence for Warming of Atlantic Water in the Southern Canadian Basin of the Arctic Ocean: Results from the Larsen-93 Expedition," *Geophysical Research Letters* 22 (9), May 1, 1995: 1061–64. Polar bears in Fred Pearce, "Too Darned Hot," *New Scientist* 159 (2146), August 8, 1998: 41–42. Icebergs in Bernice Wuethrich, "Lack of Icebergs Another Sign of Global Warming?," *Science* 285 (5424), July 2, 1999: 37. Warming confirmed in Antonio Regalado, "Listen Up! The World's Oceans May Be Starting to Warm," *Science* 268 (5216), June 9, 1995: 1436–37. Warming and circulation in Robie Macdonald, "Awakenings in the Arctic," *Nature* 380, March 28, 1996: 286–87, and Bob Dickson, "All Change in the Arctic," *Nature* 397 (6718), February 4, 1999: 389–92. Denmark Strait thinning in B. Dickson et al., "Possible Predictability in Overflow from the Denmark Strait," *Nature* 397 (6716), January 21, 1999: 243–46. Deep-water and Faeroe Islands in Rob Edwards, "Freezing Future," *New Scientist* 164 (2214), November 27, 1999: 6.

15. Wallace S. Broecker, "Thermohaline Circulation, the Achilles Heel of Our Climate System: Will Man-Made CO_2 Upset the Current Balance?," *Science* 278(5343), November 28, 1997: 1582–88. Temperature predictions in Don Kennedy, "New Climate News," *Science* 290 (5494), November 10, 2000: 1091. Global warming and rain in U.S. Office of Science and Technology Policy, "Climate Change," October 1997. Evaporation and precipitation over Atlantic in Raymond W. Schmitt, "The Ocean Component of the Global Water Cycle," *Reviews of Geophysics,* supplement, July 1995: 1395–1409, in Raymond W. Schmitt, "If Rain Falls on the Ocean, Does It Make a Sound?" *Oceanus,* Fall-Winter 1996: 4–8, and in Raymond W. Schmitt, "Mysteries of Planetary Plumbing," *Oceanus,* Summer 1992: 38–45. Historical flooding of the Mississippi in James C. Knox, "Large Increases in Flood Magnitude in Response to Modest Changes in Climate," *Nature* 361 (6411), February 4, 1993: 430–32. Europe flooding in Mick Hamer, "The Rain for Spain Fell Mainly on Luxembourg," *New Scientist* 145 (1964), February 11, 1995: 7.

16. Weakening of deep circulation in Stefan Rahmstorf, "Shifting Seas in the Greenhouse?," *Nature* 399 (6736), June 10, 1999: 523–24, in R. A. Wood et al.,

"Changing Spatial Structure of the Thermohaline Circulation in Response to Atmospheric CO_2 Forcing in a Climate Model," *Nature* 399 (6736), June 10, 1999: 572–76, in T. F. Stocker and A. Schmittner, "Influence of CO_2 Emission Rates on the Stability of the Thermohaline Circulation," *Nature* 388 (6645), August 28, 1997: 862–65, in J. L. Sarmiento and C. L. Quere, "Oceanic Carbon Dioxide Uptake in a Model of Century-Scale Global Warming," *Science* 274 (5291), November 22, 1996: 1346–50, in Stefan Rahmstorf, "Bifurcations of the Atlantic Thermohaline Circulation in Response to Changes in the Hydrological Cycle," *Nature* 378 (6553), November 9, 1995: 145–49, in S. Manabe and R. J. Stouffer, "Century-Scale Effects of Increased Atmospheric CO_2 on the Ocean-Atmosphere System," *Nature* 364 (6682), July 15, 1993: 215–17, and in Rahmstorf, "Ice-Cold in Paris."

17. Models and carbon dioxide uptake in Stocker and Schmittner, "Influence of CO_2 Emission Rates," in Sarmiento and Quere, "Oceanic Carbon Dioxide Uptake," in F. Joos, G. K. Plattner, and T. F. Stocker, "Global Warming and Marine Carbon Cycle Feedbacks on Future Atmospheric CO_2," *Science* 284 (5413), April 16, 1999: 464–67, and in K. Caldeira and P. B. Duffy, "The Role of the Southern Ocean in Uptake and Storage of Anthropogenic Carbon Dioxide," *Science* 287 (5453), January 28, 2000: 620–22.

18. Exchange of CO_2 between earth and atmosphere in Houghton, *Global Warming: The Complete Briefing,* 30. Historical temperature range and CO_2 levels in U.S. Office of Science and Technology Policy, "Climate Change," October 1997, 4, 5, 9. Isthmus of Panama and onset of ice age in G. H. Haug and R. Tiedemann, "Effect of the Formation of the Isthmus of Panama on Atlantic Ocean Thermohaline Circulation, *Nature* 393 (6686), June 18, 1998: 673–79, and in K. J. Willis et al., "The Role of Sub-Milankovitch Climatic Forcing in the Initiation of the Northern Hemisphere Glaciation," *Science* 286 (5427), July 23, 1999: 568–71. Isthmus of Panama and human evolution in Steven Stanley, *Children of the Ice Age* (New York: Harmony Books, 1996).

Chapter 6: Long-distance Swimmers

1. Visibility in *The Sailor's Handbook,* Halsey C. Herreshoff, consulting editor (Boston: Little Brown, 1983) 111. Visibility and navigational accuracy of Columbus in *The Log of Christopher Columbus,* trans. Robert H. Fuson (Camden, Maine: International Marine of TAB Books, McGraw-Hill, 1992), 48, 29, xiv. Longitude and John Harrison in Dava Sobel, *Longitude* (New York: Penguin, 1995).

2. Rigid paddle for swimming in Ren Hirayama, "Oldest Known Sea Turtle," *Nature* 392 (6678), April 16, 1998: 705–11. Ascension turtles in B. W. Bowen and J. C. Avise, "Tracking Turtles Through Time," *Natural History,* December 1994, 36–44, and personal communications, Brian Bowen, University of Florida, August and September 1998. Eight-week hatch, size of hatchlings, and age of maturity in B. W. Bowen et al., "Global Population Structure and Natural History of the Green Turtle (*Chelonia mydas*) in Terms of Matriarchal Phylogeny," *Evolution* 46 (4), 1992: 865–81. Hatching in K. J. Lohmann et al., "Orientation, Navigation, and Natal Beach Homing in Sea Turtles," in *The Biology of Sea Turtles,* ed. P. L. Lutz and J. A. Musick (Boca Raton: CRC Press, 1997), 109.

3. Turtles moving toward light and establishing orientation and bearings in K. J. Lohmann et al., "Orientation, Navigation, and Natal Beach Homing." Years at sea in

A. B. Bolten et al., "Transatlantic Developmental Migrations of Loggerhead Sea Turtles Demonstrated by mtDNA Sequence Analysis," *Ecological Applications* 8 (1) 1998: 1–7. Turning south at split in Gulf Stream in K. Lohmann and C. M. F. Lohmann, "Detection of Magnetic Field Intensity by Sea Turtles," *Nature* 380 (6569), March 7, 1996: 59–62. Strandings in "Turtle Tally," *Seabits* 14 (1), New England Aquarium, January 2000, in "Decking the Halls with Sea Turtles," *Seabits* 13 (12), New England Aquarium, December 1999, and in "Only 5600 Kilometres from Home," *New Scientist* 134 (1815), April 4, 1992: 11. Turtle navigation in Henry Fountain, "Navigation Satellites Have Rival: The Sea Turtle," *New York Times,* March 12, 1996, C4, in John Travis, "Internal Compass Guides Sea Turtles," *Science News* 48 (19), July 8, 1995: 31, in M. Goff, M. Salmon, and K. J. Lohmann, "Hatchling Sea Turtles Use Surface Waves to Establish a Magnetic Compass Direction," *Animal Behavior* 55, 1998: 59–67, in E. L. Manning, H. S. Cate, and K. J. Lohmann, "Discriminating of Ocean Wave Features by Hatchling Logggerhead Sea Turtles, *Caretta caretta,*" *Marine Biology* 127, 1997: 539–44, and in K. J. Lohmann and C. M. F. Lohmann, "Sea Turtle Navigation and the Detection of Geomagnetic Field Features," *Journal of Navigation* 5 (1), January 1998: 10–22.

4. Evolution of sea turtles in Hirayama, "Oldest Known Sea Turtle." Speciation of turtles in B. W. Bowen and S. A. Karl, "Population Genetics, Phylogeography, and Molecular Evolution," *Biology of Sea Turtles,* 31–38. Common ancestor of green turtles in Bowen et al., "Natural History of the Green Turtle."

5. Natal homing and DNA in Bowen and Karl, "Population Genetics, Phylogeography, and Molecular Evolution." Natal homing and Atol dos Rocas in Bowen et al., "Natural History of the Green Turtle."

6. Glaciers and nesting in Florida and Cyprus, and new rookeries created by one gravid female in Bowen et al., "Natural History of the Green Turtle." Warmth needed for incubation in S. E. Encalada et al., "Population Structure of Loggerhead Turtle (*Caretta caretta*) Nesting Colonies in the Atlantic and Mediterranean as Inferred from Mitochhondrial DNA Control Region Sequences," *Marine Biology* 130, 1998: 567–75.

7. Fidelity as short-term liability in Bowen et al., "Natural History of the Green Turtle."

8. X'cacel in Sam Dillon, "Voice of the Turtles Opposes a Hotel Development," *New York Times,* November 16, 1998, A4, and Brian Bowen personal communication, August 1998. Lights distracting turtles in Fred Pearce, "Return to Turtle Bay," *New Scientist* 135 (1836), August 29, 1992: 12, and in K. Katselidis and D. Dimopoulos, "The Impact of Touristic Development on Loggerhead Nesting at Daphni Beach, Zakynthos, Greece," abstract from 18th International Sea Turtle Biology and Conservation Symposium, March 3–7, 1998, Mazatlán, Sinaloa, Mexico. Dune buggies in Michael L. Weber, "Contested Coastlines," *National Parks,* January–February 1997, 30–34.

9. Turtles in Cyprus and Turkey, and gene pool down in Sandra E. Encalada, "Conservation Genetics of Atlantic and Mediterranean Green Turtles: Inferences," in Proceedings of the International Symposium on Sea Turtle Conservation Genetics, NOAA technical memorandum NMFS-SEFSC-396, December 1996, 37–38. Three to five hundred green turtles in Mediterranean in S. Hochscheid et al., "An Investigation of Green Turtle Inter-Nesting Behavior in the Eastern Mediterranean Using Data-Logging Devices," 18th International Symposium.

10. Rancho Nuevo in D. C. Rostal et al., "Nesting Physiology of Kemp's Ridley Sea Turtles, *Lepidochelys kempi*, at Rancho Nuevo, Tamaulipas, Mexico, with Observations on Population Estimates," *Chelonian Conservation and Biology* 2 (4) 1997: 538–47, in Jon R. Luoma, "Endangered Ridley Turtle Makes a Comeback on Mexican Coast," *New York Times,* November 29, 1994, C4. Nests from David Owens, University of Charleston Grice Marine Laboratory, personal communication, September 20, 1998, from *http://www.ex.ac.uk/MEDASSET/dkemps* from data presented at 18th International Symposium, and in Bernice C. Wuethrich, "Into Dangerous Waters," *International Wildlife,* March–April 1996, 44.

11. Kemp's Ridley nesting population in Luoma, "Endangered Ridley Turtle Makes a Comeback." *Arribada* in C. E. Eckrich and D. W. Owens, "Solitary versus Arribada Nesting in the Olive Ridley Sea Turtles (*Lepidochelys olivacea*): A Test of the Predator-Satiation Hypothesis," *Herpetologica* 51(3), September 1985: 349, and David Owens, personal communication, September 1998.

12. Use of hawksbill shell in Roy Pemberton, "Hawksbill: Turtle of the Reef," National Audubon Society's *Living Oceans News,* Summer 1997, and in Brian W. Bowen, "Tracking Marine Turtles with Genetic Markers," NOAA technical memorandum NMFS-SEFSC-396, 113. Hawksbill population and trends in Anne B. Meylan, "Status of the Hawksbill Turtle (*Eretmochelys imbricata*) in the Caribbean Region," *Chelonian Conservation and Biology* 3(2), April 1999: 177–84, and in A. B. Meylan and M. Donnelly, "Status Justification for Listing the Hawksbill Turtle (*Eretmochelys imbricata*) as Critically Endangered on the 1996 IUCN Red List of Threatened Animals," *Chelonian Conservation and Biology* 3 (2), 1999: 200–224.

13. Foraging and nesting sites and turtle migration in Meylan and Donnelly, in Anna L. Bass, "Genetic Analysis to Elucidate the Natural History and Behavior of Hawksbill Turtles (*Eretmochelys imbricata*) in the Wider Caribbean: A Review and Reanalyis," pp. 195–99, in Anne B. Meylan, "International Movements of Immature and Adult Hawksbill Turtles (*Eretmochelys imbricata*) in the Caribbean Region," p.189–94, both in *Chelonian Conservation and Biology* 3 (2), 1999, and from Anna L. Bass, University of Florida, personal communication, September 1998.

14. Turtle migration in Bolten et al., "Migrations of Loggerhead Sea Turtles," from David Owens, personal communication, April 2000. Half the loggerheads in the Mediterranean from southeastern U.S. in Bowen, "Tracking Marine Turtles with Genetic Markers." Half the turtles caught on longlines and almost all caught in bottom trawls originate in Mediterranean in L. Laurent et al., "Molecular Resolution of Marine Turtle Stock Composition in Fishery Bycatch: A Case Study in the Mediterranean," *Molecular Ecology* 7, 1998: 1529–42.

15. Atlantic loggerheads nesting in Encalada et al., "Population Structure of Loggerhead Turtle." Protecting juveniles in L. B. Crowder et al., "Predicting the Impact of Turtle Excluder Devices on Loggerhead Sea Turtle Populations," *Ecological Applications* 4 (3), 1994: 437–45. Loggerhead population dynamics and reproduction in National Research Council, *Decline of the Sea Turtles: Causes and Prevention* (Washington, D.C.: National Academy Press, 1990), 71, and in D. T. Crouse, L. B. Crowder, and H. Caswell, "A Stage-Based Population Model for Loggerhead Sea Turtles and Implications for Conservation," *Ecology,* 68 (5), 1987: 1412–23. Impact of shrimp trawls, potential to increase turtle populations in Crowder et al., "Predicting the Impact of Turtle Excluder Devices." Debate and dissension in Deborah Crouse,

"After TEDs: What's Next?," from 18th International Symposium. Reduction of strandings in South Carolina in Jerry Lockett, "Ecologists and Shrimpers at Loggerheads," *Geographical Magazine* 68 (4), April 1996: 9.

16. Restoration at Padre Island, and fewer strandings with shrimping ban in D. J. Shaver and C. W. Caillouet, Jr., "More Kemp's Ridley Turtles Return to South Texas to Nest," *Marine Turtle Newsletter* 82, October 1998, 1–5. Seasonal commercial fishing ban at Rancho Nuevo in R. Marquez et al., "Results of the Kemp's Ridley Nesting Beach Conservation Efforts in Mexico," *Marine Turtle Newsletter* 85, July 1999: 2–4. Turtle strandings in "More Dead Turtles Strand in Texas," *Marine Turtle Newsletter* 85, July 1999: 27. Shrimping violations of TED regulations, "NOAA Snares Shrimp Trawlers," *Marine Turtle Newsletter* 85, July 1999: 29–30.

17. Mediterranean turtles in Bowen, "Tracking Marine Turtles with Genetic Markers." Longliners in Mediterranean in Bolten et al., "Migrations of Loggerhead Sea Turtles," in R. Aguilar, J. Mas, and X. Pastor, "Impact of Spanish Swordfish Longline Fisheries on the Loggerhead Sea Turtle *Caretta caretta* Population in the Western Mediterranean," in J. I. Richardson and T. H. Richardson, compilers, Proceedings of the Twelfth Annual Workshop on Sea Turtle Biology and Conservation, NOAA technical memorandum NMFA-SEFSC-361, 1995: 1–6. Grand Banks closure in "NOAA Fisheries Closes Area in the Grand Banks to Longline Fishermen Under Emergency Rules to Protect Sea Turtles," NOAA press release, October 6, 2000. Fibropapillomatosis in A. A. Aquirre, "Fibropapillomatosis in Marine Turtles," from a workshop at the 18th International Symposium, and in *Marine Turtle Newsletter* 82, October 1998: 10–12. Papilloma tumors in Janet Raloff, "Sea Sickness: Marine Epidemiology Comes of Age," *Science News* 155 (5), January 30, 1999: 72–74.

18. Decline of turtle populations in Bowen and Avise, "Tracking Turtles Through Time," in K. A. Bjorndal et al., "Twenty-six Years of Green Turtle Nesting at Tortuguero, Costa Rica: An Encouraging Trend," *Conservation Biology,* in press. Rookeries wiped out in Bermuda and Cayman Islands in Bowen et al., "Natural History of the Green Turtle." Hawksbill nests up in Anne Meylan, "Status."

19. Bluefin migration in Lorelei Stevens, "Block's Tag Work Proves Pop-up Technology Sound," *Commercial Fisheries News,* June 1998, B2. Tuna spawning and migration in B. A. Block et al., "A New Satellite Technology for Tracking the Movements of Atlantic Bluefin Tuna," *Proceedings of the National Academy of Sciences USA* 95 (16), August 4, 1998: 9384–89. Spawning in Frank J. Mather III, "Historical Document: Life History and Fisheries of Atlantic Bluefin Tuna," NOAA technical memorandum NMFS-SEFSC-370, 1995: 79–91. Swordfish routes in "Atlantic Swordfish Overview," National Marine Fisheries Service, 1998, *http://www.nmfs.gov/sword.html.* Possible bluefin spawning area in mid-Atlantic in Workshop Recommendations from the Workshop on the Biology of Bluefin Tuna in the Mid-Atlantic, May 5–7, 2000, Hamilton, Bermuda.

20. Habits of swordfish in Carl Safina, *Song for the Blue Ocean* (New York: Henry Holt, 1997), 73–75, in Natural Resources Defense Council, "Swordfish in the North Atlantic: The Case for Conservation," January 1998, in Francis G. Carey, "Through the Thermocline and Back Again: Heat Regulation in Big Fish," *Oceanus,* Fall 1992, 81–84, in Catherine Dold, "A Mystery Explained: Why 25 Fish Species Are Warm-Blooded," *New York Times,* May 4, 1993, C4, in B. A. Block et al., "Evolution of Endothermy in Fish: Mapping Physiological Traits on a Molecular Phylogeny,"

Science 260 (5105), April 9, 1993: 210–15, and in H. B. Bigelow and W. C. Schroeder, *Fishes of the Gulf of Maine* (Washington, D.C.: U.S. Fish and Wildlife Service, 1953), 352–54.

21. Physical characteristics and habits of bluefin in Carey, "Through the Thermocline," in Safina, *Song for the Blue Ocean,* 56–60, in Block et al., "New Satellite Technology," in Jane Ellen Stevens, "Data Every 2 Minutes Are Fast Dispelling Mystery of Giant Tuna," *New York Times,* October 21, 1997, in Stevens, "Block's Tag Work," in Jonathan Beard, "Charlie the Robot Tuna," *New Scientist* 13 (1945), October 1, 1994: 22, in William K. Stevens, "Appetite for Sushi Threatens Giant Tuna," *New York Times,* September 17, 1991, C1, in Douglass Whynott, *Giant Bluefin* (New York: Farrar, Straus & Giroux, 1995), in Block et al., "Evolution of Endothermy."

22. No Cape Hatteras fishery in winter of 1998 in Stevens, "Block's Tag Work." Magnetite chips in Stevens, "Appetite for Sushi." Polarized light, motion of current, and scent of prey in Whynott, *Giant Bluefin,* 24.

23. More limited migration and needed quota reductions in Safina, *Song for the Blue Ocean,* 51, 101, 107, 110. More extensive migration in Block et al., "A New Satellite Technology," in Stevens, "Block's Tag Work," in M. Lutcavage et al., "Summary of Pop-up Satellite Tagging of Giant Bluefin Tuna in the Joint US-Canadian Program, Gulf of Maine and Canadian Atlantic," draft ICCAT working document SCRS/20/XX.

24. Block Island fishery in Van Campen Heilner, *Salt Water Fishing* (New York: Knopf, 1953), 115. Tuna fishing in Whynott, *Giant Bluefin,* in Safina, *Song for the Blue Ocean,* and in George Reiger, *Zane Grey: Outdoorsman* (Englewood Cliffs: Prentice-Hall, 1972).

25. Gulf of Mexico larval densities down, nine out of ten swordfish caught are juveniles, description of older fishermen in the Straits of Florida in Safina, *Song for the Blue Ocean,* 88, 96. Recreation bluefin catch from William Henchy, technical adviser to ICCAT, personal communication, January 20, 1999. Bluefin egg production in Whynott, *Giant Bluefin,* 20. Tagging and Gulf of Mexico breeding ground in Stevens, "Bluefin Spawning Questions," and in Lorelei Stevens, "Tags Slowly Revealing Bluefin Mysteries," *Commercial Fisheries News,* June 1999, 3B and 4B. Spring spawning in Gulf of Mexico in Block et al., "A New Satellite Technology." Size at maturity, legal catch sizes, and late closure of the fishery in Patrick A. Nickler, "A Tragedy of the Commons in Coastal Fisheries: Contending Prescriptions for Conservation, and the Case of the Atlantic Bluefin Tuna," *Boston College Environmental Affairs Law Review* 26 (3), Spring 1999: 549 (28 pp.). Percentage immature fish in catch, fishing in juvenile feeding areas, nondiscriminating gear, no transoceanic migration, and imports in Natural Resources Defense Council, "Swordfish in the North Atlantic." Swordfish catch two-thirds juveniles in "Florida Asks U.S. to Halt Commercial Swordfishing," *National Fisherman,* July 1998, 12. Two thirds of swordfish caught by longliners are juveniles in "Give Swordfish a Break Campaign Redoubles Pressure on Anniversary of Campaign Launch," Pew Charitable Trusts press release, January 21, 1999. More than 80 percent of female swordfish caught by longliners are immature in Carl Safina, "Song for the Swordfish," *Audubon,* May–June 1998. Bluefin harpooners granted less than 5 percent of the quota in "U.S. Meets Quota Requirements for 1997 on Bluefin Tuna; Announces 1998 Proposed Allocation," NOAA press release 98-R118, April 1, 1998.

26. Swordfish quota too high and imports almost half U.S. consumption in Natural Resources Defense Council, "Swordfish in the North Atlantic." Swordfish quota too high to rebuild stocks in "Draft Fishery Management Plan for Atlantic Tunas, Swordfish and Sharks," NOAA, October 1998, prepared by Office of Highly Migratory Species Management Division, Office of Sustainable Fisheries, National Marine Fisheries Service, Silver Spring, Maryland, pp. 2-2, 2-33, and in Lorelei Stevens, "Report from ICCAT: US Bluefin Quota Restored," *Commercial Fisheries News,* January 1999, B1. Eastern bluefin tuna quota twice as high as sustainable in Report of the Standing Committee on Research and Statistics, Bluefin Tuna Executive Summary, October 1999, and personal communication from William Henchy, January 20, 1999. European quota and landing violations in Lorelei Stevens, "ICCAT backs 16,000mt Eastern Bluefin Cut," *Commercial Fisheries News,* January 1999, 10B, in Lorelei Stevens, "U.S. Condemns EC Move to Hide Bluefin Overages," *Commercial Fisheries News,* January 2000, 3B, and in Caroline Raymakers and Jacqui Lynham, "Slipping the Net: Spain's Compliance with ICCAT Recommendations for Swordfish and Bluefin Tuna," World Wide Fund for Nature, November 1999, in "The Plunder of Bluefin Tuna in the Mediterranean," Greenpeace International, May 1999, and in SCRS Executive Summary. Ban on 33-pound swordfish in "Commerce Department Makes Final Import Ban on Small Swordfish," NOAA press release 99-R114, March 16, 1999. Swordfish nursery closures in Janice M. Plante, "NMFS Proposes Longline Time/Area Closures," *Commercial Fisheries News,* January 2000, 12A, 20A, in Janice M. Plante, "Industry's Pelagic Longline Plan," *Commercial Fisheries News,* January 2000, 13A, 20A, and in Janice M. Plante, "Pelagic Longliners Face NMFS Bycatch Reduction Closures," *Commercial Fisheries News,* September 2000, 12B. What the European quota should be in Safina, *Song for the Blue Ocean,* 51, 111.

27. Settling for less in "Draft Fishery Mangement Plan for Atlantic Tunas, Swordfish and Sharks," NOAA, October 1998, 2–42. Need 30 to 40 percent cut in U.S. swordfish quota in Natural Resources Defense Council, "Swordfish in the North Atlantic, " in "Report from ICCAT: U.S. Bluefin Quota Restored," *Commercial Fisheries News,* January 1999, 1A, 1B. 1994 good year class for tuna, William Henchy, personal communication, January 20, 1999, and 1997–98 good years for swordfish in Janice M. Plante, "ICCAT Commits to 10-Year Rebuilding Program," *Commercial Fisheries News,* January 2000, B1.

28. Evolution of whales in J. G. M. Thewissen, S. T. Hussain, and M. Arif, "Fossil Evidence for the Origin of Aquatic Locomotion in Archaeocete Whales," *Science* 263 (5144), January 14, 1994: 210–13, in "Ancient Whales: Thirsty at Sea," *Science News* 149 (25), June 22, 1996: 399, in R. Monastersky, "Fossil Whale Feet: A Step in Evolution," *Science News* 145 (3), January 15, 1994: 36, in R. Monastersky, "Fossil Jaw Tells Tale of Whale Evolution," *Science News* 154 (15), October 10, 1998, in Stephen Jay Gould, "Hooking Leviathan by Its Past," *Natural History,* May 1994, 8, and in John Noble Wilford, "How the Whale Lost Its Legs and Returned to the Sea," *New York Times,* May 3, 1994, C1, C11.

29. Blandford, Nova Scotia, whaling station from Phil Clapham, presentation at Right Whale Consortium, October 20, 1998, New England Aquarium. Sighting on Cape Farewell in A. R. Knowlton et al., "Long-distance Movements of North Atlantic Right Whales (*Eubalaena glacialis*)," *Marine Mammal Science* 8 (4), October 1992: 397–405.

30. Fifty and 350 whales in A. R. Knowlton, S. D. Kraus, and R. D. Kenney, "Reproduction in North Atlantic Right Whales (*Eubalaena glacialis*)," *Canadian Journal of Zoology* 72, 1994: 1297–1305.

31. Whales in calving ground in Chris Slay, "Notes from the Calving Ground," *Right Whale Research News,* Spring 1998, and in Lisa Conger, "Aerial Surveys in the Calving Grounds, Winter 1997," *Right Whale Research News,* Spring 1997. Whales in Provincetown Harbor in Marilyn K. Marx, "Cape Cod Bay, 1998," *Right Whale Research News,* Spring 1998. Whales in Great South Channel in R. D. Kenney, H. E. Winn, and M. C. Macaulay, "Cetaceans in the Great South Channel, 1979–89: Right Whale (*Eubalaena glacialis*)," *Continental Shelf Research* 15 (4–5), 1995: 385–414. 1,800-mile journey in S. D. Kraus et al., "Migration and Calving of Right Whales (*Eubalaena glacialis*) in the Western North Atlantic," *Reports of the International Whaling Commission,* special issue no. 10, 1986: 139–44. Nurseries from Ruth Waldick, presentation at Right Whale Consortium 1998. Browns Bank from Phil Clapham, presentation at Right Whale Consortium 1998. Bay of Fundy in "Bay of Fundy, 1997," *Right Whale Research News,* Fall 1997.

32. Southern versus northern right whales in C. M. Schaeff et al., "Comparison of Genetic Variability of North and South Atlantic Right Whales (*Eubalaena*), Using DNA Fingerprinting," *Canadian Journal of Zoology* 75, 1997: 1073–80. Nonsustainable growth rate in H. Caswell, M. Fujiwara, and S. Brault, "Declining Survival Probability Threatens the North Atlantic Right Whale," *Proceedings of the National Academy of Sciences USA,* 96 (6), March 16, 1999: 3308–13.

33. Calving ground in "Notes from the Calving Ground," *Right Whale Research News,* Spring 1998, and in Lisa Conger, presentation at Right Whale Consortium 1998. Tankers in Alan White, "Shippers Save Whales," *Telegraph Journal* (St. John, New Brunswick), February 14, 1998, A1, A2. Ship strikes in "Shackleton Lives," *Right Whale Research News,* Fall 1997, in Alan White, "New Brunswick Reader," *Telegraph Journal* (St. John, New Brunswick), September 20, 1997, 14–19, in "Mortalities and Entanglements," *Right Whale Research News,* Fall 1997, in "Leviathan Forensics," *Woods Hole Currents* 8 (1), 1999: 16, in Scott Allen, "One Whale's Death Spotlights Struggle of Species," *Boston Globe,* April 22, 1999, B1, in Scott Allen, "Boat Hits Rare Whale Despite Protection Efforts, Autopsy Shows," *Boston Globe,* April 28, 1999, B4, in Amy Knowlton, presentation at Right Whale Consortium 1998, in "High Speed Vessels Pose Risk to Right Whales," *Right Whale News,* May 1998. Number of calves per mother from Amy Knowlton, New England Aquarium, personal communication, November 19, 1998.

34. Lobster fishery in Scott Allen, "Lobster Supply Imperiled, Biologists Say," *Boston Globe,* July 24, 1998, A1, A24. Entanglements in "Mortalities and Entanglements," *Right Whale Research News,* Fall 1997, in David Mattila, "What Can We Learn from Entangled Whales?," presentation at Right Whale Consortium 1998, in A. R. Knowlton and S. D. Kraus, "Serious Injury and Mortality of Right Whales: 1999," abstracts of the Right Whale Consortium, October 21–22, 1999, in Cathy Quinn, "Cape Cod Bay: A Bird's Eye View," *Right Whale Research News,* Spring 1997, in Phil Hamilton, Right Whale Consortium Abstracts, 1998, in Amy Knowlton, Right Whale Consortium 1998, and in "The Demise of #2030," *Right Whale Research News,* Fall 1999.

35. Zero take in "NMFS Reduces Biological Removal Level for Right Whales to Zero," *Right Whale News,* February 2000. Decrease in female life expectancy in H. Caswell,

M. Fujiwara, and S. Brault, "Stage-Structured Demography of the North Atlantic Right Whale; Part 2: Projection Matrices, Trends in Virtual Rates, and Population Growth," Right Whale Consortium Abstracts, 1999. Dead right whale with fishing gear in January 2000 in "Another Dead Whale Reported off the Rhode Island Coast," *Commercial Fisheries News,* February 2000, 9A. Further possible restrictions for fishing gear in Lorelei Stevens, "Whale Report Triggers Alarm Among Lobstermen," *Commercial Fisheries News,* February 2000, 8A, 9A. Right whales and ships in M. W. Brown et al., "Assessing and Reducing the Risk of Collisions Between North Atlantic Right Whales and Vessels in Canadian Waters," Right Whale Consortium Abstracts, 1999, in R. Leaper and H. Clyde, "Evaluation of the Potential for Vessels to Take Avoiding Action in Response to Sightings of Right Whales, Based on Results from Simple Simulation," Right Whale Consortium Abstracts, 1999, in Ernst Frankel, "Economic Overview of the Shipping Industry," in A. R. Knowlton et al., Shipping-Right Whale Workshop, Report 97-3, New England Aquarium, April 17 and 18, 1997, 134–37, and in A. R. Knowlton, "The Regulation of Shipping to Protect North Atlantic Right Whales," master's thesis, University of Rhode Island, 1997.

36. Growth rates and reproductive success in C. A. Miller et al., "Ultrasonic Measurement of Blubber Thickness in Right Whales," International Whaling Commission Scientific Committee Meeting Document SC/M98/RW27. Shrunken gene pool, inbreeding, reproductive success, and elephant seal rebound in Schaeff et al., "Comparison of Genetic Variability." Bones in Red Bay in M. W. Brown, S. L. Cumbaa, and B. N. White, "Zooarchaeological and Molecular Perspective on Basque Whaling in 16th Century Labrador," Right Whale Consortium Abstracts, 1999, and in B. N. White et al., "The Status of the North Atlantic Right Whale as Determined from Genetic Data," Right Whale Consortium Abstracts, 1999.

37. No increase in calves with increase in mothers in "Report of the IWC/MtN Workshop on Assessing the Status of Right Whales Worldwide, Cape Town, South Africa, March 16–25, 1998," in *Right Whale News,* May 1998. 1999 calving season in Scott Kraus and Philip Hamilton, "North Atlantic Right Whales Reproduction, 1999," Right Whale Consortium Abstracts, 1999. 2000 calving season in Chris Slay, "Report from the Calving Ground: A Brief Summary of Notes from the Calving Ground, 2000," *Right Whale News,* May 2000. Copepod densities and peaks, and whale energy needs in R. C. Beardsley et al., "Spatial Variability in Zooplankton Abundance Near Feeding Right Whales in the Great South Channel," *Deep-Sea Research II* 43 (7–8), 1996: 1601–25.

38. How whales find plankton from R. D. Kenney, University of Rhode Island, personal communication, July 28, 1998, and in R. D. Kenney, C. A. Mayo, and H. E. Winn, "Migration and Foraging Strategies at Varying Spatial Scales in Western North Atlantic Right Whales," Right Whale Consortium Abstracts, 1999. *Phaeocystis* and Cape Cod Bay in Marilyn Marx, "CCS Reports from Cape Cod Bay—A Water View," *Right Whale Research News,* Spring 1997, and in Scott Allen, "Whales' Early Exit Feeds Bay Fears," *Boston Globe,* May 1997. No mother-calf pairs in 1998 in Marilyn K. Marx, "Cape Cod Bay, 1998," *Right Whale Research News,* Spring 1998. None in 1999 in Scott Kraus and Philip Hamilton, "North Atlantic Right Whales Reproduction: 1999," Right Whale Consortium Abstracts, 1999. Copepods and pteropods in Robert D. Kenney, "Anomalous 1992 Spring and Summer Right Whale (*Eubalaena glacialis*) Distributions in the Gulf of Maine: Local Effects of Global-Scale Changes," International Whaling Commission Scientific Committee Meeting

Document SC/M98/RW29, 1998. No frequenters of inshore habitat giving birth in 1998–99 from Scott Kraus and Philip Hamilton, "North Atlantic Right Whales Reproduction: 1999," Right Whale Consortium Abstracts, 1999. Reproduction and quality of food in C. Mayo, E. Lyman, and A. DeLorenzo, "A Comparison of Caloric Availability in Cape Cod Bay with North Atlantic Right Whale Calving Rates: 1984–2000," Right Whale Consortium Abstracts, 2000. Blubber studies in Miller et al., "Blubber Thickness," and presentation of Carolyn A. Miller at Right Whale Consortium, October 1998.

39. Possible toxics in copepods from Michael M. Moore, Woods Hole Oceanographic Institution, personal communication, August 7 and November 9, 1998, and in M. J. Moore et al., "Cytochrome P450 1A and Chemical Contaminants in Dermal Biopsies of Northern and Southern Right Whales," Scientific Committee of the International Whaling Commission, SC/M98/RW24. Molecular common ground in Jennifer Ackerman, "Common Ground," *Sanctuary,* January–February, 1999, 3.

40. Navigation in "A Whale of a Mission for the U.S. Navy," *Technology Review,* January 1994, 15–16. Stranded sperm whales and silenced dolphins in Fred Pearce, "Seismic Bans Silence Sensitive Dolphins," *New Scientist* 151 (2045), August 31, 1996: 10. Heard Island and Navy low-frequency sonar effects in Bob Holmes, "Noises Off," *New Scientist* 153 (2071), March 1, 1997: 30–33. Entangled and deafened whales in Trinity Bay in Malcolme B. Brown, "Human Noises in Ocean Held to Threaten Marine Mammals," *New York Times,* October 19, 1993, C1, C12. Strandings in Alexandros Frantzis, "Does Military Acoustic Testing Strand Whales?," *Nature* 391 (6671), March 5, 1998: 29, in Michelle Faul, "Sonar May Be Confusing Whales, a Biologist Says," *Boston Globe,* June 15, 2000, A29, and in Wendy Williams, "Dead Whales Underscore the Threat of Human Noise to Ocean World," *Boston Globe,* June 27, 2000, C6.

41. Blue whale population in Johann Sigurjonsson, "On the Life History and Autecology of North Atlantic Rorquals," in *Whales, Seals, Fish and Man* (New York: Elsevier Science, 1995). Sightings in Jordana Hart, "Rare Blue Whale Found Dead off Rhode Island," *Boston Globe,* March 8, 1998, B8, and in J. Sigurjonsson and T. Gunnlaugsson, "Recent Trends in Abundance of Blue (*Balaenoptera musculus*) and Humpback Whales (*Megaptera novaeangliae*) off West and Southwest Iceland, with a Note on Occurrence of Other Cetacean Species," Report to the International Whaling Commission 40, 1990: 441–48.

42. Krill company in Bruce Upbin, "Don't Tell the Whale Lovers," *Forbes,* October 20, 1997, 153. Fin whale reproductive health and food in Sigurjonsson, "Life History and Autecology of North Atlantic Rorquals."

Chapter 7: Wide, Wide Sargasso

1. Ocean productivity and nutrient recycling in P. G. Falkowski, R. T. Barber, and V. Smetacek, "Biogeochemical Controls and Feedbacks on Ocean Primary Production," *Science* 281 (5374), July 10, 1998: 200–206, in M. D. McCarthy, J. I. Hedges, and R. Benner, "Major Bacterial Contribution of Marine Dissolved Organic Nitrogen," *Science* 281 (5374) July 10, 1998: 231–34, and in C. M. Duarte and S. Agusti, "The CO_2 Balance of Unproductive Aquatic Ecosystems," *Science* 281 (5374), July 10, 1998: 234–36. Nitrate in eddies in R. G. Williams and M. J. Follows, "Eddies Make Ocean Deserts Bloom," *Nature* 394 (6690), July 16, 1998:

228–30, in A. Oschilies and V. Garcon, "Eddy-Induced Enhancement of Primary Production in a Model of the North Atlantic Ocean," *Nature* 394 (6690), July 16, 1998: 266–68, and in D. J. McGillicuddy Jr. et al., "Influence of Mesoscale Eddies on New Production in the Sargasso Sea," *Nature* 394 (6696), July 16, 1998: 263–65. Nitrates in wakes of passing animals in George A. Jackson, "Phytoplankton Growth and Zooplankton Grazing in Oligotrophic Oceans," *Nature* 284, April 3, 1980: 439–40. Genetic markers identify bacteria in S. J. Giovannoni et al., "16S rRNA Genes Reveal Stratified Open Ocean Bacterioplankton Populations Related to the Green Non-sulfur Bacteria," *Proceedings of the National Academy of Sciences USA* 93 (15), 1996: 7979–84.

2. Gelatinous animals of the open ocean in William M. Hamner, "Blue-water Plankton," *National Geographic,* October 1974, 530–45, in G. Richard Harbison, "The Gelatinous Inhabitants of the Ocean Interior," *Oceanus,* Fall 1992, 18–24, in Kate Madin, "Adrift in the Blue," *Woods Hole Currents,* Winter 1997, 7–12, in Laurence P. Madin, "Sea Jellies and Other Alien Life in the Sea," lecture at the New England Aquarium, April 13, 1995, and in Robert Kunzig, "At Home with the Jellies," *Discover,* September 1997, 64 (8 pp.).

3. Biological pump in Sallie W. Chisholm, "What Limits Phytoplankton Growth?," *Oceanus,* Fall 1992, 36–43. Open ocean as CO_2 source in Duarte and Agusti, "The CO_2 Balance of Unproductive Aquatic Ecosystems," *Science* 281 (5374), July 10, 1998: 234–36. Open ocean as CO_2 sink in P. J. leB. Williams, "The Balance of Plankton Respiration and Photosynthesis in the Open Oceans," *Nature* 394(6688), July 2, 1998: 55–58, and ocean balanced in P. J. leB. Williams and D. G. Bowers, "Regional Carbon Imbalances in the Oceans," *Science* 284 (5421), June 11, 1999: 1735B. Global warming associated with more stratification, fewer nutrients, and less export of CO_2 to the deep in Falkowski, Barber, and Smetacek, "Biogeochemical Controls."

4. Salps in John and Mildred Teal, *The Sargasso Sea* (Boston: Atlantic Monthly Press, 1975), 139, and in Paul Erickson, "Aliens from the Deep," *Aqualog,* Winter 2000, 3. Salps, krill, and penguins in V. Loeb et al., "Effects of Sea-Ice Extent and Krill or Salp Dominance on the Antarctic Food Web," *Nature* 387 (6636), June 26, 1997: 897–900. Sea ice, chlorophyll, salps, and krill in S. Nicol, T. Pauly, and N. L. Bindoff, "Ocean Circulation off East Antarctica Affects Ecosystem Structure and Sea-Ice Extent," *Nature* 406 (6795), August 3, 2000: 504–7.

5. Salps, medusae, siphonophores, and jellyfish in Madin, "Sea Jellies and Other Alien Life." 78.5 percent of earth's habitat is the deep sea in Harbison, "Gelatinous Inhabitants."

6. Shrimp, hatchetfish, and countershading in Bruce H. Robison, "Light in the Ocean's Midwaters," *Scientific American,* July 1995, 60–69. Mid-water is typical in Madin, "Sea Jellies and Other Alien Life." Viperfish and anglerfish, p.140, bioluminescence, pp.145, 408–9, in James W. Nybakken, *Marine Biology: An Ecological Approach,* 3d ed. (New York: HarperCollins, 1993). Viperfish, p.145, hatchetfish, p.149, and anglerfish, pp. 543–34, in H. B. Bigelow and W. C. Schroeder, *Fishes in the Gulf of Maine* (Washington, D.C.: U.S. Fish and Wildlife Service, 1953). Ninety percent of midwater animals emit light in G. Leisman et al., "Bacterial Origin of Luminescence in Marine Animals," *Science* 208, June 13, 1980: 1271–73. Hatchetfish, anglerfish, and bioluminescence in Peter Herring, "Lights in the Night Sea," *New Scientist* 101 (1398), February 23, 1984: 45–48. Lanternfish and Alexander's Acres, pp. 82–83,

squid ink bioluminescent, p. 146, and animals' ability to reflect, focus, control, and filter light, p. 148, in Teal and Teal, *The Sargasso Sea*. Dragonfish in S. Milius, "Red-Flashing Fish Have Chlorophyll Eyes," *Science News* 153 (23), June 6, 1998: 359. Ideas on the origin of bioluminescence in F. McCapra and R. Hart, "The Origins of Marine Bioluminescence," *Nature* 286, August 14, 1980: 660–61, in C. M. Thomson, P. J. Herring, and A. K. Campbell, "Coelenterazine Distribution and Luciferase Characteristics in Oceanic Decapod Crustaceans," *Marine Biology,* 124 (2), December 31, 1995: 197, in J.-F. Rees et al., "The Origins of Marine Bioluminescence: Turning Oxygen Defence Mechanisms into Deep-sea Communication Tools," *Journal of Experimental Biology* 201 (8), April 15, 1998: 1211, and in T. Wilson and J. W. Hastings, "Bioluminescence," *Annual Review of Cell and Developmental Biology* 14, 1998: 197–230. Sucker to photophore in the octopus in S. Johnsen, E. J. Balser, and E. A. Widder, "Light-Emitting Suckers in an Octopus," *Nature* 398 (6723), March 11, 1999: 113, and S. Milius, "Octopus Suckers Glow in the Deep, Dark Sea," *Science News* 155 (11), March 13, 1999: 167. Squid ejecting bioluminescent ink in Richard Ellis, *Deep Atlantic: Life, Death, and Exploration in the Abyss* (New York: Alfred A. Knopf, 1996), 180.

7. Deep sea unexplored in Sylvia A. Earle, *Sea Change* (New York: G. P. Putnam's Sons, 1995), 13. Diving whales in James L. Sumich, *Biology of Marine Life* (Dubuque, Iowa: William C. Brown, 1984), 304–8, in Teal and Teal, *The Sargasso Sea,* 75, and in T. M. Williams, R. W. Davis, and L. A. Fuiman, "Sink or Swim: Strategies for Cost-Efficient Diving by Marine Mammals," *Science* 288 (5463), April 7, 2000: 133–36.

8. Extensive discussion of sperm whale feeding habits, giant squid biology and encounters with sperm whales in Richard Ellis, *The Search for the Giant Squid* (New York: Lyons Press, 1998). Squid evolution in Teal and Teal, *The Sargasso Sea,* 118, and in Natalie Angier, "At Love and Play under the Sea in Octopus's Garden," *New York Times,* August 11, 1998, C1, C5. Squid eyes and vision in William J. Broad, "Biologists Closing in on Hidden Lair of Giant Squid," *New York Times,* February 13, 1996, C1, C6, in Sandra Sinclair, *How Animals See* (New York: Facts on File, 1985), 43–45, and in Peter Aldhous, "Angry Cuttlefish See One Another in a Different Light," *New Scientist* 149 (2023), March 30, 1996: 16.

9. Sperm packets in William J. Broad, "When Big Squids Mate, It's a Stab in the Dark," *New York Times,* October 21, 1997, F1, F6. Squid stomach contents in Ellis, *Search for the Giant Squid,* 124. Toxics in sperm whale blubber, J. de Boer et al., "Do Flame Retardants Threaten Ocean Life?," *Nature* 394 (6688), July 2, 1998: 28–30, and in Marlise Simons, "Whale Tissue Raises Worry on Toxic Chemicals," *New York Times,* August 30, 1998: 13. Toxics settling in the deep from personal communication, John Stegeman, Woods Hole Oceanographic Institution, March 4, 1999. PCBs, chlordanes, and dieldrin in A. H. Knap, K. S. Binkley, and W. G. Deuser, "Synthetic Organic Chemicals in the Deep Sargasso Sea," *Nature* 319 (6054), 1986: 572–74.

10. Life at the bottom in Teal and Teal, *The Sargasso Sea,* 162–68, and in Ellis, *Deep Atlantic,* 216–22. Seasonal changes in productivity felt on or near the bottom in Richard Lampitt, "Fast Living on the Ocean Floor," *New Scientist* 105, February 28, 1985: 37–40, in W. G. Deuser, E. H. Ross, and R. F. Anderson, "Seasonality in the Supply of Sediment to the Deep Sargasso Sea and the Implications for the Rapid Transfer of Matter to the Deep Ocean," *Deep Sea Research,* 28 (5A), 1981: 495–505, and in Andrew Gooday, "Epifaunal and Shallow Infaunal Foraminiferal Communities

at Three Abyssal NE Atlantic Sites Subject to Differing Phytodetritus Input Regimes," *Deep Sea Research, Part 1, Oceanographic Research Papers* 43 (9), September 1996: 1395–1432. Deep sea may be starving in K. L. Smith Jr. and R. S. Kaufman, "Long Term Discrepancy Between Food Supply and Demand in the Deep Eastern North Pacific," *Science* 284 (5417), May 14, 1999: 1174–77, in William S. Broad, "The Diverse Creatures of the Deep May Be Starving," *New York Times,* June 1, 1999, D5, and in E. R. M. Druffel and B. H. Robison, "Is the Deep Sea on a Diet?," *Science* 284 (5417), May 14, 1999: 1139–40. Flat topography, sedimentation, sand off the Sahara, and slow deposition of red clay in James Kennett, *Marine Geology* (Englewood Cliffs, N.J.: Prentice-Hall, 1982), 37, 424–33. Sedimentation in O. H. Pilkey and W. J. Cleary, "Turbidite Sedimentation in the Northwestern Atlantic Basin," *The Geology of North America,* vol. M, *The Western North Atlantic Region* (Boulder: Geological Society of America, 1986), 437.

11. Manganese nodules in Kennett, *Marine Geology,* 497–99, in Bruce C. Heezen and Charles D. Hollister, *The Face of the Deep* (New York: Oxford University Press, 1971), 423–38, and in Ellis, *Deep Atlantic,* 95. Prohibitive cost of mining in Dick Russell, "Deep Blues," *Amicus,* Winter 1998, 26.

12. Burial of radioactive wastes in Charles D. Hollister and Steven Nadis, "Burial of Radioactive Waste Under the Seabed," *Scientific American,* January 1998: 60–65.

13. Eels swimming up the Mississippi in Richard E. Sparks, "Need for Ecosystem Management of Large Rivers and Their Floodplains," *Bioscience* 45 (3), March 1995: 168–83, and personal communication from Jay Hatch, University of Minnesota, September 15, 2000.

14. Two sets of teeth and metamorphosis in Dr. James McCleave, University of Maine at Orono, personal communication, July 23, 1998. Breeding populations in T. Wirth and L. Bernatchez, "Genetic Evidence Against Panmixia in the European Eel," *Nature* 409 (6823), February 22, 2001: 1037–40. U.S. eel fishery in Steven Waterman, "The Great Eel Rush," *National Fisherman,* August 1996, 16–17, in Stephen Rappaport, "Marine Patrol Scours Local Shorelines to Enforce Elver Harvest Regulations," *Ellsworth American,* April 23, 1998, in Hoyt Childers, "Scientists Are Perplexed by Parasite Attacking Eels," *National Fisherman,* December 1998, 12, in Stephen Rappaport, "DMR Deftly Shortens Elver Fishing Season," *Ellsworth American,* May 7, 1998, in Larry Chowning, "Poached Elvers Land North Carolina Man in Hot Water—and Jail," *National Fisherman,* April 1999, 17, in Stephen Rappaport, "Asian Economies Are Melting Down but Maine Fishermen Feel Chilled," *Ellsworth American,* February 19, 1998, in Stephen Rappaport, "DMR Will Take Action to Cut Elver Fishery," *Ellsworth American,* April 30, 1998, in Paul Molyneaux, "Elver Fishermen Discard Catch in Protection of Price," *National Fisherman,* August 1999, 9, and in Paul Molyneaux, "Easter Eels," *National Fisherman,* July 1999, 20–21.

15. St. Lawrence eels in L. A. Marcogliese, J. M. Casselman, and P. V. Hodson, "Dramatic Declines in Recruitment of American eel (*Anguilla rostraata*) Entering Lake Ontario—Long-Term Trends, Causes and Effects," plenary presentation at 3d National EMAN (Ecological Monitoring and Assessment Network) meeting, Saskatoon, Saskatchewan, January 22, 1997, in J. M. Casselman, L. A. Marcogliese, and P. V. Hodson, "The American Eel, *Anguilla rostrata,* Stock of the Upper St. Lawrence River and Lake Ontario: Long-Term Trends, Decreasing Abundance, Cause and Effect," presented at 1998 American Fisheries Society Annual Meeting,

August 23–27, 1998, and in J. M. Casselman et al., "Status of the Upper St. Lawrence River and Lake Ontario American Eel Stock—1996," in R. H. Peterson, ed., *The American Eel in Eastern Canada: Stock Status and Management Strategies,* Proceedings of Eel Workshop, January 13–14, 1997, Quebec City, Canadian Technical Report of Fisheries and Aquatic Sciences, no. 2196, 106–20. Organochlorines in European rivers in J. de Boer et. al., "8-Year Study on the Elimination of PCBs and Other Organochlorine Compounds from Eel (*Anguilla anguilla*) Under Natural Conditions," *Environmental Science and Technology* 28 (13), 1994: 2242–48. Belugas in S. DeGuise et al., "Possible Mechanisms of Action of Environmental Contaminants on St. Lawrence Beluga Whales," *Environmental Health Perspectives* 103 (4), May–June 1995: 73–77, in Pierre Beland, "The Beluga Whales of the St. Lawrence River," *Scientific American,* May 1996, 74–81, and in Steve Kemper, "The 'Sea Canary' Sings the Blues," *Smithsonian,* November 1999, 86. Spawning migration in "Maintaining Biodiversity," National Water Research Institute, Environment Canada, digest no. 18, February 2, 1997, in Backlund, "The American Eel," *South Dakota Conservation Digest,* July–August 1997. Fat needed for journey and gonads in V. J. T. van Ginneken and G. E. E. J. M. van den Thillart, "Eel Fat Stores Are Enough to Reach the Sargasso," *Nature* 403 (6766), January 13, 2000: 156–57. Transmission of pollutants in eggs in T. Hesman, "DDT Treatment Turns Male Fish into Mothers," *Science News* 157 (6), February 5, 2000: 87.

16. *Anguillicola crassus* in G. Peters and F. Hartmann, "*Anguillicola,* a Parasitic Nematode of the Swim Bladder Spreading Among Eel Populations in Europe," *Diseases of Aquatic Organisms* 1, October 15, 1986: 229–30, in J. Wurtz, K. Knopf, and H. Taraschewski, "Distribution and Prevalence of *Anguillicola crassus* (*Nematoda*) in Eels *Anguilla anguilla* of the Rivers Rhine and Naab, Germany," *Diseases of Aquatic Organisms* 32, March 5, 1998: 137–43, in L. T. Fries, J. Williams, and S. K. Johnson, "Occurrence of *Anguillicola crassus,* an Exotic Parasitic Swim Bladder Nematode of Eels, in the Southeastern United States," *Transactions of the American Fisheries Society* 125, 1996: 794–97, in Eel Nematode information sheet from Nonindigenous Aquatic Species information list of United States Geological Survey and Florida Caribbean Science Center, in J. Wurtz, H. Taraschewski, and B. Pelster, "Changes in Gas Composition in the Swim Bladder of the European Eel (*Anguilla anguilla*) Infected with *Anguillicola crassus* (*Nematoda*)," *Parasitology* 112, 1996: 233–38. Infection in the U.S. in Hoyt Childers, "Scientists Are Perplexed by Parasite Attacking Eels," *National Fisherman,* December 1998, 12, in E. R. Johnston Jr., M. C. Moser, and J. R. Hall, "Infection of *Anguilla rostrata* by *Anguillicola crassus* in the Cape Fear River Basin," Final Report to the North Carolina Sea Grant, Raleigh, North Carolina, 1999, in personal communication, Mary Moser, University of North Carolina at Wilmington, April 16, 1999, in A. Barse and D. H. Secor, "Widespread Invasion by the Nonindigenous Nematode Parasite, *Anguillicola crassus,* in Mid-Atlantic Bight American Eels," *Fisheries* 24(2), 1999: 6–10, in personal communication, David Secor, University of Maryland Center for Environmental Science, April 14, 1999, in D. Secor, J. Barker, and A. Barse, "Estuarine Gradients in Contamination and *Anguillicola crassus* Infestation in Hudson River American Eels," presentation at American Fisheries Society Annual Meeting, August 23–27, 1998, Hartford, Connecticut.

17. Chemicals sink with everything else in Deuser, Ross, and Anderson, "Seasonality in the Supply of Sediment." Traces of PCBs, chlordane, and dieldrin on Sargasso floor

in Knap, Brinkley, and Deuser, "Synthetic Chemicals in the Deep Sargasso Sea." Turtles, tar, and plastic in Blair E. Witherington, "Habitats and Bad Habits of Young Loggerhead Turtles in the Open Ocean," Proceedings of the 18th Annual Symposium on Sea Turtle Biology and Conservation, March 1998, and in Blair E. Witherington, "Flotsam, Jetsam, Post-Hatchling Loggerheads, and the Advecting Surface Smorgasbord," Proceedings of the 14th Annual Symposium on Sea Turtle Biology and Conservation, March 1994. Illegal dumping in Douglas Frantz, "Cruise Line Is Indicted in Dumping," *New York Times,* February 26, 1999, A10. *Sargassum* declining in "Time to Check Disappearance of the Wide Sargasso Weed," *National Fisherman,* November 1998, 6–7. Restricting the take of *Sargassum* in "Fishery Council Votes to Ban the Harvest of *Sargassum* Sea Weed," Sea Turtle Survival League/Caribbean Conservation Corp press release, December 14, 1998. *Sargassum* "stretching to the horizon . . . so thick that it actually held back the ships . . . so much weed that the sea seemed to be a solid mat . . . if we had not seen it was weed, we would have thought it was shoals" in *The Log of Christopher Columbus,* trans. Robert H. Fuson (Camden, Maine: International Marine, Tab Books, McGraw-Hill, 1992), 65–66, 182.

18. Community of *Sargassum* in Teal and Teal, *The Sargasso Sea,* 28–36, in Bigelow and Schroeder, *Fishes in the Gulf of Maine,* 541–42, in Gunnar Thorson, *Life in the Sea* (New York: McGraw-Hill, 1971), 59–63, in Sir Alister Hardy, *The Open Sea: Its Natural History, Part 1: The World of Plankton* (Boston: Houghton Mifflin, 1965), 36, in South Atlantic Fishery Management Council, "Fishery Management Plan for Pelagic *Sargassum* Habitat of the South Atlantic Region," December 1998, 14–23, 42–43, and in L.Coston-Clements et al., "Utilization of the *Sargassum* Habitat by Marine Invertebrates and Vertebrates—A Review," NOAA technical memorandum NMFS-SEFSC-296, October 1991. Iron from the Sahara in J. Wu et al., "Phosphate Depletion in the Western North Atlantic Ocean," *Science* 289 (5480), August 4, 2000: 759–62. *Trichodesmium* as nitrogen source in D. G. Capone et al., "*Trichodesmium,* a Globally Significant Marine Cyanobacterium," *Science* 276 (5316), May 23, 1997: 1221–29, and in Lawrence R. Pomery, "The Microbial Food Web," *Oceanus,* Fall 1992, 28–36.

Part III: The Empty Basin

Chapter 8: The Moving Earth Beneath the Sea

1. History of early sounding and mapping in Susan Schlee, *Edge of an Unfamiliar World* (New York: Dutton, 1973), and in Anthony Rice, "Finding Bottom," *Sea Frontiers,* March–April, 1991, 28–33. Finding volcanic basalt in Jean Francheteau, "The Ocean Crust," *Scientific American,* September 1983, 114–29.

2. Mid-Atlantic Ridge volcanoes in Jean-Louis Cheminée et al., *Arcyana FAMOUS: Photographic Atlas of the Mid-Atlantic Ridge* (Paris: Gauthier-Villars), 1978.

3. Spreading rates in Ken MacDonald, "A Slow but Restless Ridge," *Nature* 348 (6297), November 8, 1990: 108, in Ken C. MacDonald, "The Crest of the Mid-Atlantic Ridge: Models for Crustal Generation Processes and Tectonics," in *The Geology of North America,* vol. M, *The Western North Atlantic Region* (Boulder, Colo.: Geological Society of America, 1986), 52, and in David A. Ross, *Introduction to Oceanography* (Englewood Cliffs, N.J.: Prentice-Hall, 1997), 142.

4. Spreading rates of Europe and North America in Macdonald, "The Crest of the Mid-Atlantic Ridge," 59. Magnetic stripes and seafloor drift in J. R. Heirtzler and W. B. Bryan, "The Floor of the Mid-Atlantic Rift," *Scientific American,* August 1975, 78–90.

5. Iceland eruptions in *The Voyage of Saint Brendan,* trans. John J. O'Meara (Mountrath, Portlaoise, Ireland: Dolmen Press, 1976), 52–54, from Nordic Volcanological Institute, University of Iceland, *http://www.norvol.hi.is,* and in Katherine Scherman, *Daughter of Fire: A Portrait of Iceland* (Boston: Little Brown, 1978), 134–57.

6. Rate of Iceland eruptions in "Workshop Reports," *RIDGE Events Newsletter,* July 1998, 25. Rate of rift widening in Haraldur Sigurdsson, "Dyke Injection in Iceland: A Review," *Mafic Dyke Swarms, Geological Association of Canada, Special Paper 34,* edited by H. C. Halls, and W. F. Fahrig, 1987.

7. Myvatn and Krafla in Sigurdsson, "Dyke Injection in Iceland," in Sigurdur Thorainsson, "The Postglacial History of the Myvatn Area," *Oikos* 32, 1979: 17–28, in M. Bamlett and J. F. Potter, *Iceland* (London: Geologists' Association, 1994), 38–39, in Nordic Volcanological Institute, University of Iceland, *http://norvol.hi.is,* in P. Einarsson, and B. Prandsdottir, "Seismological Evidence for Lateral Magma Intrusion During July 1978 Deflation of the Krafla Volcano in NE Iceland," *Journal of Geophysics* 47, 1980: 160–65, in Haraldur Sigurdsson, *www.volcanotours.com/iceland/fieldguide.htm.*

8. Laki in Haraldur Sigurdsson, "Volcanic Pollution and Climate: The 1783 Laki Eruption," *EOS* 63 (32), August 10, 1982: 601–2. Flood under ice cap in M. T. Gudmundsson, F. Sigmundsson, and H. Bjornsson, "Ice-Volcano Interaction of the 1996 Gjalp Subglacial Eruption, Vatnojokull, Iceland," *Nature* 389 (6654), October 30, 1997: 954–57, and R. Monastersky, "Volcanoes Under Ice: Recipe for a Flood," *Science News* 150 (21), November 23, 1996: 327. Norse myths in Virginia Hamilton, *In the Beginning: Creation Stories from Around the World* (London: Harcourt Brace Jovanovich, 1988), and in Ingri and Edgar D'Aulaire, *Norse Gods and Giants* (Garden City, N.Y.: Doubleday, 1967).

9. Birth of Surtsey and eruptions on Heimay in Scherman, *Daughter of Fire,* 3–10, 45–60, and in Haraldur Sigurdssson, *http://www.volcanotours.com/iceland/fieldguide.htm.*

10. Volcanic activity on ridge in Ken C. MacDonald, "Exploring the Global Mid-Ocean Ridge," *Oceanus* 41 (1), 1998: 2–8, and in Donald W. Forsyth, "The Big MELT," *Oceanus* 41 (2), 1998, 27–31. Helium isotopes for mantle plume in J.-G. Schilling, "Iceland Mantle Plume: Geochemical Evidence Along the Reykjanes Ridge," *Nature* 242, April 27, 1973: 565–71, and in M. Kurz, P. S. Meyer, and H. Sigurdsson, "Helium Isotopic Systematics Within the Neovolcanic Zones of Iceland," *Earth and Planetary Science Letters* 74 1985: 291–305. Depth of mantle plume in Richard A. Kerr, "Iceland's Fires Tap the Heart of the Planet," *Nature* 284 (5417), May 14, 1999: 1095–97, in Cecily J. Wolfe, "Prospecting for Hotspot Roots," *Science* 396 (6708), November 19, 1998: 212–13, and in D. V. Helmberger, L. Wen, and X. Ding, "Seismic Evidence That the Source of the Iceland Hotspot Lies at the Core-Mantle Boundary," *Nature* 396 (6708), November 19, 1998: 251–55. Hot-spot heat and Azores/Iceland fueling Mid-Atlantic Ridge in Jian Lin, "Hitting the Hotspots," *Oceanus* 41 (2), 1998, 34–37.

11. Rift valley in J. R. Delaney et al., "The Quantum Event of Oceanic Crustal Accretion: Impacts of Diking at Mid-Ocean Ridges," *Science* 281 (5374), July 10, 1998: 222–30, in D. K. Smith and J. R. Cann, "Mid-Atlantic Ridge Volcanic Processes," *Oceanus* 41 (1), 1998, 11–14, in D. K. Smith, S. E. Humphris, and W. B. Bryan, "A Comparison of Volcanic Edifices at the Reykjanes Ridge and the Mid-Atlantic Ridge at 24°–30°N," *Journal of Geophysical Research* 100 (B11), November 10, 1995: 22, 485–98, in Heirtzler and Bryan, "Floor of the Mid-Atlantic Rift," and in J.-L. Cheminée et al., *Arcyana FAMOUS: Photographic Atlas of the Mid-Atlantic Rift and Transform Fault at 3000 Meters* (Paris: Gauthier-Villars), 1978.

12. Ridge structure in G. M. Purdy et al., "Bathymetry of the Mid-Atlantic Ridge, 24°–31°N: A Map Series," *Marine Geophysical Researches* 12, 1990: 247–52, in J. Lin et al., "Evidence from Gravity Data for Focused Magmatic Accretion Along the Mid-Atlantic Ridge," *Nature* 344 (6267), April 12, 1990: 627–32, in A. J. Calvert, "Seismic Evidence for a Magma Chamber beneath the Slowspreading Mid-Atlantic Ridge," *Nature* 377 (6548), October 5, 1995: 410–13, in Macdonald, "The Crest of the Mid-Atlantic Ridge," 56, in Brian E. Tucholke, "Discovery of 'Megamullions' Reveals Gateways into Ocean Crust and Upper Mantle," *Oceanus* 41 (1), 1998, 15–17, and in J. Pilot et al, "Palaeozoic and Proterozoic Zircons from the Mid-Atlantic Ridge," *Nature* 393 (6686), June 18, 1998: 676–79. Underwater avalanche in "Rock Slide Under the Waves," *Science News* 141 (11), March 14, 1992: 175, and in Willard Bascom, *Waves and Beaches* (Garden City, N.Y.: Anchor Doubleday, 1980), 115.

13. Lucky Strike in C. Langmuir et al., "Hydrothermal Vents near a Mantle Hotspot: The Lucky Strike Vent Field at 37°N on the Mid-Atlantic Ridge," *Earth and Planetary Science Letters* 148, 1997: 69–91, and in C. L. Van Dover, D. Desbruyères, and M. Segonzac, "Biology of the Lucky Strike Hydrothermal Field," *Deep-Sea Research, Part I, Oceanographic Research Papers* 43 (9), September 1996: 1509–30.

14. Plumbing of hot springs in S. E. Humphris and T. McCollom, "The Cauldron Beneath the Seafloor," *Oceanus* 41 (2), 1998, 18–21, in R. Kerrich, "Nature's Gold Factory," *Science* 284 (5423), June 25, 1999: 2101–02, in S. E. Humphris et al., "The Internal Structure of an Active Sea-floor Massive Sulphide Deposit," *Nature* 377 (6551), October 26, 1995: 713–16, and in M. C. Kleinrock and S. E. Humphris, "Structural Control on Sea-floor Hydrothermal Activity at the TAG Active Mound," *Nature* 382 (6587), July 11, 1996: 149–53. Smokers in M. K. Tivey, "How to Build a Black Smoker Chimney," *Oceanus* 41 (2), 1998, 22–26, in J. M. Edmond and K. V. Damm, "Hot Springs on the Ocean Floor," *Scientific American*, April 1983, 78–93, and in Peter A. Rona, "Mineral Deposits from Sea-Floor Hot Springs," *Scientific American*, January 1986, 84–92.

15. Deep-sea mining and Law of the Sea in William J. Broad, *The Universe Below* (New York: Simon & Schuster, 1997), 262, 274–75, in Georg Breuer, "A Strategy for the Sea Floor," *New Scientist* 132 (1791), October 12, 1991: 34–37, in David E. Pitt, "U.S. Seeks to Fix Mining Provisions of Sea Treaty," *New York Times*, August 28, 1993, 3, in William J. Broad, "Plan to Carve Up Ocean Floor Riches Nears Fruition," *New York Times*, March 29, 1994, C1, C8, in J. L. Jacobson and A. Rieser, "The Evolution of Ocean Law," *Scientific American Presents the Oceans*, Fall 1998, 100–105, in Dick Russell, "Deep Blues," *Amicus Journal*, Winter 1998, 25–29, in William J. Broad, "Undersea Treasure and Its Odd Guardians," *New York Times*, December 30, 1997, F1, F5, in B. Wagner and P. Lofrumento, "It's Time to Ratify the

Law of the Sea Treaty," *Washington Quarterly* 22 (3), Summer 1999: 17–21, in K. Iizasa et al., "A Kuroko-Type Polymetallic Sulfide Deposit in a Submarine Silicic Caldera," *Science* 283 (5404), February 12, 1999: 975–77, and in G. P. Glasby, "Lessons Learned from Deep-Sea Mining," *Science* 289 (5479), July 28, 2000: 551–53.

16. Location of vents in C. R. German and L. M. Parson, "Distribution of Hydrothermal Activity Along the Mid-Atlantic Ridge: Interplay of Magmatic and Tectonic Controls," *Earth and Planetary Science Letters* 160, 1998: 327–41, in C. S. Chin, G. P. Klinkhammer, and C. Wilson, "Detection of Hydrothermal Plumes on the Northern Mid-Atlantic Ridge: Results from Optical Measurements," *Earth and Planetary Science Letters* 162, 1998: 1–13, in J. Scholten et al., "Hydrothermal Activity Along the Tjoernes Fracture Zone, North of Iceland: Initial Results of R/V *Poseidon* Cruises 252 and 253," *InterRidge News,* Fall 1999, 28–32, in "Deep South," *New Scientist* 164 (2213), November 20, 1999: 27, and personal communication, Roy Livermore, British Antarctic Survey, September 2000. Manganese and methane concentrations in C. R. German, M. D. Rudnicki, and G. P. Klinkhammer, "A Segment-Scale Survey of the Broken Spur Hydrothermal Plume," *Deep-Sea Research, Part I* 46, 1999: 701–14. Balance of salt in the sea in Ross, *Introduction to Oceanography,* 180–81, in James Trefil, *A Scientist at the Seashore* (New York: Charles Scribner's Sons, 1984), 11–23. And in Richard A. Kerr, "A Cooler Way to Balance the Sea's Salt," *Science* 285 (5428), July 30, 1999: 657–58. Volume of water passing through hot springs in Humphris and McCollom, "The Cauldron Beneath the Seafloor."

17. Sulfur chemosynthesis in Holgar W. Jannasch, "Microbial Interactions with Hydrothermal Fluids," in S. E. Humphris et al., eds., *Seafloor Hydrothermal Systems: Physical, Chemical, Biological and Geological Interactions,* geophysical monograph 91, (Washington, D.C.: American Geophysical Union, 1995), 273–96. Shrimp in William J. Broad, "Floodlight of Divers May Blind Shrimp," *New York Times,* March 16, 1999, D3. Mussel densities in C. L. Van Dover, "Ecology of Mid-Atlantic Ridge Hydrothermal Vents," in I. M. Parsons, C. L. Walker, and D. R. Dixon, eds., *Hydrothermal Vents and Processes,* special publication no. 87 (London: Geological Society, 1995), 257–94. Toxicity and mussels in C. R. Fisher, "Temperature and Sulphide Tolerance of Hydrothermal Vent Fauna," pp. 283–86, and F. Geret et al., "Metal Compartmentalization and Metallothionein Isoforms in Mussels from the Mid-Atlantic Ridge; Preliminary Approach to the Fluid-Organism Relationship," pp. 291–93, in *Cahiers de Biologie Marine* 39 (3–4), 1998. Bacteria in Lucky Strike mussels in Langmuir et al., "Lucky Strike Vent Field." DNA damage in mussels from The InterRidge Workshop: Long-Term Monitoring of the Mid-Atlantic Ridge (MOMAR) October 28–31, 1998, Lisbon, and in D. R. Dixon, J. T. Wilson, and L. R. J. Dixon, "Toxic Vents and DNA Damage," *InterRidge News,* Spring 2000, 13–15.

18. Prevalence of mussels versus shrimp at some fields, personal communication, Daniel Desbruyères, IFREMER, July 16, 1999. Habits of shrimp in Van Dover, "Ecology of Mid-Atlantic Ridge Hydrothermal Vents," in R. N. Jinks et al., in "Sensory Adaptations in Hydrothermal Vent Shrimps from the Mid-Atlantic Ridge," pp. 309–12, and D. R. Dixon, L. R. J. Dixon, and D. W. Pond, "Recent Advances in Our Understanding of the Life History of Bresiliid Vent Shrimps on the MAR," pp. 383–86, in *Cahiers de Biologie Marine* 39 (3–4), 1998. Dim light at the vents in S. N. White and A. D. Chave, "ALISS in Wonderland," *Oceanus* 41 (2), 1998, 14–17,

in Ruth Flanagan, "The Light at the Bottom of the Sea," *New Scientist* 156 (2112), December 13, 1997: 42–46, and in Carl Zimmer, "The Light at the Bottom of the Sea," *Discover,* November 1996, 62–73.

19. Extinct chimneys and empty shells at Lucky Strike in Langmuir et al., "Lucky Strike Vent Field." Larvae in eddies in Jon Copley, "Going for a Spin," *New Scientist* 160 (2164), December 12, 1998: 31–33. Symbionts through the generations in J. L. Trask and C. L. Van Dover, "Site-Specific and Ontogenetic Variations in Nutrition of Mussels (*Bathymodiolus* sp.) from the Lucky Strike Hydrothermal Vent Field, Mid-Atlantic Ridge," *Limnology and Oceanography* 44 (2), March 1999: 334–44. Dispersal and metabolism in L. S. Mullineaux and S. C. France, "Dispersal Mechanisms of Deep-sea Hydrothermal Vent Fauna," in *Seafloor Hydrothermal Systems,* 408–24. DNA in mussels in D. Jollivet et al., "Unexpected Relationship Between Dispersal Strategies and Speciation within the Association *Bathymodiolus* (*Bivalvia*)—*Branchipolynoe* (*Polychaeta*) Inferred from the rDNA Neutral ITS2 Marker," pp. 359–69, and mussel species and worm feeding in P. Chevaldonne et al., "Commensal-Scale Worms of the Genus *Branchipolynoe* (*Polychaeta: Polynoidae*) at Deep-sea Hydrothermal Vents and Cold Seeps," pp. 347–50, in *Cahiers de Biologie Marine,* 39 (3–4), 1998.

20. Broken Spur field in J. T. P. Copley et al., "Spatial and Interannual Variation in the Faunal Distribution of the Broken Spur Vent Field (29°N, Mid-Atlantic Ridge), *Marine Biology* 129 (4), December 1997: 723–33, and in Van Dover, "Ecology of Mid-Atlantic Ridge Hydrothermal Vents." Movements of larval and juvenile shrimp in P. J. Herring and D. R. Dixon, "Extensive Deep-sea Dispersal of Postlarval Shrimp from a Hydrothermal Vent," *Deep-Sea Research, Part I* 45 (12), December 1998: 210–18, in P. J. Herring, "North Atlantic Midwater Distribution of the Juvenile Stages of Hydrothermal Vent Shrimps (*Decapoda: Bresiliidae*)," *Cahiers de Biologie Marine* 39 (3–4), 1998: 387–90, and from The InterRidge Workshop: Long-Term Monitoring of the Mid-Atlantic Ridge (MOMAR), October 28–31, 1998, Lisbon. Photosynthethic shrimp and egg size in Dixon, Dixon, and Pond, "Life History of Bresiliid Vent Shrimps." Oxygen-poor environment of the vents in Chevaldonne et al., "Commensal-Scale Worms." Mussel worms packed with hemoglobin in Daniel Desbruyères, INFREMER, personal communication, July 16, 1999. Ancient fossil hot springs in R. M., "Deep Ocean Is No Place to Hide," *Science News* 151 (6), February 8, 1997: 91.

21. Vent microbes in Jannasch, "Microbial Interactions with Hydrothermal Fluids," in M. T. Madigan and B. L. Narrs, "Extremophiles," *Scientific American,* April 1997, 82–88, in William J. Broad, "Strange Oases in Sea Depths Offer Map to Riches," *New York Times,* November 16, 1993, C1, C15, and in Don A. Cowan, "Hyperthermophilic Enzymes; Biochemistry and Biotechnology," in *Hydrothermal Vent Processes,* 351–63.

22. Deep biosphere and methanogenesis in D. C. Nelson, "Recent Progress in the Microbiology of Deep-sea Hydrothermal Vents and Seeps," *Cahiers de Biologie Marine* 39 (3–4), 1998: 373, 378, in Delaney et al., "Oceanic Crustal Accretion," in Holger W. Jannasch, "The Discovery of a 'Deep-Surface Biosphere': Remarks by a Microbiologist," *InterRidge News,* November 1997, and methanogenesis in Jannasch, "Microbial Interactions with Hydrothermal Fluids." Deep-sea biosphere as hope for life on Europa in John R. Delaney, "Life on the Seafloor and Elsewhere in the Solar System," *Oceanus* 41 (2), 1998: 10–13.

23. Redrawing the tree of life in W. Ford Doolittle, "Phylogenetic Classification and the Universal Tree," *Science* 284 (5423), June 25, 1999: 2124–28, in John Travis, "Third Branch of Life Bares Its Genes," *Science News* 150 (8), August 24, 1996: 116, in Nicholas Wade, "Tree of Life Turns Out to Have Surprisingly Complex Roots," *New York Times,* April 14, 1998, C1, C6, in Elizabeth Pennisi, "Is It Time to Uproot the Tree of Life?," *Science* 284 (5418), May 21, 1999: 1305–7, in Elizabeth Pennisi, "Genome Data Shake Tree of Life," *Science* 280 (5364), May 1, 1998: 672–74, and in Nicholas Wade, "Life's Origins Get Murkier and Messier," *New York Times,* June 13, 2000, D1, D2. Swapping genes in Elizabeth Pennisi, "Borrowing—Genes—from Microbial Neighbors," *Science* 284 (5418), May 21, 1999: 1318–22.

24. Early life in Richard A. Kerr, "Early Life Thrived Despite Earthly Travails," *Science* 284 (5423), June 25, 1999: 2111–113, in Delaney, "Life on the Seafloor," in Nicholas Wade, "Evidence Emerges for Placing Life's Origins Around Volcano," *New York Times,* April 11, 1997, A27, in Nicholas Wade, "Amateur Shakes Up Ideas on Recipe for Life," *New York Times,* April 22, 1997, C1, C8, in Sarah Simpson, "Life's First Scalding Steps," *Science News* 55 (2), January 9, 1999: 24–26, and in Wade, "Life's Origins Get Murkier and Messier."

Chapter 9: Atlantic's Primordial Ancestors

1. L'Anse aux Meadows in E. Kaye Fulton, "The Secrets on the Shore: Artifacts Revealed the Viking Past," *Maclean's,* August 5, 1991, 41, in Canadian Parks Service, "L'Anse aux Meadows National Historic Park," and in John Noble Wilford, "Ancient Site Offers Clues to Vikings in America," *New York Times,* May 9, 2000, D1, D4.

2. Crab nebula in R. Cowen, "Shedding Light on an Ancient Supernova," *Science News* 156 (6), August 7, 1999: 87. New stars in Nigel Henbest, "Birth of the Planets," *New Scientist* 131 (1783), August 24, 1991: 30–35, and in P. Smaglik, "Dust Sheds New Light on Planetary Birth," *Science News* 151 (10), March 8, 1997: 141. Hydrogen cloud as possible star in J. Bland-Hawthorn, "Clues to Galaxy Formation," *Nature* 400 (6741), July 15, 1999: 220–21. Cooling into planetesimals in Ron Cowen, "In the Footsteps of Descartes," *Science News* 147 (16), April 22, 1995: 251–53. Origin of universe and planets in R. Osborne and D. Tarling, eds., *The Historical Atlas of the Earth* (New York: Henry Holt, 1996), 18–23, in S. Liebes, E. Sahtouris, and B. Swimme *A Walk Through Time* (New York: John Wiley and Sons, 1998), 8–28, and in Govert Schilling, "From a Swirl of Dust, a Planet Is Born," *Science* 286 (5437), October 1, 1999: 66–68.

3. Origin of the moon, liquid sea and melting in A. N. Halliday and M. J. Drake, "Origin of Earth and Moon: Colliding Theories," *Science* 283 (5409), March 19, 1999: 1861–63, in S. R. Taylor and S. M. McLennan, "The Evolution of Continental Crust," *Scientific American,* January 1997, 76–80, and in Osborne and Tarling, *Historical Atlas of the Earth,* 26.

4. Possible sources of earth's water in R. Meier et al., "A Determination of the HDO/H_2O Ratio in Comet C/199501 (Hale-Bopp)," *Science* 279 (5352), February 6, 1998: 842–43, in William J. Broad, "Spotlight on Comets in Shaping of Earth," *New York Times,* June 3, 1997, C1, C6, and in C. F. Chyba, "Response to Organic Shielding of Greenhouse Gases on Early Earth," *Science* 259 (5352), February 6, 1998: 779a. Young sun 30 percent less warmth in Richard A. Kerr, "Early Life Thrived Despite Earthly Travails," *Science* 284 (5423), June 25, 1999: 2111–13. Amino acids

in Murchison meteor in Christopher Chyba, "Origins of Life: Buried Beginnings," *Nature* 395 (6700), September 24, 1998: 329–30, and in Richard A. Kerr, "Making New Worlds with a Throw of the Dice," *Science* 286 (5437), October 1, 1999. 68–69.

5. Making of the continents, volcanics on other planets in Taylor and McLennan, "Evolution of Continental Crust," in "William J. Broad, "How the First Land Arose from the Waters," *New York Times,* July 25, 1995, C1, C5, in Carl Zimmer, "Ancient Continent Opens Window on Early Earth," *Science* 286 (5448), December 17, 1999: 2254–56, and in Richard A. Kerr, "Looking Back to Early Mars, Deep into Earth," *Science* 287 (5451), January 14, 2000: 218–19.

6. Isua in Stephen Moorbath, "The Oldest Rocks and the Growth of Continents," *Scientific American,* March 1977, 92–104, and in Minik T. Rosing, "^{13}C-Depleted Carbon Microparticles in >3700-Ma Sea-Floor Sedimentary Rocks from West Greenland," *Science* 283 (5402), January 29, 1999: 674–76. Adaptation of vent bacteria to surface photosynthesis in E. G. Nisbet and C. M. R. Fowler, "Some Liked It Hot," *Nature* 382 (6590), August 1, 1996: 404–5, and in Euan Nisbet, "The Realms of Archaean Life," *Nature* 405 (6787), June 8, 2000: 625–26.

7. Fossils of hot-springs bacteria in Birger Rasmussen, "Filamentous Microfossils in a 3,235-Million-Year-Old Volcanogenic Massive Sulphide Deposit," *Nature* 405 (6787), June 8, 2000: 676–79. Trace metals in Nisbet, "Realms of Archaean Life." Brine in L. Paul Knauth, "Salinity History of the Earth's Early Ocean," *Nature* 395 (6702), October 8, 1998: 545–55. Banded iron formations in D. E. Canfield, "A Breath of Fresh Air," *Nature* 400 (7644), August 5, 1999: 503–4, in Brian F. Windley, *The Evolving Continents,* 2d ed. (New York: John Wiley, 1984), and in D. E. Canfield, "A New Model for Proterozoic Ocean Chemistry," *Nature* 396 (6710), December 3, 1998: 450–52.

8. Ancient continents in John J. W. Rogers, "A History of the Continents in the Past Three Billion Years," *Journal of Geology* 104, January 1996: 91–107. Stromatolites in Svalbard in Richard Fortey, *Life: A Natural History of the First Four Billion Years of Life on Earth* (New York: Knopf, 1998), 52. Molecular marker in R. E. Summons et al., "2-Methylhopanoids as Biomarkers for Cyanobacterial Oxygenic Photosynthesis," *Nature* 400 (6744), August 5, 1999: 554–57, and in J. J. Brocks et al., "Archean Molecular Fossils and the Early Rise of Eukaryotes," *Science* 285 (5430), August 13, 1999: 1033–36. Banded iron formations in Canfield, "Breath of Fresh Air." Cyanobacteria and stromatolites in J. William Schopf, *Cradle of Life: The Discovery of Earth's Earliest Fossils* (Princeton, N.J.: Princeton University Press), 1999: 76–98, 184–93.

9. Origin and rise of eukaryotes in Lynn Margulis, *Symbiotic Planet* (New York: Basic, 1998), in William Martin, "A Powerhouse Divided," *Science* 287 (5456), February 18, 2000: 1219, in Geoff McFadden, "Ever Decreasing Circles," *Nature* 400 (6740), July 8, 1999: 119–21, in J. J. Brocks et al., "Archean Molecular Fossils and the Early Rise of Eukaryotes," *Science* 285 (5430), August 13, 1999: 1033–36, and in A. H. Knoll, "A New Molecular Window on Early Life," *Science* 285 (5430), August 13, 1999: 1025–26.

10. Grenville mountains in Ian W. D. Dalziel, "Earth Before Pangea," *Scientific American,* January 1995, 58–64, in Rogers, "A History of the Continents," in Ian W. D. Dalziel, "Neoproterozoic-Paleozoic Geography and Tectonics: Review, Hypothesis,

Environmental Speculation," *GSA Bulletin* 109 (1), January 1997: 16–42, in E-an Zen, "Exotic Terranes in the New England Appalachians—Limits, Candidates, and Ages: A Speculative Essay," in R. D. Hatcher, H. Williams, and I. Zietz, eds., *Contributions to the Tectonics and Geophysics of Mountain Chains,* memoir 158 (Boulder, Colo.: Geological Society of America, 1983), 55–76, and in Harold Williams, "Tectonic Lithofacies Map of the Appalachian Orogen," no. 1A (St. John's, Newfoundland: Memorial University, 1978). Greenland into Torridon in R. G. Park, *Geological Structures and Moving Plates* (Glasgow: Blackie, 1988), 245, and in Steven Stanley, *Earth and Life Through Time,* 2d ed. (New York: Johns Hopkins University and W. H. Freeman, 1989), 289, 298. Birth of Iapetus in P. J. A. McCausland and J. P. Hodych, "Paleomagnetism of the 550Ma Skinner Cove Volcanics of West Newfoundland and the Opening of the Iapetus Ocean," *Earth and Planetary Science Letters* 163 (1–4), November 1998, 15–29, in B. Bingen, D. Demaiffe, and O. Breemen, "The 616Ma Old Egersund Basaltic Dike Swarm, SW Norway and the Late Neoproterozoic Opening of the Iapetus Ocean," *Journal of Geology* 105 (5), September 1998: 565–74, and in T. H. Torsvik et al., "Continental Break-up and Collision in the Neoproterozoic and Palaeozoic—A Tale of Baltica and Laurentia," *Earth-Science Reviews* 40 (3–4), June 1996: 229–58.

11. Failed graben in P. Robinson et al., "Paleozoic Orogens in New England, USA," *GFF,* 120, part 2, June 1998: 119–48. New Madrid in Richard A. Kerr, "From Eastern Quakes to a Warming's Icy Clues," *Science* 283 (5398), January 1, 1999: 28–29, and in "Eastern Quakes Recycle Old Faults," *Science News* 147 (9), March 4, 1995: 143. Rise of acritarchs in Gonzalo Vidal, "The Oldest Eukaryotic Cells," *Scientific American,* February 1984, 48–57, and in Andrew H. Knoll, "End of the Proterozoic Eon," *Scientific American,* October 1991, 64–73. Earth's earliest animals in S. Xiao, Y. Zhang, and A. H. Knoll, "Three-Dimensional Preservation of Algae and Animal Embryos in a Neoproterozoic Phosphorite," *Nature* 391 (6667), February 5, 1998: 553–59, in C.-W. Li, J.-Y. Chen and T.-E. Hua, "PreCambrian Sponges with Cellular Structures," *Science* 279 (5352), February 6, 1998: 879–82, in S. Bengston, "Animal Embryos in Deep Time," *Nature* 391 (6667), February 5, 1998: 529–30, and in Richard A. Kerr, "Pushing Back the Origins of Animals," *Science* 279 (5352), February 6, 1998: 603–4.

12. Snowball earth in Richard Monastersky, "Popsicle Planet: The King of All Ice Ages May Have Spurred Animal Evolution," *Science News* 54 (9), August 29, 1998: 137, in Richard A. Kerr, "Did an Ancient Deep Freeze Nearly Doom Life?," *Science* 281 (5381), August 28, 1998: 1251–61, in P. F. Hoffman et al., "A Neoproterozoic Snowball Earth," *Science* 281 (5381), August 28, 1998: 1342–46, in Richard A. Kerr, "An Appealing Snowball Earth That's Still Hard to Swallow," *Science* 287 (5459), March 10, 2000: 1734–36, in W. T. Hyde et al., "Neoproterozoic 'Snowball Earth' Simulations with a Coupled Climate/Ice-Sheet Model," *Nature* 405 (6785), May 25, 2000: 425–29, in Bruce Runnegar, "Loophole for Snowball Earth," *Nature* 405 (6785), May 25, 2000: 403–4, and in P. Weiss, "Warm Band May Have Girdled Snowball Earth," *Science News* 157 (22), May 27, 2000: 343.

13. Ediacara in R. Monastersky, "Living Large on the Precambrian Planet," *Science News* 149 (20), May 18, 1996: 308, in M. A. S. McMenamin and D. L. S. McMenamin, *The Emergence of Animals: The Cambrian Breakthrough* (New York: Columbia University Press, 1990), and in Mark A. S. McMenamin, *The Garden of Ediacara* (New York: Columbia University Press, 1998), in E. Landing, G. M. Narbonne, and

P. Myrow, eds., "Trace Fossils, Small Shelly Fossils and the Precambrian-Cambrian Boundary," Proceedings, bulletin 463, University of the State of New York, New York State Museum/Geological Survey, May 1988, 29, in M. W. Martin, D. V. Grazhdankin, and S. A. Bowring, "Age of Neoproterozoic Bilatarian Body and Trace Fossils, White Sea, Russia: Implications for Metazoan Evolution," *Science* 288 (5467), May 5, 2000: 841–45, and in Richard A. Kerr, "Stretching the Reign of Early Animals," *Science* 288 (5467), May 5, 2000: 789.

14. Isle of Islay worm in Jonathan Knight, "Gutsy Ancestors," *New Scientist* 156 (2112), December 13, 1997: 20. Fortune Head in Landing, Narbonne, and Myrow, "Precambrian-Cambrian Boundary." An environment conducive to life in R. Monastersky, "Eruptions Spark Explosions of Life," *Science News* 148 (1), July 1, 1995: 4. Conodonts in Douglas Palmer, "First Vertebrates Went in for the Kill," *New Scientist* 146 (1975), April 29, 1995: 16, in R. J. Aldridge and D. E. G. Briggs, "A Soft Body of Evidence," *Natural History,* May 1989, 6–9, and in Carl Zimmer, "In Search of Vertebrate Origins: Beyond Brain and Bone," *Science* 287 (5458), March 3, 2000: 1576–79. *Anomalocaris* in R. Monastersky, "The First Monsters," *Science News* 146 (9), August 27, 1994: 138–39, in Tim Thwaites, "When Monster Shrimps Ruled the World," *New Scientist* 143 (1936), July 30, 1994: 16, and in McMenamin and McMenamin, *The Emergence of Animals,* 111. No new body plans in Simon Conway Morris, "Showdown on the Burgess Shale: The Challenge," and Stephen Jay Gould, "Showdown on the Burgess Shale: The Reply," *Natural History,* December 1998–January 1999, 48–55. 0.1 percent of the dead made into fossils in Schopf, *Cradle of Life.*

15. Archaeocyanthans' own body plan in McMenamin and McMenamin, *The Emergence of Animals,* 60. Joyce R. Richardson, "Brachiopods," *Scientific American,* September 1996, 100–106. Lottery for survival in Stephen Jay Gould, *Wonderful Life: The Burgess Shale and the Nature of History* (New York: W. W. Norton, 1989), and in Richard Fortey, "Shock Lobsters," *London Review of Books,* October 1, 1998.

16. Dimensions of Iapetus, Scotland in North America, and the Precordillera in Dalziel, "Neoproterozoic-Paleozoic Geography." The Precordillera in Richard A. Kerr, "Missing Chunk of North America Found in Argentina," *Science* 270 (5242), December 8, 1995: 1567–68, in W. A. Thomas and R. A. Astini, "The Argentine Precordillera: A Traveler from the Ouachita Embayment of North American Laurentia," *Science* 273 (5276), August 9, 1996: 752–58, and in Tim Appenzeller, "Travels of America," *Discover,* September 1996, 80–88.

17. Baltica and reach of larvae in L. R. M. Cocks and R. A. Fortey, "The Lower Palaeozoic Margins of Baltica," *GFF* 120, part 2, June 1998: 173–79. Gondwana in Osborne and Tarling, *Historical Atlas of the Earth,* 52. Gondwana, Avalonia, Amazonia, and Florida in L. R. M. Cocks, W. S. McKerrow, and C. R. van Staal, "The Margins of Avalonia," *Geological Magazine* 135 (5), September 1997: 627–36. Gondwana in Torsvik et al., "Continental Break-up and Collision."

18. Carbonate bank in Robert D. Hatcher Jr., "Tectonic Synthesis of the U.S. Appalachians," in R. D. Hatcher Jr., W. A. Thomas, and G. W. Viele, eds., *The Geology of North America,* vol. F-2, *The Appalachian-Ouachita Orogen in the United States* (Boulder, Colo.: Geological Society of America, 1989), 514, and in Stanley, *Earth and Life Through Time.* Long arc in C. MacNiocaill, B. A. van der Pluijm, and R. Vandervoo, "Ordovician Paleography and the Evolution of the Iapetus Ocean,"

Geology 25 (2), February 1997: 159–62. Black and red shale, the Taconics, and Stark's Knob in Ed Landing, "Depositional Tectonics and Biostratigraphy of the Western Portion of the Taconic Allochthon, Eastern New York State," in Ed Landing, ed., *The Canadian Paleontology and Biostratigraphy Seminar,* bulletin 462, Albany: New York State Museum, March 1988, 96–110. Taconic and Humber Arm in Harold Williams, "Taconic Lithofacies Map of the Appalachian Orogen." Crown Point in Stephen E. Speyer, "Stratigraphy and Sedimentology of the Chazy Group (Middle Ordovician)—Lake Champlain Valley," in Landing, *Canadian Paleontology and Biostratigraphy Seminar,* 135–47. Marble and slate in Vermont and New York in Bradford B. Van Diver, *Roadside Geology of Vermont and New Hampshire* (Missoula, Mont.: Mountain Press Publishing, 1987), 121, 125. Thetford mine ophiolite in R. Laurent and B. Baldwin, "Thetford Mines Ophiolite, Quebec," in David C. Roy, ed., *Centennial Field Guide,* vol. 5, *Northeastern Section of the GSA* (Boulder, Colo.: Geological Society of America, 1987).

19. Island arc in Williams, "Taconic Lithofacies Map of the Appalachian Orogen," in Robinson et al., "Paleozoic Orogens in New England, USA," and in L. R. M. Cocks, W. S. McKerrow, and C. R. van Staal, "Margins of Avalonia." Scotland and the Caledonides in John F. Dewey, "Plate Tectonics and the Evolution of the British Isles," *Journal of the Geological Society of London* 139, 1982: 371–412, and in Osborne and Tarling, *Historical Atlas of the Earth,* 60. Katahdin and Monadnock in Van Diver, *Roadside Geology of Vermont and New Hampshire*. Acadian orogeny in Robinson et al., "Paleozoic Orogens in New England, USA," 119–48. Trilobites in Newfoundland and Shropshire, in Cocks, McKerrow, and van Staal, "Margins of Avalonia." Avalon trilobites in Torsvik et al., "Continental Break-up and Collision."

20. Geology of Newfoundland in Canadian Parks Service, *Rocks Adrift: The Geology of Gros Morne National Park,* 1990, and in A. R. Berger et al., "Geology, Topography and Vegetation, Gros Morne National Park, Miscellaneous Report 54," Ottawa, Geological Survey of Canada, 1992.

21. Soapstone in Vermont in R. Clements and D. Robinson, "The Carlton Quarry," *Rocks and Minerals* 71 (4), July–August 1996, 231–36. Ophiolites along Baie-Verte–Brompton Line in Robinson et al., "Paleozoic Orogens in New England." Soapstone at Fleur-de-Lys, and maritime archaic people in John Erwin, "Fleur de Lys Archaeological Project," *http://www.acs.ucalgary.ca/~jcerwin/introduction.htm,* and James A. Tuck, "The Maritime Archaic Tradition," museum notes, Newfoundland Museum, Fall 1991. Atlantic and Pacific realm trilobites in Niles Eldredge, *Life Pulse: Episodes from the Story of the Fossil Record* (New York: Facts on File, 1987), 57.

22. Beothuks in Ralph T. Pastore, "The Beothuks," museum notes, Newfoundland Museum, Fall 1991. Archaeology at Avalon at *http://www.heritagenf.ca/avalon* and in Tuck, "The Maritime Archaic Tradition."

Chapter 10: The Birth of Atlantic

1. River in Pangea in Bernice Wuethrich, "Long Ago, a River Ran Through It," *Science* 273 (5271), July 5, 1996: 31. Permian climate and evaporite basins in A. M. Ziegler, M. L. Hulver, and D. B. Rowley, "Permian World Topography and Climate," in I. P. Martini, ed., *Late Glacial and Postglacial Environmental Changes* (New York and Oxford: Oxford University Press, 1997), 111–46. Crinoids in Steven M. Stanley,

Earth and Life Through Time (New York: Johns Hopkins University and W. H. Freeman, 1989), 322–24, 386–89. Sea life in Tethys, and a unified world is a reduced world in Richard Fortey, *Life: A Natural History of the First Four Billion Years of Life on Earth* (New York. Knopf, 1998), 166–209. Sealife in Niles Eldredge, *The Miner's Canary: Unraveling the Mysteries of Extinction* (New York: Prentice-Hall, 1991), 49–78, and in Douglas H. Erwin, *The Great Paleozoic Crisis* (New York: Columbia University Press, 1993), 15–41.

2. Climate models versus botanical data, and warm polar current ameliorates coast but interior still cools in P. McA. Rees, et al., "Permian Climates: Evaluating Model Predictions Using Global Paleobotanical Data," *Geology* 27 (10), October 1999: 891–94. *Glossopteris* forest in Permian Antarctica in E. L. Taylor, T. N. Taylor, and R. Cuneo, "The Present Is Not the Key to the Past: A Polar Forest from the Permian of Antarctica," *Science* 257 (5077), September 18, 1992: 1675–78. Climate models versus paleobotany, and mature forest in Antarctica in E. L. Taylor, T. N. Taylor, and R. Cuneo, "Permian and Triassic High Latitude Paleoclimates: Evidence from Fossil Biotas," in B. T. Huber, K. G. MacLeod, and S. L. Wing, eds., *Warm Climates in Earth History* (Cambridge: Cambridge University Press, 2000), 321–50.

3. Disappearance of spore-bearing plants in C. Raymo and M. E. Raymo, *Written in Stone* (Chester, Conn: Globe Pequot, 1989), 91–93. *Lystrosaurus* diversity of mammals as result of breakup in Walter Sullivan, *Continents in Motion,* 2d ed. (New York: American Institute of Physics, 1991), 205–12. *Lystrosaurus* in Gillian King, "When the Desert Was Green," *Natural History* 105 (3), March 1996, 50 (6 pp.). Dearth of reptiles and diversity of mammals as result of continental drift in Bjorn Kurten, "Continental Drift and Evolution," *Scientific American,* March 1969, reprinted in *Continents Adrift* (San Francisco: W. H. Freeman), 114–23. Homogeneity of fauna in Permian in R. H. Dott and R. L. Batten, *Evolution of the Earth,* 4th ed. (New York: McGraw-Hill, 1988), 445. Evolution of reptiles and gymnosperms, change from provincial to cosmopolitan biotic provinces as Iapetus closed, continental fragments increasing biotic diversity, fast mammal evolution, and faunal diversity lowest in single continent in Brian F. Windley, *The Evolving Continents,* 3d ed. (Chichester: Wiley, 1995), 29–48. Number of marine provinces in Erwin, *The Great Paleozoic Crisis.* Marine provinces today in J. W. Valentine and E. M. Moores, "Plate Tectonics and the History of Life in the Oceans," *Scientific American,* April 1974, reprinted in *Continents Adrift and Continents Aground* (San Francisco: W. H. Freeman). David Jablonski, "Plate Tectonics and Evolution," in J. Scotchmoor and D. A. Springer, *Evolution: Investigating the Evidence,* special publication, vol. 9 (Pittsburgh: Paleontological Society, 1999), 283–92.

4. Extinction rates in S. A. Bowring, D. H. Erwin, and Y. Isozaki, "The Tempo of Mass Extinction and Recovery: The End-Permian Example," *Proceedings of the National Academy of Sciences USA* 96 (16), August 3, 1999: 8827–28, and in Y. G. Jin et al., "Pattern of Marine Mass Extinction near the Permian-Triassic Boundary in South China," *Science* 289 (5478), July 21, 2000: 432–36. Polar climates in Ziegler, Hulver, and Rowley, "Permian World Topography and Climate." Anoxia and similar temperature at Poles and equator in P. B. Wignall and R. J. Twitchett, "Oceanic Anoxia and the End Permian Mass Extinction," *Science* 272 (5265), May 24, 1996: 1155–59. Sluggish circulation and anoxia in R. Hotinski and L. Kump, Penn State University, personal communication, April 2000. Red and black chert and ocean suffocation in Yukio Isozaki, "Permo-Triassic Boundary Superanoxia and Stratified Superocean: Records from a Lost Deep Sea," *Science* 276 (5310), April 11, 1997: 235–39.

5. Gas release in Africa in Haraldur Sigurdsson, *Melting the Earth* (New York: Oxford Univeristy Press, 1999), 32–33. Possible crater in William J. Broad, "Newfound Crater Could Explain Worst Mass Extinction," *New York Times,* April 25, 2000, D12, and Richard A. Kerr, "Whiff of Gas Points to Impact Mass Extinction," *Science* 291 (5508), February 23, 2001: 1469–70. Massive lava from Siberian traps in Bowring, Erwin, and Isozaki, "Tempo of Mass Extinction and Recovery." Synchronicity of Permian extinction with massive Siberian traps in V. Courtillot et al., "On Causal Links Between Flood Basalts and Continental Breakup," *Earth and Planetary Science Letters* 166 (3–4), March 15, 1999: 177–95, and Paul E. Olsen, "Giant Lava Flows, Mass Extinctions, and Mantle Plumes," *Science* 284 (5414), April 23, 1999: 604–5. Volume of flood basalts versus Laki volcano, and coincidence of extinctions with flood basalts in Vincent Courtillot, *Evolutionary Catastrophe: The Science of Mass Extinction* (Cambridge: Cambridge University Press, 1999). Theories of extinction in Erwin, *The Great Paleozoic Crisis.*

6. Homogeneity in Pangea part of the equation in Fortey, *Life,* 186–209. Theories in Douglas H. Erwin, "The Mother of Mass Extinctions," *Scientific American,* July 1996, 72 (7 pp.), in Erwin, *The Great Paleozoic Crisis.* Suddenness in S. Perkins, "Was It Sudden Death for the Permian Period?," *Science News* 158 (3), July 15, 2000: 39, and in Jin et al., "Pattern of Marine Mass Extinction." Quick extinction, long recovery in Bowring, Erwin, and Isozaki, "The Tempo of Mass Extinction and Recovery."

7. Three quarters of reptile families and two thirds of amphibians extinct in Erwin, "The Mother of Mass Extinctions." Who made it and who didn't in A. H. Knoll et al., "Comparative Earth History and Late Permian Mass Extinction," *Science* 273 (5274), July 26, 1996: 452–58. Extinctions selective but unpredictable in David Raup, "Extinction: Bad Genes or Bad Luck?," *New Scientist* 131 (1786), September 14, 1991: 46–49, in D. M. Raup, "The Role of Extinction in Evolution," *Proceedings of the National Academy of Sciences USA* 91 (15), July 19, 1994: 6758–63. Resulting radiation of species not predictable in Erwin, *The Great Paleozoic Crisis.*

8. African plume in J. Ritsema, H. J. van Helijst, and J. H. Woodhouse, "Complex Shear Wave Velocity Structure Imaged Beneath Africa and Iceland," *Science* 286 (5446), December 3, 1999: 1925–28, in Joachim R. R. Ritter, "Rising Through Earth's Mantle," *Science* 286 (5446), December 3, 1999: 1865–66, and in Richard A. Kerr, "The Great African Plume Emerges as a Tectonic Player," *Science* 285 (5425), July 9, 1999: 187–88. Timing of basalt floods, Red Sea, Aden, and spreading and volcanism in C. J. Ebinger and N. H. Sleep, "Cenozoic Magmatism Throughout East Africa Resulting from Impact of a Single Plume," *Nature* 395 (6704), October 22, 1998: 788–91, and in Geoff Davies, "A Channeled Plume Under Africa," *Nature* 395 (6704), October 22, 1998: 743–44. Ethiopian plateau sends Nile north in Kevin Burke, "The African Plate," *South African Journal of Geology* 99 (4), December 1996: 341 (69 pp.).

9. Sadiman and Laetoli footprints in N. Agnew and M. Demas, "Preserving the Laetoli Footprints," *Scientific American,* September 1998, 44 (12 pp.). Forest and woodland in Terry Hardaker, "Beyond the Bones," *Geographical Magazine* 71 (1), January 1999: 17 (5 pp.), and in Carl Zimmer, "Kenyan Skeleton Shakes Ape Family Tree," *Science* 285 (5432), August 27, 1999: 1335–37. *Homo habilis* and tools at Olduvai Gorge, and stone tools at Turkana in James Steele, "Palaeoanthropology: Stone Legacy of Skilled Hands," *Nature* 399 (6731), May 6, 1999: 24–25. Oldest stone tools at Turkana in H.

Roche et al., "Early Hominid Stone Tool Production and Technical Skill 2.34 Myr Ago in West Turkana, Kenya," *Nature* 399 (6731), May 6, 1999: 57–60. Butchering at the Awash in J. de Heinzelin et al., "Environment and Behavior of 2.5-Million-Year-Old Bouri Hominids," *Science* 284 (5414), April 23, 1999: 625–29, and in Elizabeth Culotta, "A New Human Ancestor?," *Science* 284 (5414), April 23, 1999: 572–73. Most early traces of man traces in rifts in Chet Raymo, *The Crust of Our Earth* (Englewood Cliffs, N.J.: Prentice-Hall, 1983), 96–103. Separation rate between Nubia and Somalia in D. Chu and R. G. Gordon, "Evidence for Motion Between Nubia and Somalia Along the Southwest Indian Ridge," *Nature* 398 (6722), March 4, 1999: 64–67, and in Fred F. Pollitz, "From Rifting to Drifting," *Nature* 398 (7622), March 4, 1999: 21–22.

10. Description of Afar in Haroun Tazieff, "The Afar Triangle," *Scientific American,* February 1970, in *Continents Adrift,* 133–41. Red Sea tectonics and mineral deposits in Enrico Bonatti, "The Rifting of Continents," *Scientific American,* March 1987, 97–102, and in Peter A. Rona, "Hydrothermal Mineralization at Slow-Spreading Centers: Red Sea, Atlantic Ocean, and Indian Ocean," *Marine Mining* 5 (2), 1985: 117–43.

11. Helium$_3$ in William J. Broad, "Heavy Volcanic Eras Were Caused by Plumes from the Earth's Core," *New York Times,* August 21, 1995, C1, C10. CAMP plume in Olsen, "Giant Lava Flows, Mass Extinctions, and Mantle Plumes." Rift basins in Paul E. Olsen, "Stratigraphic Record of the Early Mesozoic Breakup of Pangea in the Laurasia-Gondwana Rift System," *Annual Review of Earth and Planetary Science* 25, 1997: 337–401, in J. S. Schlee, W. Manspeizer, and S. R. Riggs, "Paleoenvironments: Offshore Atlantic U.S. Margin," and in Warren Manspeizer and Harold L. Cousminer, "Late Triassic-Early Jurassic Synrift Basins of the U.S. Atlantic Margin," both from R. E. Sheridan and J. A. Grow, eds., *Geology of North America,* vol. I-2, *The Atlantic Continental Margin: United States* (Boulder, Colo.: Geological Society of America, 1988.

12. Flood basalts in A. Marzoli et al., "Extensive 200-Million-Year-Old Continental Flood Basalts of the Central Atlantic Magmatic Province," *Science* 284 (5414), April 23, 1999: 616–18. Triassic-Jurassic extinction in Steven M. Stanley, *Extinction* (New York: Scientific American Library, 1987), 115–16. Plant extinctions in J. C. McElwain, D. J. Beerling, and F. I. Woodward, "Fossil Plants and Global Warming at the Triassic-Jurassic Boundary," *Science* 285 (5432), August 27, 1999: 1386–90. Connecticut geology in Michael Bell, *The Face of Connecticut* (Hartford: Connecticut Geological and Natural History Survey, 1985), 13–26, 113–15, 157–64. Plumes and mass extinctions in Olsen, "Giant Lava Flows, Mass Extinctions, and Mantle Plumes." Extinction and evolution in Eldredge, *The Miner's Canary,* 49–78, and Fortey, *Life,* 186–209.

13. Florida and south Georgia part of Africa in "Hidden Graft," *Scientific American,* September 1985: 67–68. Mississippi River valley as false start in Dott and Batten, *Evolution of the Earth,* 463–64. Evaporite deposits in Schlee, Manspeizer, and Riggs, "Paleoenvironments: Offshore Atlantic U.S. Margin," and in Manspeizer and Cousminer, "Late Triassic-Early Jurassic Synrift Basins." Single salt basin split in two in Nova Scotia and Morocco in Lubomir F. Jansa, "Paleoceanography and Evolution of the North Atlantic Basin During the Jurassic," in P. R. Vogt and B. E. Tucholke, eds., *The Geology of North America,* vol. M, *The Western North Atlantic Region*

(Boulder, Colo.: Geological Society of America, 1986), 603–16. Ammonites in Peter Ward, "The Extinction of the Ammonites," *Scientific American,* October 1983, 136–47, and in Fortey, *Life,* 230–31. Ammonites show Tethys–Pacific connection in Brian E. Tucholke and Floyd W. McCoy, "Paleogeographic and Paleobathymetric Evolution of the North Atlantic Ocean," in Vogt and Tucholke, eds., *Western North Atlantic,* 589–603.

14. White cliffs of Dover and coccoliths in Roger Osborne and Donald Tarling, *The Historical Atlas of the Earth* (New York: Henry Holt, 1996), 106, and in Raymo, *Crust of Our Earth,* 70–71. Coccolith blooms, from the *Emiliania huxleyi* home page, *http://www.soc.soton.ac.uk/SUDO/tt/eh/.* Circulation, seawater, and sediments in Tucholke and McCoy, "Paleogeographic and Paleobathymetric Evolution," and in Jansa, "Paleoceanography and Evolution." Diversification of plankton, poor circulation, and black shale in James P. Kennett, *Marine Geology* (Englewood Cliffs, N.J.: Prentice-Hall, 1982), 696–745.

15. Rivers flowing west in Robert S. White, "Ancient Floods of Fire," *Natural History,* April 1991, 51–60. Fossil plume in J. C. VanDecar, D. E. James, and M. Assumpção, "Seismic Evidence for a Fossil Mantle Plume Beneath South America and Implications for Plate Driving Forces," *Nature* 378 (6652), November 2, 1995: 25–31, and in Norman H. Sleep, "A Wayward Plume?," *Nature* 378 (6652), November 2, 1995: 19–20. Serral Geral linked to Kaoka in Namibia in Osborne and Tarling, *Historical Atlas of the Earth.* Lava flowed for one million years in Coutillot et al., "Flood Basalts and Continental Breakup." Opening of South Atlantic in Kennett, *Marine Geology,* 696–82. North-south connection in W. A. Berggren, and R. K. Olsson, "North Atlantic Mesozoic and Cenozoic Paleobiogeography," in Vogt and Tucholke, eds., *Western North Atlantic,* 565–87. Cretaceous seas in Stanley, *Earth and Life Through Time,* 477–97, in Fortey, *Life,* 210–35, and in Raymo and Raymo, *Written in Stone,* 115–22.

16. Deccan traps in India in Courtillot et al., "Flood Basalts and Continental Breakup." Three quarters of marine species extinct and no reason in David Jablonski and David M. Raup, "Selectivity of End-Cretaceous Marine Bivalve Extinctions," *Science* 268 (5209), April 21, 1995: 389–92. Blake Nose description in "Critical Boundaries in Earth's History—and the K-T Boundary," *Joint Oceanographic Institutions for Deep Earth Sampling (JOIDES) Journal,* Spring 1997. Core in John Lauerman, "Anatomy of a Disaster," *Woods Hole Currents,* Fall 1997, 3–6, and in R. D. Norris, B. T. Huber, and J. Self-Trail, "Synchroneity of the K-T Oceanic Mass Extinction and Meteorite Impact: Blake Nose, Western North Atlantic," *Geology* 27 (5), May 1999: 419–22.

17. K-T recovery in Stanley, *Earth and Life Through Time,* 524–34. Evergreens in Kennett, *Marine Geology,* 696–82. Methane in Richard A. Kerr, "When Climate Twitches, Evolution Takes Great Leaps," *Science* 257 (5077), September 18, 1992: 1622-25, in Richard A. Kerr, "A Smoking Gun for an Ancient Methane Discharge," *Science* 286 (5444), November 19, 1999: 1465, in M. E. Katz et al., "The Source and Fate of Massive Carbon Input During the Latest Paleocene Thermal Maximum," *Science* 286 (5444), November 19, 1999: 1531–33, in Richard Norris and Ursula Rohl, "Carbon Cycling and Chronology of Climate Warming During the Palaeocene/Eocene Transition," *Nature* 401 (6755), October 21, 1999: 775–78, in Gerald R. Dickens, "The Blast in the Past," *Nature* 401 (6755), October 21, 1999: 752–55, and William C. Clyde, University of New Hampshire, personal communication, November 1999.

18. Temperature differential between ridge and hot spot, and Cape Verde Island in White, "Ancient Floods of Fire." Greenland traps in Courtillot et al., "Flood Basalts and Continental Breakup." Yellowstone in Sandra Blakeslee, "Below Yellowstone, Earth Is on the Boil," *New York Times,* April 7, 1998, C1, C5. Giant's Causeway, Rockall and Orphan Knoll in Raymo, *Crust of Our Earth,* 66–77, and in K. O. Emery and Elazar Uchupi, *The Geology of the Atlantic Ocean* (New York: Springer-Verlag, 1984), 144.

19. New ridge in Courtillot et al., "Causal Links Between Flood Basalts and Continental Breakup."

20. Thermohaline circulation associated with isthmus and ice age in Gerald H. Haug and Ralf Tiedemann, "Effect of the Formation of the Isthmus of Panama on Atlantic Ocean Thermohaline Circulation," *Nature* 393 (6686), June 18, 1998: 673–76. Animals dwelling in island South America in A. Wyss, J. Flynn, and R. Charrier, "Fire, Ice, and Fossils," *Natural History,* June 1999, 38 (4 pp.). Isthmus of Panama and evolution in John F. Ross, "A Few Miles of Land Arose from the Sea—and the World Changed," *Smithsonian,* December 1996, 112 (9 pp.), in N. Knowlton et al., "Divergence in Proteins, Mitochondrial DNA, and Reproductive Compatibility across the Isthmus of Panama," *Science* 260 (5114), June 11, 1993: 1629–33, in Kathy A. Svitil, "Oceans Divided," *Discover,* November 1993, 38 (2 pp.), and in J. B. J. Jackson et al., "Diversity and Extinction of Tropical American Mollusks and Emergence of the Isthmus of Panama," *Science* 260 (5114), June 11, 1993: 1624–27.

Part IV: Full Circle

Chapter 11: Fraying Edges

1. Hog farms and waste in Peter J. Kilborn, "Storm Highlights Flaws in Farm Law in North Carolina," *New York Times,* October 17, 1999, A1, A26, in U.S. Senate Committee on Agriculture, "Animal Waste Pollution in America: An Emerging Problem," 1998, and in Natural Resources Defense Council, "America's Animal Factories: How States Fail to Prevent Pollution from Livestock Waste," 1998. *Pfiesteria* life cycle, fish kills, kleptochloroplastidy, and nutrient enrichment in JoAnn M. Burkholder, "Harmful Microalgae and Heterotrophic Dinoflagellates in Management of Sustainable Marine Fisheries," *Ecological Applications* 8 (1), supplement, February 1998: S37–62. *Pfiesteria* on North Carolina State University Aquatic Biology Laboratory Web site, *www.ncsu.edu,* on University of Maryland Web site, *www.mdsg.umd.edu,* in Todd Spangler, "Chesapeake Mystery: Fish and People Fall Ill," *Boston Globe,* September 4, 1997, A3, and in "Fish-Killing Microbe Is Found to Cause Serious Harm to People," *New York Times,* August 14, 1998, A14. Poultry production in Natural Resources Defense Council, "America's Animal Factories," in Dick Russell, "Underwater Epidemic," *Amicus,* Spring 1998, 28–32, and in Francis X. Clines, "Perdue Offers a Plan to Fight Odor and Pollution," *New York Times,* October 19, 1999, A14.

2. *Pfiesteria* fish kill and nutrients in Burkholder, "Harmful Microalgae and Heterotrophic Dinoflagellates." *Gymnodinium breve* off Florida in D. F. Boesch et al., "Harmful Algal Blooms in Coastal Waters: Options for Prevention, Control and Mitigation," National Oceanic and Atmospheric Administration (NOAA) Coastal

Ocean Program, 1997, 1–20. Increase in number and persistence of toxic species in HEED (Health Ecological and Economic Dimensions of Global Change Program), "Marine Ecosystems: Emerging Diseases as Indicators of Change," Washington, D.C.: NOAA, 1998, 1–18. Blooms in Europe in European Environment Agency, "Environment in the European Union at the Turn of the Century," 1999, 357–75, in Common Wadden Sea Secretariat, "The Wadden Sea Quality Status Report," 1999, in Robert Koening, "Black Spots Blot German Coastal Flats," *Science* 273 (5271), July 5, 1996: 25, in Tara Patel, "Toxins Bloom Round French Coast," *New Scientist* 139 (1883), July 23, 1993: 6, in Roger Milne, "North Sea Algae Threaten British Coasts," *New Scientist* 122 (1668), June 10, 1989: 28, in "Mystery Toxin in French Shellfish," *New Scientist* 137 (1861), February 20, 1993: 7, in Commission for the Protection of the Marine Environment in the North-East Atlantic (OSPAR), "OSPAR Quality Status Report 2000: Region IV—Bay of Biscay and Iberian Coast," June 2000, and in S. Gallacher et al., "The Occurrence of Amnesic Shellfish Poisons in Scottish Waters," abstract from HAB 2000, the Ninth International Conference on Harmful Algal Blooms, Tasmania, February 2000. Prince Edward Island mussels in National Research Council, *From Monsoons to Microbes: Understanding the Ocean's Role in Human Health* (Washington, D.C.: National Academy Press, 1999), 59–70. Scallops in Peconic Bay, Long Island Sound in Clifford Krauss, "Trying to Turn the Tide," *New York Times,* July 23, 1997, B1, B6. Chesapeake, North Carolina, and Florida in Christine Mlot, "The Rise in Toxic Tides," *Science News* 152 (13), September 27, 1997: 202–5. Numbers of toxic species and sublethal effects in Burkholder, "Harmful Microalgae and Heterotrophic Dinoflagellates." New shellfish toxic syndrome from Irish mussels in K. J. James et al., "Azaspiracid Poisoning (AZP): A New Shellfish Toxic Syndrome in Europe," HAB 2000. New toxic *Pfiesteria* in H. Glasgow et al., "A New Species of Toxic *Pfiesteria*," HAB 2000. "Fish-Killing Microbe Is Found to Cause Serious Harm to People," *New York Times,* August 14, 1998, A14.

3. Michael L. Parsons and Quay Dortch, "Sedimentological Evidence of an Increase in *Pseudo-Nitzschia* (*Bacillariophycecae*) Abundance in Response to Coastal Eutrophication," from HAB 2000. *Cryptosporidium* in Chesapeake Bay in R. Fayer et al., "*Cryptosporidium parvum* in Oysters from Commercial Harvesting Sites in the Chesapeake Bay," *Emerging Infectious Diseases* 5 (5), September–October 1999: 706–13. Cholera in Rita R. Colwell, "Global Climate and Infectious Disease: The Cholera Paradigm," *Science* 274 (5295), December 20, 1996: 2025–31. Cholera in the Chesapeake in S. C. Jiang et al., "Genetic Diversity of *Vibrio cholerae* in Chesapeake Bay Determined by Amplified Fragment Length Polymorphism Fingerprinting," *Applied and Environmental Microbiology* 66 (1), January 2000: 140–48. Nutrients and red tides in HEED, "Marine Ecosystems: Emerging Diseases," and in Burkholder, "Harmful Microalgae and Heterotrophic Dinoflagellates." Algal blooms and Po River nutrients in J. A. Downing et al., "Gulf of Mexico: Land and Sea Interactions," Council for Agricultural Science and Technology, Task Force Report no. 134, June 1999.

4. Effects of hypoxia on benthic communities in Robert J. Diaz and Rutger Rosenberg, "Marine Benthic Hypoxia: A Review of Its Ecological Effects and Behavior Responses of Benthic Macrofauna," *Oceanography and Marine Biology: an Annual Review* 33, December 1995: 245–303, and in Donald E. Harper Jr. and Nancy N. Rabalais, "Responses of Benthonic and Nektonic Organisms, and Communities to

Severe Hypoxia on the Inner Shelf of Louisiana and Texas," *Proceedings of the First Gulf of Mexico Hypoxia Management Conference,* December 5–6, 1995, Kenner, La., pp. 41–45. Effect on shrimp migration in NOAA, National Ocean Service, "Draft Integrated Assessment of Hypoxia in the Northern Gulf of Mexico," October 1999, and in Downing et al., "Gulf of Mexico: Land and Sea Interactions."

5. Nitrogen and crops in Robert H. Socolow, "Nitrogen Management and the Future of Food: Lessons from the Management of Energy and Carbon," *Proceedings of the National Academy of Sciences USA* 96 (11), May 1999: 6001–8. Nitrogen, fertilizer, and wetland loss in NOAA, "Assessment of Hypoxia." Fertilizer in Downing et al., "Gulf of Mexico: Land and Sea Interactions." Wetlands loss in Gulf of Mexico in Ken Kelley, "Fish Habitat: Who's Minding the Store?," *National Fisherman,* October 1994, 14–15. Mississippi River drainage in Tim Beardsley, "Death in the Deep," *Scientific American,* November 1997, 17 (2 pp.). Effect of channel size in R. B. Alexander, R. A. Smith, and G. E. Schwarz, "Effect of Stream Channel Size on the Delivery of Nitrogen to the Gulf of Mexico," *Nature* 403 (6771), February 17, 2000: 758–61. Wetlands loss in Louisiana in Joel Bourne, "Louisiana's Vanishing Wetlands: Going, Going . . . ," *Science* 289 (5486), September 15, 2000: 1860–63.

6. Tripling of nitrogen flux in NOAA, "Assessment of Hypoxia." *Pseudo-nitzschia* and fertilizer in Gulf of Mexico in Parsons and Dortch, "Increase in *Pseudo-nitzschia,*" HAB 2000. Carbon production in Downing et al., "Gulf of Mexico: Land and Sea Interactions." Effect of hypoxia and anoxia on benthic communities in Diaz and Rosenberg, "Marine Benthic Hypoxia." Shift in energy pulses and shrimp migration in NOAA, "Assessment of Hypoxia." Energy shift in R. E. Turner et al., "Fluctuating Silicate: Nitrate Ratios and Coastal Plankton Food Webs," *Proceedings of the National Academy of Sciences USA* 95 (22), October 27, 1998: 13048–51, and in Janet Pelley, "Scientists Elucidate Role Elements Play in Eutrophication," *Environmental Sciences and Technology News* 34 (1), January 1, 2000, 14A.

7. Baltic geology, fisheries, nitrogen discharges, and phytoplankton increases in "Environmental Conditions in the Baltic Sea Region," Baltic Sea Joint Comprehensive Environmental Action Programme of Baltic Marine Environment Protection Commission (HELCOM), Baltic Environment Proceedings No. 48, 1993, 2-I–3-20. Extent of blooms in "The State of the Baltic Marine Environment in 1999," Environment Committee of HELCOM, October 1999, and in E. Rantajarvi et al., "Phytoplankton Blooms in the Baltic Sea in 1997," Finnish Institute of Marine Research, *http://www2.fimr.fi.* Anoxia, hypoxia, lobsters, and sulfur mats in Diaz and Rosenberg, "Marine Benthic Hypoxia." Baltic residence time in T. D. Jickells, "Nutrient Biogeochemistry of the Coastal Zone," *Science* 281 (5374), July 10, 1998: 217–22. Baltic communities, cod, watershed population, and community change in Bengt-Owe Jansson and Kristina Dahlberg, "The Environmental Status of the Baltic Sea in the 1940s, Today and in the Future," *Ambio* 28 (4), June 1999: 312–19. Cod collapse in Anne Simon Moffat, "Global Nitrogen Problem Grows Critical," *Science* 279 (5353), February 13, 1998: 988–89. Millions of dead bivalves in Kattegat in "Algae Make Ecological Waves," *New Scientist* 104 (1490), January 9, 1986: 31.

8. Nitrogen associated with harmful algal blooms in HEED, "Marine Ecosystems: Emerging Diseases," 4–18. Population predictions in Nina V. Federoff and Joel E. Cohen, "Plants and Population: Is There Still Time?," *Proceedings of the National Academy of Sciences USA* 96 (11), May 1999: 5903–7. Agricultural production and

nitrogen usage increasing in David Tilman, "Global Environmental Impacts of Agricultural Expansion: The Need for Sustainable and Efficient Practices," *Proceedings of the National Academy of Sciences USA* 96 (11), May 1999: 5995–6000, and in Scott W. Nixon, "Enriching the Sea to Death," *Scientific American Presents the Oceans,* Fall 1998, 48–53. Anthropogenic input, doubling of fixed nitrogen, and deleterious effects in Socolow, "Nitrogen Management and the Future of Food." Anthropogenic and natural nitrogen, and deleterious effects in P. M. Vitousek et al., "Human Alterations of the Global Nitrogen Cycle: Sources and Consequences," *Ecological Applications* 7 (3), 1997: 737–50.

9. Eutrophication in half U.S. estuaries in S. B. Bricker, C. G. Clement, and D. E. Pirhalla, "National Eutrophication Assessment: Effects of Nutrient Enrichment on the Nation's Estuaries," NOAA, National Ocean Service, September 1999.

10. Anoxia and hypoxia in Diaz and Rosenberg, "Marine Benthic Hypoxia." Nitrogen discharge into Norway's estuaries in "Norwegian Fish Farms Now Major Polluters," Environmental News Daily press release, April 4, 2000. Norway salmon kills in "Fjords Provide Shelter from Toxic Algae," *New Scientist* 118 (1616), June 9, 1988: 36. Cod larvae killed and older cod forced offshore in Hans Chr. Eilertsen and Tim Wyatt, "Ecosystem Effects of *Phaeocystis pouchetii* and Its Toxins," HAB 2000. *Phaeocystis* clogging nets in Burkholder, "Harmful Microalgae and Heterotrophic Dinoflagellates." *Phaeocystis* blooms in the southern North Sea in Jickells, "Nutrient Biogochemistry of the Coastal Zone." Great Britain eutrophication in Milne, "North Sea Algae." Blooms and black spots in Koening, "Black Spots."

11. Swedish fjord in Diaz and Rosenberg, " Marine Benthic Hypoxia." Nitrogen has not declined in Wadden Sea Secretariat, "Wadden Sea 1999," and in OSPAR, "OSPAR 2000."

12. Atlantic drainage in David A. Ross, *Introduction to Oceanography,* 4th ed., (Englewood Cliffs, N.J.: Prentice-Hall, 1988), 121. Nitrogen flux into North Atlantic in R. W. Howarth et al., "Regional Nitrogen Budgets and Riverine N & P Fluxes for the Drainages to the North Atlantic Ocean: Natural and Human Influences," *Biogeochemistry* 35, 1996: 75–139.

13. Mussels in World Wide Fund for Nature, "The Common Future of the Wadden Sea," November 1991. Mussels, artificial drainage, and diking in Wadden Sea Secretariat, "Wadden Sea 1999." Two thirds of stocks fished at unsustainable levels in OSPAR, "OSPAR 2000," chap. 6, Overall Assessment, June 30, 2000. Flooding as alternative to diking in "Water Management: Dutch Rivers to Be Given Space Again After Six Centuries of Dyke Building," Netherlands Ministry of Transport, Public Works and Water Management press release, March 3, 2000.

14. Pollution of North Sea, including tributyl tin (TBT) in OSPAR, "2000," Overall Assessment.

15. Reproduction rates in seals fed Wadden Sea herring in T. M. Crisp et al., "Environmental Endocrine Disruption: An Effects Assessment and Analysis," *Environmental Health Perspectives,* supplements, 106 (1), February 1998: 11–56, as recounted in P. J. H. Reijnders, "Reproductive Failure in Common Seals Feeding on Fish from Polluted Coastal Waters," *Nature* 324, 1986: 456–57. Also in Peter S. Ross et al., "PCBs Are a Health Risk for Humans and Wildlife," *Science* 289 (5486), September 15, 2000: 1878–79. Viruses increasing in marine mammals and pollutants

depressing immune systems in HEED, "Marine Ecosystems: Emerging Diseases," 24–30. Viruses increasing in A. Motluk, "Deadlier than the Harpoon?," *New Scientist* 147 (1984), July 1, 1995: 12–13. North sea seals' distemper from harp seals in C. D. Harvell et al., "Emerging Marine Diseases—Climate Links and Anthropogenic Factors," *Science* 285 (5433), September 3, 1999: 1505–10. Harp seals' distemper from huskies in David Dickson, "Canine Distemper May Be Killing North Sea Seals," *Science* 241 (4871), September 9, 1988: 1284. Polluted Baltic herring and lowered immunity in R. L. de Swart, P. S. Ross, and J. G. Vos, "Impaired Immunity in Harbour Seals (*Phoca vitulina*) Exposed to Bioaccumulated Environmental Contaminants: Review of a Long-Term Feeding Study," *Environmental Health Perspectives* 104, supplement 4, August 1996, and in P. S. Ross et al., "Contaminant-Related Suppression of Delayed-Type Hypersensitivity and Antibody Responses in Harbor Seals Fed Herring from the Baltic Sea," *Environmental Health Perspectives* 103 (2), February 1995. PCB levels higher in diseased porpoises than in porpoises caught in fishing nets in Rob Edwards, "Sea Sickness," *New Scientist* 164 (2217), December 18, 1999: 12. Human flu in seals in A. D. M. E. Osterhaus et al., "Influenza B Virus in Seals," *Science* 288 (5468), May 12, 2000: 1051–52.

16. Lessepsian migrants in Alexandre Meinesz, *Killer Algae* (Chicago: University of Chicago Press, 1999) and in B. Galil, "The Silver Lining: The Economic Impacts of Red Sea Invaders in the Mediterranean," abstracts from the First National Conference on Biomarine Invasions, MIT Seagrant, Cambridge, Mass., January 24–27, 1999, 58.

17. Conditions in the Mediterranean in European Environment Agency, "State and Pressures of the Marine and Coastal Mediterranean Environment," 1999, 1–26, 47–75, and in European Environment Agency, "Environment in the European Union," 357–75. Blooms along the Adriatic in European Environment Agency, "Marine and Coastal Mediterranean Environment," 76–104, and in Fred Pearce, "Dead in the Water," *New Scientist* 145 (1963), February 4, 1995: 26–31. Adriatic algal blooms in Robert Koenig, "Adriatic Nations Team Up to Explore Spreading Marine Mystery," *Science* 290 (5492), October 27, 2000: 694, and in Robinson Shaw, "Woes of Venice Lagoon Tackled in U.S.," ENN News, September 23, 1999.

18. *Caulerpa* in Josep-Maria Gili, "Frontline View of an Invasion," *Science* 287 (5459), March 10, 2000: 1762, in O. Jousson et al., "Invasive Alga Reaches California," *Nature* 408 (6809), November 9, 2000: 157–58, in Meinesz, *Killer Algae,* in "Rogue Algae May Harm Mediterranean Fish," *Science News* 155 (22), May 29, 1999: 343, and in Janet Roloff, "Rogue Algae: The Mediterranean Floor Is Being Carpeted with a Shaggy, Aggressive Invader," *Science News* 154 (1), July 4, 1998: 8–11. Mediterranean sewage in Fred Pearce, "Dead in the Water," *New Scientist* 145 (1963), February 4, 1995: 26–31. Toxicity in M. Uchimura, R. Sandeaux, and C. Larroque, "The Enzymatic Detoxifying System of a Native Medterranean Scorpio Fish Is Affected by *Caulerpa taxifolia* in Its Environment," *Environmental Science and Technology* 33 (10), 1999: 1671–74. *Caulerpa* in Florida in Hillary Mayell, "Spread of Seaweed Threatens Fragile Ecosystem of Southern Florida Coastline," Environmental News Network, June 11, 2000.

19. Nitrogen in Chesapeake in Chesapeake Bay Foundation, "State of the Bay 2000," *www.savethebay.cbf.org/state-of-the-bay* and *www.chesapeakebay.net/indicators.htm,* in Nixon, "Enriching the Sea to Death," and in Michael Hirshfield, Chesapeake Bay

Foundation, personal communication, September 2000. Eelgrass and nitrogen in Janet Raloff, "Sea Sickness: Marine Epidemiology Comes of Age," *Science News* 155 (5), January 30, 1999: 72–74. Chesapeake seagrass and oysters from Chesapeake Bay Foundation, "State of the Bay 2000," in Chesapeake Bay Foundation press release, May 25, 2000, "CBF Calls Increase in Underwater Grasses Encouraging but Far from Success." Previous oyster harvests in Peter S. Goodman, "Sowing Oysters to Reap a Cleaner Chesapeake," *Washington Post,* August 21, 1997, A1, A10. 2010 goals for the Chesapeake from Governor Gilmore of Virginia, Governor Glendening of Maryland, Governor Ridge of Pennsylvania et al., "Chesapeake 2000," Chesapeake Bay Program, June 2000.

20. Seagrass, mangrove, and coral habitat requirements in Jeremy B. C. Jackson and Luis d'Croz, "The Ocean Divided," in Anthony Coates, ed., *Central America: A Natural and Cultural History* (New Haven, Conn.: Yale University Press, 1997), 38–71. Life cycle of spiny lobster in Mark J. Butler, "Salinity Changes and Model Predictions: Will Spiny Lobster Tolerate Our Environmental Monkey-Business?," abstract from the 1999 Florida Bay and Adjacent Marine Systems Science Conference. Mangrove, and seagrass coral dependencies in William L. Kruczynski, "Water Quality Concerns in the Florida Keys: Sources, Effects, and Solutions," Florida Keys National Marine Sanctuary, Water Quality Protection Program, September 1999. Nighttime on the reef in Matthew S. Grober, "Starlight on the Reef," *Natural History,* October 1989, 73–80, and in John F. Ross, "The Miracle of the Reef," *Smithsonian,* February 1998, 88–96. Night habits of parrot fish in "New Blue Debut," *Seabits,* August 2000, New England Aquarium. Coral reefs rich in Peter F. Sale, "Coral Reefs: Recruitment in Space and Time," *Nature* 397 (6714), January 7, 1999: 25–27. Life and competition on the reef, and coral symbiosis in T. F. Goreau, N. I. Goreau, and T. J. Goreau, "Corals and Coral Reefs," *Scientific American,* August 1979, 124–36. Spawning in "Reef Romance: Coral Spawning Coming Up August 18–26," NOAA press release, International Year of the Reef, Week 32, August 14, 1997.

21. Geology of Florida Keys, sewage treatment, viruses and bacteria, and destruction of seagrass in Kruczynski, "Water Quality in the Florida Keys." Florida Keys sewage and pathogens in Rick Bragg, "Crowded Florida Keys: A Paradise in Trouble," *New York Times,* September 28, 1999, A14. Sewage treatment plants fail to remove nitrogen and viruses completely in National Research Council, *Monsoons to Microbes,* 43–58, and in HEED, "Marine Ecosystems: Emerging Diseases," 18–23. Florida sewage in D. W. Griggin et al., "Detection of Viral Pathogens by Reverse Transcriptase PCR and of Microbial Indicators by Standard Methods in the Canals of the Florida Keys," *Applied and Environmental Microbiology* 65 (9), September 1999: 4118–4216.

22. Water from Florida Bay to reef in Kruczynski, "Water Quality in the Florida Keys." Lake Okeechobee in K. E. Havens et al., "Rapid Ecological Changes in a Large Subtropical Lake Undergoing Cultural Eutrophication," *Ambio* 25 (3), May 1996: 150–55. Okeechobee, nitrogen, algal blooms, and possible problems with replumbing the Everglades in Larry Brand and Maiko Suzuki, "Nutrient Bioassays and the Redfield Ratio in Florida Bay," from abstracts of 1999 Florida Bay Conference. Nitrogen inputs into Florida Bay in D. T. Rudnick et al., "Phosphorus and Nitrogen Inputs to Florida Bay: The Importance of the Everglades Watershed," *Estuaries* 22 (2B), June 1999: 398–416, and in William K. Stevens, "A Bay Is Sick; Will the Cure Make it Worse?" *New York Times,* April 15, 1997, C1, C6.

23. Coral reef diseases in HEED, "Marine Ecosystems: Emerging Diseases," 50–57. *Aspergillus* in D. M. Geiser et al., "Cause of Sea Fan Death in the West Indies," *Nature* 394 (6689), July 9, 1998: 137–39, and in Bob Holmes, "Ravaged Reefs," *New Scientist* 161 (2171), January 30, 1999: 11. Star coral in L. L. Richardson et al., "Florida's Mystery Coral-Killer Identified," *Nature* 392 (6676), April 9, 1998: 557–58. Florida Keys study in P. Dustan, J. W. Porter, and W. C. Jaap, "Coral Reef Monitoring Project Analysis of Trends, 1996–1999, Executive Summary to Steering Committee," March 30, 2000. Decline on Carysfort Reef from Phil Dustan, College of Charleston, personal communication, June 2000. Staghorn, elkhorn eradication and no similar past record in C. Harvell et al., "Emerging Marine Diseases."

24. Rate of reef growth in Kruczynski, "Water Quality in the Florida Keys: Sources." Urchin decline and multiple redundancies in Elizabeth Pennisi, "Coral Reefs Dominate Integrative Biology Meeting," *Science* 279 (5352), February 6, 1998: 897–99, and in T. P. Hughes and J. H. Cornell, "Multiple Stressors on Coral Reefs: A Long-term Perspective," *Limnology and Oceanography* 44 (3, part 2), 1999: 932–40, and in Jackson and d'Croz, "The Ocean Divided." Jamaican coral cover in Global Coral Reef Monitoring Network, "Status of Coral Reefs of the World, 2000," October 2000, 1–6. Two thirds of reefs at risk, but sediment expands nonreef area in D. Bryant et al., "Reefs at Risk," World Resources Institute, 11–22. Sea urchin die-off in K. B. Ritchie et al., "A Tetrodotoxin-Producing Marine Pathogen," *Nature* 404 (6776), March 23, 2000: 354. Problems near population centers from NOAA International Year of the Reef press release, Week 39, September 29, 1997. Resurgence of fish in no-take zone in J. A. Bohnsack et al., "Initial Responses of Exploited Organisms to No-Take Protection Zones in the Florida Keys National Marine Sanctuary," abstracts of 1999 Florida Bay Conference.

25. Corals and carbon dioxide in J. A. Kleypas et al., "Geochemical Consequences of Increased Atmospheric Carbon Dioxide on Coral Reefs," *Science* 285 (5411), April 2, 1999: 118–20, in R. W. Buddemeier and Stephen V. Smith, "Coral Adaptation and Acclimatization: A Most Ingenious Paradox," *American Zoologist* 39 (1), February 1999: 1–6, and in Elizabeth Pennisi, "New Threat Seen from Carbon Dioxide," *Science* 279 (5353), February 13, 1998: 989. Coccolithophores and carbon dioxide in Jean-Pierre Gattuso and Robert W. Buddemeier, "Calcification and CO_2," *Nature* 407 (6800), September 21, 2000: 311–13. Coral bleaching associated with warm water in HEED, "Marine Ecosystems: Emerging Diseases," in Peter W. Gynn, "Widespread Coral Mortality and the 1982–83 El Niño Warming Event," *Environmental Conservation* 11 (2), Summer 1984: 132–44, in Tom Goreau and Tim McClanahan, "Conservation of Coral Reefs After the 1998 Global Bleaching Event," *Conservation Biology* 14 (1), February 2000: 5–15, in Peter Pockley, "Global Warming Could Kill Most Coral Reefs by 2100," *Nature* 400 (6740), July 8, 1999: 98, in R. B. Aronson et al., "Coral Bleach-Out in Belize," *Nature* 405 (6782), May 4, 2000: 36. Tropical water warming in "Tropical Waters in Northern Hemisphere Heating at an Accelerated Rate," National Oceanic and Atmospheric Administration (NOAA) press release, July 28, 2000.

Chapter 12: Landfall

1. Local migrants in Brad Chase, "Tracking Coastal Invasion," *Boston Globe,* North Section, February 20, 2000, 1, 14. Periwinkles in Michael and Deborah Berrill, *The*

North Atlantic Coast: A Sierra Club Naturalist's Guide (San Francisco: Sierra Club Books, 1981), 89. Periwinkle shapes rocky shore in Cory Dean, "Tiny Snail Is Credited as a Force Shaping the Coast," *New York Times,* August 23, 1988, C1, C6. Green crab, clam, and veined rapa whelk in Sue Robinson, "Hostile Takeover," *National Fisherman,* March 1999, 22–24. Pacific shore crab in M. D. Brandhagen et al., "Geographic Differentiation of an Introduced Crab Species (*Hemigrapsus sanguineus*) on the Atlantic Coast of North America," in D. Brousseau, P. Kurchari, and C. Pflug, "Flood Preference Studies of the Japanese Shore Crab (*Hemigrapsus sanguineus*) from Western Long Island Sound," and in T. Casanova, "The Ecology of the Japanese Shore Crab (*Hemigrapsus sanguineus*) and Its Niche Relationship to the Green Crab (*Carcinus maenas*) along the Coast of Connecticut, U.S.A.," in abstracts of First National Conference on Biomarine Invasions, MIT Seagrant, Cambridge, Mass., January 24–27, 1999, 31, 32, 36. Veined rapa whelk in Scott Harper, "Predators Are Breeding in the Bay," *Virginian-Pilot* (Norfolk, Va.), September 22, 1998, and in J. Harding and R. Mann, "Habitat and Prey Preferences of Veined Rapa Whelks (*Rapana venosa*) in the Chesapeake Bay," abstracts of Conference on Biomarine Invasions, 66.

2. Acceleration of introduced species and its significance in P. M. Vitousek et al., "Biological Invasions as Global Environmental Change," *American Scientist* 84, September–October 1996: 468–78. Rate of ballast water release in "Alien Aquatic Species Will Be Turned Away at U. S. Borders," National Oceanic and Atmospheric Administration (NOAA) press release, April 30, 1999. Chesapeake invaders in G. M. Ruiz, P. Fofonoff, and A. H. Hines, "Nonindigenous Species as Stressors in Estuarine Marine Communities: Assessing Invasion Impacts and Interactions," *Limnology and Oceanography* 44 (3, part 2), May 1999: 950–72, and personal communication from Paul Fofonoff, Smithsonian Environmental Research Center, July 2000. Content of ballast water and introduced species in German ports in Deborah MacKenzie, "Alien Invaders," *New Scientist* 162 (2183), April 24, 1999: 18–19. Pathogens in T. A. McCollin, J. P. Hamer, and I. A. N. Lucas, "Transport of Marine Organisms Via Ships' Ballast into Ports Around England and Wales," p. 94, and L. A. Drake et al., "Inventory of Microbes in Ballast Water of Ships Arriving in Chesapeake Bay," p. 53, in abstracts of Conference on Biomarine Invasions. Shrimp farming in Ken Kelley, "The Viral Threat," *National Fisherman,* November 1997, 20–23, 92, and in Maureen Milne Donald, "Deadly Virus Threatens South Carolina Shrimp Farms," *National Fisherman,* January 1999, 12.

3. Diversity and invasion, and value of redundancies in J. J. Stachowicz, R. B. Whitlach, and R. W. Osman, "Species Diversity and Invasion Resistance in a Marine Ecosystem," *Science* 286 (5444), November 19, 1999: 1577–79. Stability in complexity in Gary A. Polis, "Stability Is Woven by Complex Webs," *Nature* 745 (6704), October 22, 1999: 744–45. Number of mammals earth can support, and invasions promote extinctions in Vitousek et al., "Biological Invasions as Global Environmental Change."

4. Atlantic seafloor descending in Paul Mann, "Caribbean Sedimentary Basins: Classification and Tectonic Setting from Jurassic to Present," in Paul Mann, ed., *Sedimentary Basins of the World, Caribbean Basins* (Amsterdam: Elsevier, 1999), 3–31. Pelée eruption in Haraldur Sigurdsson and Steven Carey, *Caribbean Volcanoes: A Field Guide* (Toronto: Geological Association of Canada, 1990), and Peter Francis, *Volcanoes* (Middlesex: Penguin, 1976), and in *Volcano* (Alexandria, Va.: Time-Life,

1982). Thousands of earthquakes in Guadeloupe's volcano in Michael J. Carr and Richard E. Stoiber, "Volcanism," in G. Dengo and J. E. Case, eds., *The Caribbean Region,* vol. H, *The Geology of North America* (Boulder, Colo.: Geological Society of America, 1990). Boiling river in "Natural Park of Guadeloupe," (Saint-Jorioz, France: SPEA Press), 1973. Ash from St. Vincent to Haiti and Dominica's Valley of Desolation in Sigurdsson and Carey, *Caribbean Volcanoes*. Health hazard in Montserrat in P. J. Baxter et al., "Cristobalite in Volcanic Ash and the Soufrière Hills Volcano, Montserrat, British West Indies," *Science* 283 (5405), February 19, 1999: 1142–45. Seven thousand out of eleven thousand evacuated in "Caribbean Follies," *Economist,* August 3, 1997, 41 (2 pp.). Three-year eruption in "Montserrat Volcano Erupts Again, Surprising Scientists," *Boston Globe,* March 22, 2000, A4.

5. Salt water causing melting at subduction zones in Terry Plank, "The Brine of the Earth," *Nature* 380 (6571), March 21, 1996: 202–3, and in J. Brendan Murphy and R. Damian Nance, "Mountain Belts and the Supercontinent Cycle," *Scientific American,* April 1992, 84–91.

6. Barbados in G. K. Westbrook, A. Mascle, and B. Biju-Duval, "Geophysics and the Structure of the Lesser Antilles Forearc," in B. Biju-Duval et al., *Initial Reports, Deep Sea Drilling Project,* vol. 78A (Washington, D.C.: U.S. Government, 1984), and in G. K. Westbrook and W. R. McCann, "Subduction of the Atlantic Lithosphere Beneath the Caribbean," in *The Geology of North America,* vol. M, *The Western North Atlantic Region* (Boulder, Colo.: Geological Society of America, 1986). Geology of Cuba contains history of Caribbean, ancient rifts of proto-Caribbean, and oceanic plateau in A. C. Kerr et al., "A New Plate Tectonic Model of the Caribbean: Implications from a Geochemical Reconnaissance of Cuban Mesozoic Volcanic Rocks," *Geological Society of America Bulletin* 111 (11), November 1999: 1581–99. Caribbean as part of Atlantic in David A. Ross, Introduction to Oceanography, 4th ed., (Englewood Cliffs, N.J.: Prentice-Hall, 1988), 14. History of the Caribbean in Kevin Burke, "Tectonic Evolution of the Caribbean," *Annual Review of Earth and Planetary Sciences* 16, 1988: 201–30, and in Anthony G. Coates, "The Forging of Central America," in Anthony Coates, ed., *Central America: A Natural and Cultural History* (New Haven, Conn.: Yale University Press, 1997), 1–37, in Philippe Bouysse, "Opening of the Grenada Back-arc Basin and Evolution of the Caribbean Plate During the Mesozoic and Early Paleogene," *Tectonophysics* 149, 1988: 121–43, and in Mann, "Caribbean Sedimentary Basins." Port Royal and Kingston in Jon Erickson, *Volcanoes and Earthquakes* (Blue Ridge Summit, Pa.: Tab Books, 1988). Reshaping of Lesser Antilles in R. C. Maury et al., "Geology of the Lesser Antilles," in Dengo and Case, eds., *The Caribbean Region*.

7. Cold slabs sinking to mantle in Rob D. van der Hilst and Hrafnkell Kárason, "Compositional Heterogeneity in the Bottom 1000 Kilometers of Earth's Mantle: Toward a Hybrid Convectional Model," *Science* 283 (5409), March 19, 1999: 1885–87, in L. H. Kellog, B. H. Hager, and R. D. van der Hilst, "Compositional Stratification in the Deep Mantle," *Science* 283 (5409), March 19, 1999: 1881–84, in Satoshi Kaneshima and George Helffrich, "Dipping Low-Velocity Layer in the Mid-Lower Mantle: Evidence for Geochemical Heterogeneity," *Science* 283 (5409), March 19, 1999: 1888–90, in Richard A. Kerr, "A Lava Lamp Model for the Deep Earth," *Science* 283 (5409), March 19, 1999: 1826–27, in R. Monastersky, "A Stirring Tale from Inside Earth," *Science News* 155 (12), March 20, 1999: 180, in Richard A. Kerr, "Deep Sinking Slabs Stir the Mantle," *Science* 276 (5300), January

31, 1997: 613–15, and in R. Monastersky, "Global Graveyard," *Science News* 152 (3), July 19, 1997: 46–47. Recycled ocean crust at Mid-Atlantic Ridge in J. M. Eiler et al., "Oxygen-isotope Evidence for Recycled Crust in the Sources of Mid-Ocean Ridge Basalts," *Nature* 403 (6769), February 3, 2000: 530–34, and personal communication from John Eiler, California Institute of Technology, March 3, 2000.

8. Rift in Europe from plume, and extension ceasing because of subduction in S. Goes, W. Spakam, and H. Bijwaard, "A Lower Mantle Source for Central European Volcanism," *Science* 286 (5446), December 3, 1999: 1928–30, in Joachim R. R. Ritter, "Rising Through Earth's Mantle," *Science* 286 (5446), December 3, 1999: 1865–66, and in K. Hoernle, Y.-S. Zhang, and D. Graham, "Seismic and Geochemical Evidence for Large-Scale Mantle Upwelling Beneath the Eastern Atlantic and Western and Central Europe," *Nature* 374 (6517), March 2, 1995: 34–38. Source of Roermond earthquake in T. van Eck and L. Ahorner, "The Earthquake of the Century in Northwestern Europe: The Roermond, the Netherlands, Earthquake," *Earthquakes and Volcanoes* 24 (1), January 1993: 15 (7 pp.).

9. Back and forth between Europe and Africa, Zagros Mountains, and Alps in Roger Osborne and Donald Tarling, *The Historical Atlas of the Earth* (New York: Henry Holt, 1996). Africa and Europe approaching at four millimeters per year in J. Morales, I. Serriano, and A. Jabaloy, "Active Continental Subduction Beneath the Betic Cordillera and the Alboran Sea," *Geology* 27 (8), August 1999: 735–38. Troodos in Ian G. Gass, "Ophiolites," *Scientific American,* August 1982, 122–31, and in "Report on the InterRidge/Bridge Field Trip to the Troodos Ophiolite, Cyprus, July 11–17, 1999," *InterRidge News,* Fall 1999, 8–10.

10. Cleopatra and the sewage in Douglas Jehl, "Down Among the Sewage, Cleopatra's Storied City," *New York Times,* October 29, 1997, A4, and in Angela M. H. Schuster, "Uncovering Cleopatra's Egypt Underwater," *New York Times,* Art and Architecture, March 14, 1999, 41–42. Herakleion in Vijay Joshi, "Ancient Cities' Ruins Found on Sea Bottom, Archeolologists Say," *Boston Globe,* June 4, 2000, A3.

11. Mediterranean salinity crisis in W. Krijgsman et al., "Chronology, Causes and Progression of the Messinian Salinity Crisis," *Nature* 400 (6745), August 12, 1999: 652–55, and in Judith A. McKenzie, "From Desert to Deluge in the Mediterranean," *Nature* 400 (6745), August 12, 1999: 613–14.

12. Thera eruption 1650 B.C. possibly related to plagues in Exodus, pyroclastic flows for Pompeii, and Virgil, Dante, and the Phlegraean Fields in Haraldur Sigurdsson, *Melting the Earth* (New York: Oxford University Press, 1999). Temperature increases in Campi Flegrei caldera, in "Campi Flegrei," Volcanic Activity Reports, Global Volcanism Program, Smithsonian Institution, November 1997 and October 1999. State of Vesuvius in "Vesuvius," Volcanic Activity Reports, Global Volcanism Program, Smithsonian Institution, June 1996, August 1996, April 1997, and October 1999. Solid dike below Vesuvius summit in Grant Heiken, "Will Vesuvius Erupt? Three Million People Need to Know," *Science* 286 (5445), November 26, 1999: 1675–87. Twenty thousand people died in Pompeii in Robert Etienne, *Pompeii: the Day a City Died* (New York: Harry N. Abrams, 1992), 43. Future of Atlantic from Christopher R. Scotese, PALEOMAP Project, personal communication, September 2000, and Kevin Burke, Carnegie Institution of Washington, personal communication, August 2000.

Acknowledgments

I could not have researched and written this book alone. Many people provided essential help and guidance for which I am deeply appreciative.

A number of scientists thoroughly reviewed long portions of the manuscript, generously giving their time and expertise, and providing detailed and thoughtful suggestions: Kevin J. Burke, JoAnn Burkholder, Sallie Chisholm, Robert Diaz, Cynthia Ebinger, Douglas Erwin, Neil Glickstein, Peter Herring, Michael Hirshfield, Susan Humphris, Amy Knowlton, Molly Lutcavage, James McCleave, David Owens, James Price, Eugene Turner, and Cindy Lee Van Dover. The remaining errors are my own.

Peg Brady, John McMahon, and the Sea Education Association made it possible for me to share the first leg of the Fall Sea Semester, and Captain Terry Hayward and his crew welcomed me aboard the *Corwith Cramer*.

Barbara Kelley, Harriet Webster, Diana Peck, and Jackie Vaccarello offered insightful comments on early drafts; Judith Kildow, Ken Mallory, and Susan and Frank Shepard opened doors into the scientific community; and Nubar Alexanian and Peg Anderson provided unending moral support. Many scientists described their research, answered my questions, and explained how their particular piece fit into a larger whole. The list of names is far too long to print, but the time these scientists gave provided the book with whatever breadth, depth, and cohesion it might have.

Writing this book was hard. The subject often seemed overwhelming, but I have been lucky. Rob Cowley was there at the very beginning, and Peter Davison and Merloyd Lawrence provided key pieces of advice along the way. It was a pleasure to have Amy Robbins's copyediting and Stefanie Diaz's editorial assistance at Norton. When time was running out, Barbara Gale provided additional copyediting. No one could ask for an agent more kind and skilled than Jonathan Matson or an editor more gracious and wise than Angela von der Lippe at Norton. I am grateful for their faith in the idea of *Great Waters,* and in me.

Protecting the integrity of the sea is a never-ending job. Over the years, Frank Mirarchi, Joe Brancaleone, and Steve Murawski have labored to protect New England's fisheries. Their tenacity and vision, along with that of Robert "Stubby" Knowles, Bill Eichbaum, and Robert Buchsbaum, continues to inspire. Ellie Dorsey, another staunch advocate for the sea, died while I was completing *Great Waters*. Though we miss her, perhaps there is some comfort seeing her spirit live on in the work of others.

To Dan, whose loyalty and generosity have been way above and beyond the call of love or duty. To him, and to Abby and Susannah, all of whose enthusiasm, kindness, and generosity mean more to me than I can ever say.

Index